W0227457

recent advances in phytochemistry

volume 23

Plant Nitrogen Metabolism

RECENT ADVANCES IN PHYTOCHEMISTRY

Proceedings of the Phytochemical Society of North America
General Editor: Eric E. Conn, *University of California, Davis, California*

Recent Volumes in the Series

Volume 15 **The Phytochemistry of Cell Recognition and Cell Surface Interactions**
Proceedings of the First Joint Meeting of the Phytochemical Society of North America and the American Society of Plant Physiologists, Pullman, Washington, August, 1980

Volume 16 **Cellular and Subcellular Localization in Plant Metabolism**
Proceedings of the Twenty-first Annual Meeting of the Phytochemical Society of North America, Ithaca, New York, August, 1981

Volume 17 **Mobilization of Reserves in Germination**
Proceedings of the Twenty-second Annual Meeting of the Phytochemical Society of North America, Ottawa, Ontario, Canada, August, 1982

Volume 18 **Phytochemical Adaptations to Stress**
Proceedings of the Twenty-third Annual Meeting of the Phytochemical Society of North America, Tucson, Arizona, July, 1983

Volume 19 **Chemically Mediated Interactions between Plants and Other Organisms**
Proceedings of the Twenty-fourth Annual Meeting of the Phytochemical Society of North America, Boston, Massachusetts, July, 1984

Volume 20 **The Shikimic Acid Pathway**
Proceedings of the Twenty-fifth Annual Meeting of the Phytochemical Society of North America, Pacific Grove, California, June, 1985

Volume 21 **The Phytochemical Effects of Environmental Compounds**
Proceedings of the Twenty-sixth Annual Meeting of the Phytochemical Society of North America, College Park, Maryland, July, 1986

Volume 22 **Opportunities for Phytochemistry in Plant Biotechnology**
Proceedings of the Twenty-seventh Annual Meeting of the Phytochemical Society of North America, Tampa, Florida, June, 1987

Volume 23 **Plant Nitrogen Metabolism**
Proceedings of the Twenty-eighth Annual Meeting of the Phytochemical Society of North America, Iowa City, Iowa, June 1988

A Continuation Order Plan is available for this series. A continuation order will bring delivery of each new volume immediately upon publication. Volumes are billed only upon actual shipment. For further information please contact the publisher.

recent advances in phytochemistry

volume 23

Plant Nitrogen Metabolism

Edited by

Jonathan E. Poulton

University of Iowa
Iowa City, Iowa

John T. Romeo

University of South Florida
Tampa, Florida

and

Eric E. Conn

University of California, Davis
Davis, California

PLENUM PRESS • NEW YORK AND LONDON

Proceedings of the Twenty-eighth Annual Meeting of the
Phytochemical Society of North America on Plant Nitrogen Metabolism,
held June 26–30, 1988, at the University of Iowa, Iowa City, Iowa

ISBN-13: 978-1-4612-8106-1 e-ISBN: 978-1-4613-0835-5
DOI: 10.1007/978-1-4613-0835-5

© 1989 Plenum Press, New York
Softcover reprint of the hardcover 1st edition 1989
A Division of Plenum Publishing Corporation
233 Spring Street, New York, N.Y. 10013

All rights reserved

No part of this book may be reproduced, stored in a retrieval system, or transmitted
in any form or by any means, electronic, mechanical, photocopying, microfilming,
recording, or otherwise, without written permission from the Publisher

PREFACE

This volume is based on papers presented by invited speakers at a symposium entitled "Plant Nitrogen Metabolism" held in conjunction with the 28th Annual Meeting of the Phytochemical Society of North America. The meeting took place on the campus of the University of Iowa at Iowa City during June 26-30, 1988, and attracted 110 participants from 11 countries. The goal of the symposium was to trace the pathway by which nitrogen passes from soil and atmosphere into both primary and secondary nitrogenous metabolites, focusing upon areas which were felt to be most rapidly expanding.

From nodulines (nodule specific proteins) and GS/GOGAT mutants to sugar mimics (polyhydroxyalkaloids) and herbicide inhibitors of amino acid metabolism, research in nitrogen metabolism has expanded into areas barely envisioned only a few years ago. Both the nitrogen specialist and the general plant biochemist will be pleased by the range of topics covered here. Following an overview in Chapter 1 of plant nitrogen metabolism, the remaining chapters are loosely organized into three groups. Chapters 2-6 deal primarily with the biochemistry and molecular biology of nitrogen assimilation and transport, Chapters 7-9 with amino acid metabolism, and Chapters 10-12 with secondary metabolites.

The Symposium was organized by Jonathan E. Poulton (University of Iowa) and John T. Romeo (University of South Florida), who gratefully acknowledge advice received from fellow PSNA members Eric E. Conn, Ann Oaks, and James A. Saunders concerning choice of speakers. The organizers also wish to extend their gratitude to other parties who contributed greatly to the success of the meeting. Generous financial support was provided by the U.S. Department of Agriculture, UI Graduate College, Cone Foundation (UI Botany Department), E.I. duPont de Nemours and Company, Monsanto Chemical

Company, and the Phytochemical Society of North America. The University of Iowa is thanked for permission to hold the meeting on campus and the Botany Department for providing secretarial assistance. Special recognition is due to David Dawes and his colleagues at the UI Center for Conferences and Institutes, who assisted by Paul Durham, Tom Glendening, Hua-Cheng Wu and Gary Kuroki, ensured the smooth running of the conference. We thank Billie Gabriel (University of California, Davis) for her skillful preparation of the camera-ready manuscript.

Jonathan E. Poulton
John T. Romeo

February, 1989

CONTENTS

Chapter One

AN OVERVIEW OF NITROGEN METABOLISM IN HIGHER PLANTS

DALE G. BLEVINS

Department of Agronomy
University of Missouri
Columbia, Missouri 65211

NITROGEN SOURCES

The predominant sources of nitrogen for higher plants are NH_4^+, NO_3^- and N_2. In most forests, where soils are acidic and low in nitrogen, NH_4^+ is the major nitrogen source. In agricultural soils, NO_3^- is the primary source of nitrogen for plant growth and development, even though NH_4^+ or urea fertilizers may have been applied. Soil microorganisms, under normal conditions, oxidize the NH_4^+ or urea-N to NO_3^- by the process of nitrification. In this review, the metabolism of nitrogen obtained by higher plants from NH_4^+, NO_3^- or N_2 will be discussed.

AMMONIUM METABOLISM

The availability of nitrification inhibitors, like nitrapyrin, has increased interest in NH_4^+ uptake and metabolism. Plants adapted to acid soils,[1] i.e., many tree species, or those adapted to low soil redox potentials such as rice, have a preference for NH_4^+. The uptake and metabolism of NH_4^+ may change plant metabolism in several ways; it can (1) alter uptake of other cations like Mg^{2+}, (2) increase root respiration, (3) lower soluble carbohydrate in roots, (4) stimulate root exudation, and (5) increase putrescine synthesis.[1] The growth and development of plants grown under NH_4^+ nutrition may be quite different from plants grown with NO_3^- nutrition.

Ammonium nutrition causes a reduction in root elongation that results in short, thick roots[2,3] with increased lateral root formation.[4] This stimulation of lateral root formation may be responsible for the increase in cytokinin concentration in the xylem sap of apple root stocks and in the number of new spurs produced with NH_4^+ nutrition versus NO_3^- nutrition.[5] By judicious use of NH_4^+ and nitrification inhibitors, economic advantages might be gained in the growth and development of certain agronomic or horticultural plants as compared to the more common NO_3^- nutrition.

There are several enzymes potentially capable of incorporating ammonia into organic forms as amino acids and amides. These include alanine dehydrogenase, aspartate dehydrogenase, asparagine synthetase (AS), glutamate dehydrogenase (GDH) and glutamine synthetase (GS) (Table 1);[6] of these, only the last three have been reported in higher plants. Although most higher plants absorb very little NH_4^+ through their root systems, they are equipped with and use enzymes designed for NH_4^+ assimilation. All forms of nitrogen used by the plant are ultimately reduced to NH_4^+ and incorporated into organic combination. In addition, large quantities of NH_4^+ are assimilated daily through the pathway called the photorespiratory nitrogen cycle[7] (Fig. 1). In the intensive nitrogen metabolism involved in photorespiration, two molecules of glyoxylate are converted to glycine.[7] One molecule of glycine is decarboxylated, deaminated and the remaining methylene carbon is

Table 1. Reactions of Enzymes that Incorporate NH_4^+ into Amino Acids or Amides in Higher Plants.

(1)	Pyruvate + NH_4^+ + NADH = Alanine + H_2O + NAD^+ Alanine Dehydrogenase
(2)	L-Aspartate + NH_4^+ + ATP = L-Asparagine + H_2O + AMP + PP_i Asparagine Synthetase
(3)	Oxaloacetate + NH_4^+ + NADH = L-Aspartate + H_2O + NAD^+ Aspartate Dehydrogenase
(4)	α-Ketoglutarate + NH_4^+ + NAD(P)H = L-Glutamate + H_2O + $NAD(P)^+$ Glutamate Dehydrogenase
(5)	L-Glutamate + ATP + NH_4^+ = L-Glutamine + ADP + P_i Glutamine Synthetase

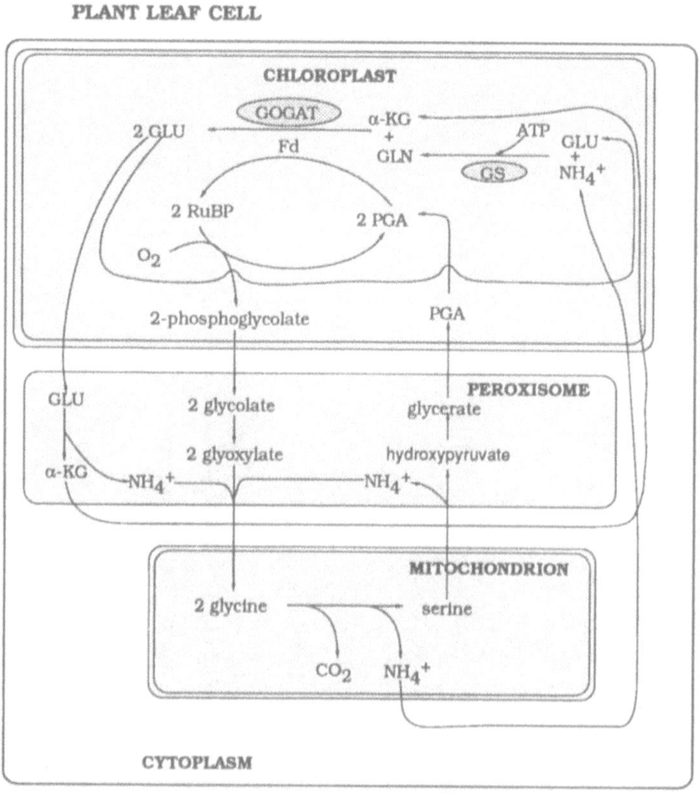

Fig. 1. Glutamine synthetase and glutamate synthase involvement in the photorespiratory nitrogen cycle. Adapted from Keys _et al_.[7] and from Chapter 5, this volume.

transferred to the other molecule of glycine to form serine, which is deaminated. One amino group for the two molecules of glycine comes from transamination with serine, while the other group comes from the cellular pool of ammonia. Only 7% of the nitrogen in the photorespiratory nitrogen cycle comes from asparagine,[8] so the glutamine synthetase/glutamate synthase (GS/GOGAT) system is probably of major importance in the recycling of NH_4^+ through photorespiration. This

$$\text{GLN} + \alpha\text{-Kg} \xrightarrow[\text{Ferredoxin}_{\text{red}} \text{ or NAD(P)H}]{\text{GOGAT}} 2\ \text{GLU}$$

was shown by Somerville and Ogren[9] who observed that *Arabidopsis* mutants lacking this enzyme system could only survive in an O_2 free atmosphere, in which the oxidative steps in photorespiration could not take place. Ammonium fixation must be very rapid during normal, active photosynthesis, attaining a rate of at least one-third that of net CO_2 fixation.[6] There is probably more nitrogen assimilated through NH_4^+ in the photorespiratory nitrogen cycle than through any other area of nitrogen metabolism (see Chapter 5, this volume).

NITRATE METABOLISM

Most agronomic, nonleguminous plants utilize NO_3^- as a nitrogen source; consequently nitrate metabolism and its effect on general plant metabolism have been extensively studied. Contrasting NO_3^- nutrition with NH_4^+ nutrition is of special interest in this discussion. The biochemical pH stat is a topic that illustrates the response of plants to NO_3^- vs NH_4^+ uptake and assimilation.

During NO_3^- reduction, OH^- is produced and the pH of the tissue increases. The increase in pH stimulates

$$NO_3^- + 8H^+ + 8e^- \xrightarrow[\text{Reduction}]{\text{Nitrate}} NH_3 + 2H_2O + OH^-$$

phosphoenolpyruvate (PEP) carboxylation by the enzyme PEP carboxylase, producing oxalacetate (OAA), which is rapidly converted to malate by malate dehydrogenase.[10] This sequence of events has been termed "the biochemical pH stat" by D.D. Davies.[11] Labelling leaves with $^{14}CO_2$ resulted in $^{14}CO_2$ (as HCO_3^-) efflux from roots of plants fed NO_3^-.[12] This indicated that an increase in alkalinity of the rhizosphere with NO_3^- nutrition was due to exchange or efflux of HCO_3^- from the roots. The HCO_3^- was produced when malate, which had descended from the shoot in the phloem, was decarboxylated by malic enzyme in the root.[13] The biochemical pH stat, which operates in conjunction with NO_3^- uptake and reduction, involves organic acid synthesis and degradation much the same as that found in C-4 photosynthesis. In most plants NO_3^- reduction is carried out in leaves, which in corn or forage sorghum may be in aerial leaves 3 m above the soil surface. This pH stat eliminates possible alkalinization of the tissues by utilizing OH^- and CO_2 to synthesize malate. Malate is subsequently transported to the root, decarboxylated and HCO_3^- effluxed into the soil. This is an effective means of eliminating into the rhizosphere a potentially dangerous situation in the leaf tissue, i.e., buildup of high pH.

Under NH_4^+ nutrition the pH stat reacts very differently -- the rhizosphere become acid (i.e., pH down to 3.8 in unbuffered solutions).[1] Apparently, NH_4^+ uptake involves H^+ exchange. With NH_4^+ inside the root, keto acids, i.e., OAA or α-ketoglutarate (αKg) are quickly aminated, depleting the supply of carbon in the root. Ammonium ion, the dominant cation in terms of uptake, is thus consumed along with organic anions in the synthesis of amino acids and amides. Mulder[14] and Kirkby and Mendel[15] found that NH_4^+ nutrition caused a lower tissue pH than did NO_3^- nutrition for leaves and roots of pea and tomato, respectively. This lower pH reduced the accumulation of organic anions (limited PEP carboxylation) and resulted in a low concentration of organic anions especially in root tissue![15] An understanding of the metabolism associated with the pH stat and NO_3^- vs NH_4^+ nutrition is important for discussion of the metabolism involved in root nodules in a later section.

In most higher plant tissues, nitrate reduction to nitrite is catalyzed by NADH-specific nitrate reductase

$$NO_3^- + NAD(P)H + 2e^- + H^+ \xrightarrow[\text{Reductase}]{\text{Nitrate}} NO_2^- + NAD(P)^+ + H_2O$$

(NR).[16-18] This step is regarded as rate-limiting and regulated.[17,19] Nearly 25 years following the discovery of NR,[16] methods for rapid purification of this unstable protein were developed.[20] Purification of the native NADH:NR from both mono- and dicots has shown that it is a dimeric enzyme composed of two identical subunits, each subunit containing one 110- to 115,000 polypeptide, one FAD, one heme-Fe (Cyt b) and one Mo-containing factor (Mo-pterin).[20-24] Nitrate reductases that utilize NADPH have been detected in both roots and leaves. Soybean leaves contain substantial amounts of the NADPH-dependent NR, while the predominant form in roots is NADH-dependent.[25] In barley a bispecific NAD(P)H form has been found as well.[26] Understanding the forms of nitrate reductase gets more complicated when one discovers that there are assimilatory and dissimilatory (respiratory) nitrate reductases. For example, many root nodule bacteria contain a constitutive, dissimilatory nitrate reductase.[27] Apparently soybean leaves contain a small quantity of dissimilatory NADPH-NR which may cause the loss of substantial gaseous nitrogen oxides during the *in vitro* NR assay.[25] This enzyme in root nodules may also catalyze NO_3^- formation and the loss of gaseous nitrogen.[28]

Hageman and Flesher[29] first discovered that light and nitrate influenced NR activity in corn seedlings. Shaner and Boyer[30] found that NR activity was maintained in leaves by NO_3^- flux into the tissue. Recent evidence has shown that both barley and corn NADH:NRs are synthesized *de novo* in the light and degraded in the dark or in the absence of NO_3^-.[31,32] The availability of purified NR and specific antibodies should facilitate our understanding of the regulation and control of this enzyme.

Campbell's laboratory[20] found that metal chelating columns and affinity chromatography were effective in the quick purification of the labile NR. The availability of purified NR has led to the development of monoclonal and polyclonal antibodies which have been used in localization studies. Early research had indicated that nitrate reductase was a soluble, cytoplasmic protein;[33] however, a few studies had found NR localized near membranes or in vesicles near the chloroplast.[34,35] Recently, use of specific antibodies and the immunogold technique have shown a chloroplastic localization for NR in spinach leaves.[36] If additional experiments confirm these results, the source of reductant for NO_3^- would be conveniently supplied and the product, NO_2^-, would be conveniently near nitrite reductase. Nitrite reductase, an iron protein, was shown to be chloroplastic

$$NO_2^- + \text{Ferredoxin}_{red} + 6H^+ + 6e^- \xrightarrow[\text{Reductase}]{\text{Nitrite}} NH_3 + \text{Ferredoxin}_{ox} + OH^- + H_2O$$

many years ago.[33] In roots this enzyme is localized in plastids.[37] The source of reductant for nitrite reductase in leaves is ferredoxin. Recently, a ferredoxin-like protein, which may be the reductant for nitrite reductase, was isolated from root tissue.[18] This ferredoxin-like protein in the roots was also localized in plastids. The isolation of an enzyme that can transfer electrons to the root ferredoxin from NADPH completes our understanding of the mechanism for NO_2^- reduction in root tissues.[18]

NITROGEN FIXATION

Nodulation of Leguminous Plants

Rhizobium and *Bradyrhizobium* infect roots of plants in the family Leguminosae. This symbiosis is the result of a complex sequence of events which requires signal molecule recognition between the host and symbiont. A few bacterial and plant genes required for the symbiosis have been identified, e.g., *Rhizobium* nodulation genes *nod* DABC. The exact functions of these genes other than being responsible for stimulating the earliest detectable responses of the host (cortical cell division and root hair curling)[39,40] have not been determined. These genes are induced by molecules in the exudates of host plants.[41-43] It has been determined that certain phenolics are signal molecules for the rhizobial infection process (Table 2). Peters *et al.*[44] identified 3',4'5,7-tetrahydroxyflavone (luteolin) as the inducer molecule for *nod* ABC expression in *Rhizobium meliloti*. Therefore, luteolin from alfalfa roots may serve to control *nod* ABC expression during infection and nodule formation. Redmond *et al.*[45] found that flavones in root washings of clover induced *nod* gene expression in *Rhizobium trifolii*.

The expression of *nod* ABC is dependent on the interaction of the *nod* D gene product and the plant-secreted phenolic molecules.[46,47] These plant-secreted molecules have now been isolated from white clover and peas as well as the alfalfa mentioned earlier. In most cases, the stimulatory molecules are hydroxylated flavones or flavanones [e.g., 7,4'-dehydroxyflavone, 4'7-dihydroxy-3'-methoxyflavone (geraldone) and 4'-hydroxy-7-methoxyflavone, clovers;

Table 2. Examples of Signal Molecules that Stimulate Nodulation Gene Expression in Symbionts of Leguminous Plants.

Host						
Common Name	Genus	Species	Bacterium	Phenolic	Type	Reference
Alfalfa	*Medicago*	*sativum*	*R. meliloti*	luteolin	flavone	44
White Clover	*Trifolium*	*repens*	*R. trifolia*	7,4'-dihydroxyflavone	flavone	45
				4',7-dihydroxy-3'-methoxy-flavone	flavone	45
Pea	*Pisum*	*sativum*	*R. leguminosarum*	eriodictyol	flavone	52
				apigenin-7-O-glucoside	flavone	52
Soybean	*Glycine*	*max*	*B. japonica*	daidzein	iso-flavone	55

luteolin, alfalfa; apigenin-7-O-glycoside, pea (Table 2)]. The stimulatory hydroxyflavones are released from the zone of emerging root hairs and the bacterial nod genes are expressed within minutes.[47] Breakdown products of flavonoid molecules (i.e., hydroxybenzoic acids) have no stimulatory effect, and other than apigenin-7-O-glycoside, most glycosylated flavonoids have no stimulatory effect on nodulation.[47]

An exciting practical extension of this signal molecule research has come from Phillips' laboratory.[48] Hairy Peruvian 32 alfalfa was selected in two generations for increased N_2 fixation and growth. These increases in the selected line were the result of greater numbers of root nodules and greater N_2 fixation than the parental Hairy Peruvian population.[49] Addition of 10 μM luteolin to the rhizosphere of parental Hairy Peruvian seedlings increased nodulation, N_2 fixation, total N and total dry weight to levels of the selected lines. Perhaps the selection process was selected for enhanced production of the signal molecule, luteolin, in Hairy Peruvian 32.

Pankhurst and Jones[49-51] published a series of three papers in 1979 describing flavolins (condensed tannins) in clover nodules. One of the most interesting aspects of their work was that roots treated with NH_4NO_3 were devoid of flavolins and nodulation was inhibited, compared to normal flavolin concentrations and normal nodulation in inoculated roots lacking NH_4NO_3.[51] With our current state of knowledge, one could propose that NH_4NO_3 thereby eliminated flavonoid-flavone biosynthesis and eliminated the signal molecule necessary for expression of nod ABC. Further, the biosynthesis of flavonoid-flavones involves the utilization of PEP (Fig. 2) and it was pointed out earlier how important PEP carboxylation is to NO_3^- assimilation, in terms of ionic balance. The PEP is consumed as NO_3^- reduction stimulates OAA and malate synthesis or as the OAA is formed for amination reactions with the NH_4^+ produced from NO_3^- reduction. This utilization of PEP may shut down phenolic biosynthesis and could be a variation on the carbohydrate deprivation hypothesis proposed as one mechanism for NO_3^- inhibition of nodulation and N_2 fixation. This variation would lead to a much more subtle control of nodulation than total carbohydrate depletion, since the synthesis of a signal molecule is

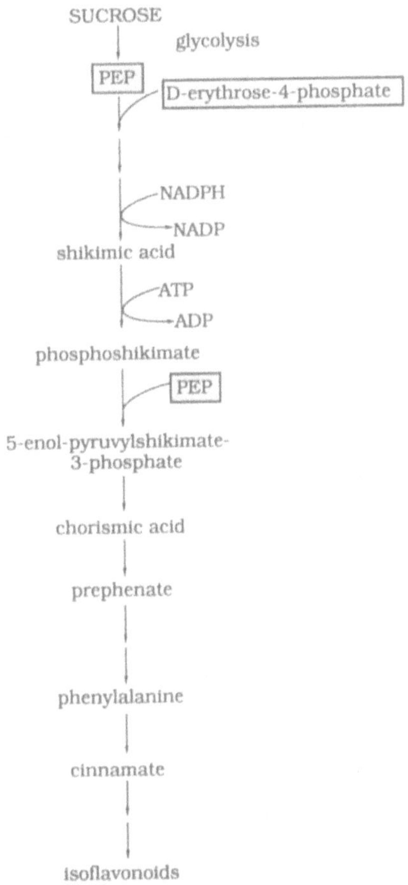

Fig. 2. The important role of phosphoenolpyruvate in the synthesis of phenolics.

involved. Note in Figure 2 that two PEP molecules are required in the pathway of phenolic biosynthesis.

Inhibitory Molecules

Although certain plant-derived signal molecules activate *nod* genes, other plant-derived molecules antagonize their induction. Certain flavonols and isoflavonoids, as well as other phenolics (e.g., acetosyringone), inhibit transcription of nod genes.[52,54] Inhibitors of nod gene expression derived from clover roots were coumarin, umbelliferone and the isoflavonoid, formononetin. The isoflavonoids, among the most potent antagonists, are characteristic of leguminous plants and are rare in other plants.[55]

In the future, seed inoculant containing phenolics as signal molecules could be used to stimulate infection events, nodule numbers and nitrogen fixation. Certain signal molecules could be used to allow an advantage to genetically engineered or selected "super" rhizobia in the nodulation of leguminous crop plants.

Nitrogenase

N_2 is a very stable molecule and requires an iron catalyst, 300 atm pressure and 300°C for industrial conversion to NH_3.[56] Nitrogenase performs this reaction at room temperature and at atmospheric pressure. Nitrogenase is composed of two iron-sulfur proteins, dinitrogenase reductase and dinitrogenase. Since both proteins are irreversibly inactivated by O_2, nodules on leguminous plants apparently protect the proteins with a variable O_2 diffusion barrier[57] and by use of leghemoglobin.[58]

Dinitrogenase is a tetramer of two identical subunits of ca 50 kDa each plus two identical subunits of ca 60 kDa each, with a combined molecular weight of 218 to 245 kDa.[59] This polypeptide contains approximately 30 iron atoms, approximately 32 acid-labile sulfides and two molybdenum atoms, possibly arranged in four Fe_4S_4 centers, two Fe_8S_6Mo cofactor centers, plus two "S" centers, each containing two iron atoms.[59] The active substrate reducing sites of nitrogenase are believed to be the Fe_8S_6Mo centers called FeMoco.[60] Recently, Shah *et al.*[61] reported that homocitrate (nif V product) was required for synthesis of active FeMoco.

The energy for fixation of N_2 by dinitrogenase is provided through the second component of nitrogenase, dinitrogenase reductase. This protein is composed of two identical subunits and has a molecular weight near 60 kDa. Dinitrogenase reductase has a single Fe_4S_4 center for

Table 3. Characteristics of Soybean Root Nodules Containing Hup^+ or Hup^- Bradyrhizobium.

Hup^+	Hup^-	References
Low nodule nitrate reductase	High nodule nitrate reductase	67
- serinol	+ serinol	66
- rhizobitoxine	+ rhizobitoxine	66
Greater CO_2 fixation	Less CO_2 fixation	68
Greater aspartate formation	Less aspartate formation	70

transfer of a single electron[62] and two binding sites for MgATP.[63]

Nitrogenase catalyzes the reduction of several substrates including H^+, N_2, C_2H_2, H_2O, CN^-, N_3^-, isonitriles and cyclopropene.[59] When functioning normally, nitrogenase catalyzes the fixation of one mole of N_2; simultaneously one mole of H_2 is evolved and this is apparently obligatory in the fixation of N_2.[56] Some Rhizobium strains possess a second enzyme, hydrogenase, which recycles H_2 evolved by nitrogenase so that part of the energy in H_2 is recovered as reducing equivalents or ATP.[64] Strains that possess the uptake hydrogenase (Hup^+) and recycle or recover H_2 in root nodules may increase the nitrogen concentration and yield of soybeans[64,65] when compared with soybeans inoculated with Hup^- strains.

The search for root nodule bacteria possessing Hup has led to further analysis and comparison of these strains with Hup^- strains. At least five other characteristics appear different: Hup^+ strains have no detectable serinol,[66] no rhizobitoxine,[66] very low nitrate reductase activities,[67] more rapid CO_2 fixation[68] and greater aspartate synthesis[69] as compared to Hup^- strains (Table 3). Earlier work indicated that at least some rhizobitoxine-producing strains of *B. japonicum* can infect non-nodulated (rj_1) soybeans.[70] Finding the significance, if any, of these different characteristics of Hup^+ and Hup^- strains will require further research.

Energy and C-Skeletons

The actual carbon sources involved in the supply of energy and C-skeletons necessary to support nitrogen fixation remain a mystery. Several reports indicate that the tricarboxylic acid (TCA) cycle acids, malate and succinate, are the most likely molecules supplied to the bacteroid.[71-73] More recently Kahn *et al.*[74] has proposed that amino acids in conjunction with a malate-aspartate shuttle (Fig. 3) could provide energy or reducing equivalents for bacteroid nitrogen fixation. Labelling and enzymological experiments[75-77] also seem to support the idea that amino acids may be involved. Additionally, Salminen and Streeter[78] have evidence that amino acids can be utilized by bacteroids and state that perhaps both organic acids and amino acids are used. Kohl *et al.*[79] have localized enzymes of proline metabolism in nodule cytosol and bacteroids, and support proline as one of the molecules shuttling reducing equivalents to bacteroids. Enzymes of the pentose phosphate pathway were relatively more active than enzymes of glycolysis in cytosolic extracts of nodules;[77] consequently these enzymes were emphasized in the carbon flow scheme presented in Figure 3.

Labeling studies, those previously mentioned and others,[80-84] where glutamate, malate, succinate and alanine were rapidly labelled with ^{14}C, ^{13}C or ^{15}N, indicate that these molecules are actively involved in nodule metabolism. However, a shuttle mechanism might preclude rapid labeling of C or N in the molecules actually involved in the shuttle.[68]

Anaerobic metabolism occurs in the plant cytosol of root nodules.[85-87] Suganuma *et al.*[88,89] have shown that ethanol production is more active in uninfected cells and cortical cells than in infected cells, and that mitochondria in the infected cells of soybean nodules are capable of active respiration in the presence of leghemoglobin. The TCA cycle in mitochondria of the infected

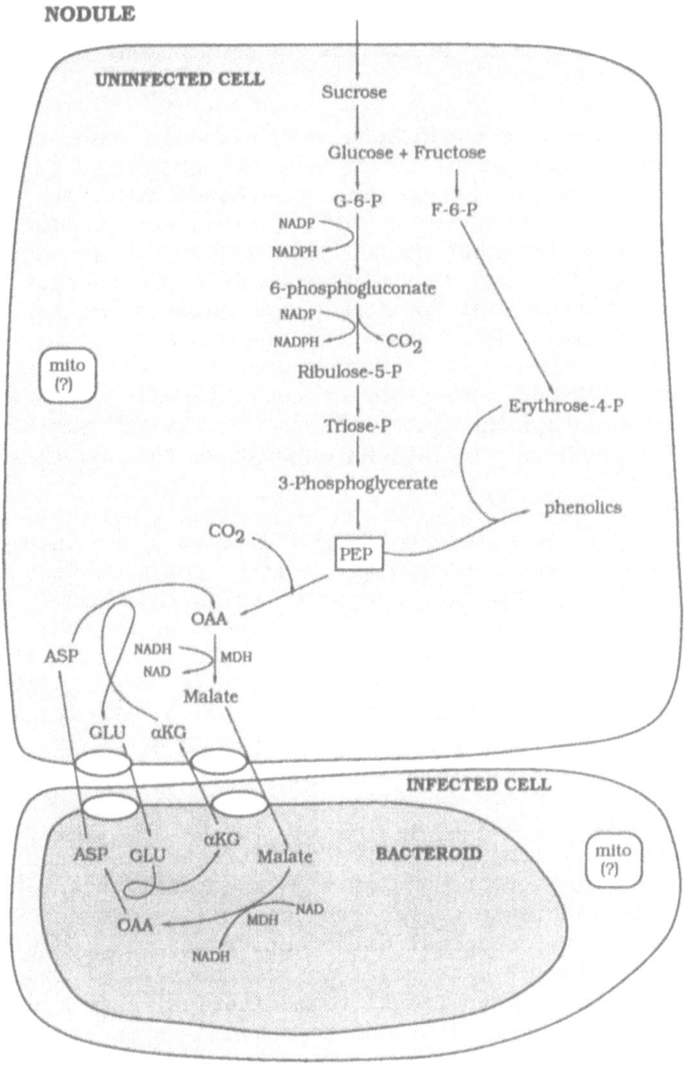

Fig. 3. A scheme for root nodule carbon metabolism supporting nitrogen fixation.

cells of soybean nodules is capable of supplying energy and carbon skeletons for assimilation of NH_4^+, metabolism of purines and initial steps in ureide synthesis. The central question may be whether or not these mitochondria are fully functional in the low O_2 environment of the plant cells in nodules.

The anaerobic conditions in the nodule, caused by the O_2 diffusion barrier,[57] affect metabolism as verified by the availability of enzymes of ethanol and aldehyde fermentation in bacteroids.[77,86,89] However, Tajima and LaRue[86] suggested that carbon flow through these products is minimal. The low O_2 status in nodule plant cells may inhibit mitochondrial function, especially oxidative phosphorylation. This shut-down of mitochondria may be a key in the symbiosis, since several studies show the importance of PEP carboxylation and malate in supporting nitrogen fixation in bacteroids.[90,91] The PEP may be shunted toward carboxylation rather than through the TCA cycle.

It is interesting that higher plants under anaerobic conditions accumulate alanine, γ-amino butyric acid, serine and glycine, the same amino acids that are rapidly labelled in ^{14}C, ^{13}C or ^{15}N studies in root nodules![92] Malate and succinate are also products of higher plant anaerobiosis[92] and succinate effectively supports nitrogen fixation by bacteroids.[71,72] The ability of a bacterium to utilize organic acids is related to symbiotic effectiveness,[93] and mutant strains of rhizobia that are defective in the uptake and metabolism of succinate and malate (J. Waters and D. Emerich, personal communication) are ineffective in nitrogen fixation (Fix^-) in legume root nodules.[94-99] Gardiol et al.[99] found that succinate dehydrogenase activity was necessary in *R. meliloti* for normal differentiation, maintenance and functioning of bacteroids in root nodules of alfalfa. Waters and Emerich (personal communication) have found that bacteroids devoid of malate dehydrogenase are Fix^-, while bacteroids with a doubled amount of malate dehydrogenase have doubled nitrogenase activity compared to controls. Thus there is a large body of evidence that suggests that succinate and malate play a central role in the bacteroid metabolism supporting nitrogen fixation. Phosphoenolpyruvate carboxylation is also an integral part of this carbon metabolism. Recent studies have shown that NH_4^+ from nitrogen fixation was assimilated primarily through the GS/GOGAT pathway in soybean nodules[80,82,100] and that asparagine was derived from glutamine via aspartate aminotransferase.[100] What is the C-source for the aspartate skeleton? Since the glyoxylate cycle, which could contribute TCA intermediates, does not operate in nodules,[101] the oxalacetate for aspartate formation must be formed by PEP carboxylation.[90,102]

A recent study with an inhibitor of glutamine synthetase that stimulated nitrogen fixation may relate back to PEP carboxylation and malate synthesis. Pseudomonas is a common soil bacterium; some strains live on the root surfaces of many plants[103-105] and release tabtoxine-β-lactam (2-amino-4-[3-hydroxy-2-oxoazacyclobutan-3-yl]-butanoic acid). This toxin can be taken up by the roots[104] and is an irreversible inhibitor of glutamine synthetase (GS) in planta.[104,105] This inhibition of GS in roots and subsequent NH_4^+ accumulation leads to plant death.[104,105] A strain of Pseudomonas syringae pv. tabaci which releases tabtoximine-β-lactam caused an approximate doubling of alfalfa growth, total plant nitrogen, nodulation and nitrogen fixation. The toxin inactivated the root-specific form of GS in the roots and nodules, but had little effect on the nodule-specific or bacteroid GS. Nodule glutamine levels were lowered, while NH_4^+ and glycine concentrations were elevated by treatment with the toxin.[106] These results imply that one form of GS, or the concentration of a metabolite influenced by the functioning of the GS, is important in the regulation of nodulation and nitrogen fixation. Peterson and Evans[107] found that NH_4^+ inhibited pyruvate kinase in soybean nodule cytosol. Therefore in the infected cells of root nodules (Fig. 3), NH_4^+ accumulated as a result of GS inhibition would inhibit pyruvate kinase and more PEP would be available for carboxylation to OAA. The important enzyme, malate dehydrogenase, would convert OAA to malate for transport to the bacteroid. Thus, dark CO_2 fixation would be higher than ethanol fermentation when NH_4^+ is accumulated following nitrogen fixation. Perhaps, NH_4^+ accumulation and its inhibition of pyruvate kinase in the tabtoxine experiments are the key to unleashing nitrogen fixation and stimulating growth of the host plant.

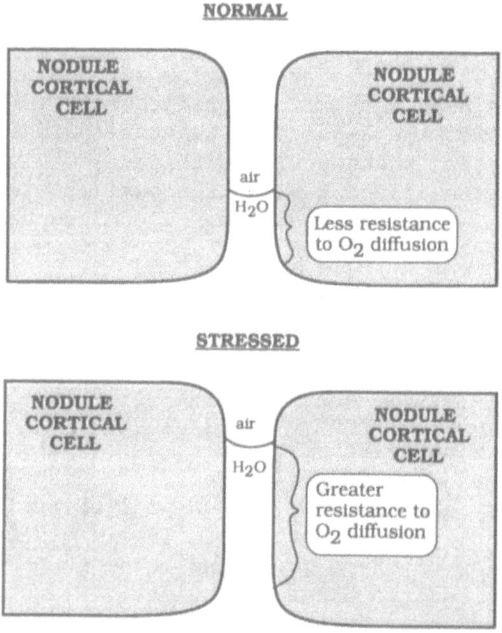

Fig. 4. Diagram of root nodule cells with water between cells serving as the variable O_2 barrier.

O_2 Diffusion to Bacteroids

Root nodule bacteria require O_2 for utilization of carbon supplies in the production of energy to support nitrogen fixation. This dependence on O_2 poses a dilemma since the nitrogenase complex within the bacteroid is very sensitive to O_2 and breaks down in its presence. The outer cortex of the root contains a barrier to O_2 diffusion that regulates the O_2 supply to the interior of the nodules.[57,108,109] The idea currently receiving much support in this area is based on the fact that O_2 diffusion in air is 10^4 times higher than in H_2O. In the outer cortex of a root nodule with a very active rate of nitrogen fixation, air fills the spaces between cells. A change in the water status of the host plant (drought, defoliation, cutting, transpiration shutdown)[110 111] may cause H_2O to fill the air space between cells[112] (Fig. 4). The finding that water columns between cells effectively limits O_2 diffusion into the nodule, and thereby energy production and subsequent nitrogen fixation by the bacteroids, could explain a major component in the regulation of nitrogen fixation.[110-113] Once O_2 gets to the infected cells, leghemoglobin facilitates the supply of O_2 to bacteroids.[58] This situation relates to the previous section, since bacteroids are dependent on O_2 to metabolize organic or amino acids as energy sources.

Ammonium Assimilation in Root Nodules

Although a shuttle system like the aspartate-malate shuttle may operate between bacteroids and the plant cytosol, many reports suggest that enzymes of NH_4^+ assimilation are repressed in bacteroids.[114-116] Therefore, most fixed nitrogen may be excreted from the bacteroid as NH_4^+ and assimilated in the plant cytosol.[115,116] Results from labelling, inhibitor and enzymological studies in the nodule cytosol are consistent with NH_4^+ assimilation via GS/GOGAT. Recent studies with $[^{13}N]N_2$ show that NH_4^+ is first incorporated into the amide position of glutamine by GS.[117] GOGAT utilizes glutamine and α-ketoglutarate to form two molecules of glutamate. Both GS and GOGAT have been isolated from legume nodules.[118-121] The GS was cytoplasmic, while GOGAT was found in both the cytoplasm and in plastids.[122-125]

GDH was originally suggested as the first enzyme for incorporation of fixed nitrogen into amino acid form.[126] Later studies showed that the Km for NH_4^+ was rather high and it was suggested that perhaps GDH was involved only in very young or very old nodules, but that GS was relatively more important throughout the life of the nodules.[114] Schmidt's laboratory[127] recently found multiple forms of GDH in *Chlorella* and one form (the α isoenzyme) had high affinity for NH_4^+ (0.02 to 3.5 mM). It remains to be seen whether or not these forms of GDH also exist in root nodules.

ASPARAGINE SYNTHESIS

It is interesting that much of the research on asparagine synthesis has been done with soybean nodules. Prior to the mid-70's, asparagine was thought to be the major transport form of nitrogen, even in xylem sap from soybeans.[128] Asparagine constitutes only 10% of the total nitrogen in the xylem sap of soybeans fixing nitrogen, while up to 90% of the total nitrogen is ureide-N.[129,130] Nevertheless, soybean nodule research with $^{14}CO_2$, $^{15}NH_4$ and [amide-^{15}N]glutamine is consistent with asparagine synthesis by asparagine synthetase (AS).[128,131,132] The $^{14}CO_2$ is incorporated into aspartate and subsequently asparagine via PEP carboxylase.[133,134] Asparagine synthetase has been detected and purified from soybean nodule extracts.[135]

Asparagine synthetase requires a source of aspartate, which may be provided by aspartate aminotransferase (AAT), an interesting nodule enzyme which is receiving much attention currently.[136] The AAT in lupine and soybean nodules exists in two forms, one of which is specifically induced in response to nitrogen fixation and localized within the plastid in nodule cytosol.[136-138]

UREIDE METABOLISM

In leguminous plants that transport ureides (allantoin and allantoic acid), extensive metabolism takes place following transformation of the fixed nitrogen into glutamine.[128] The synthesis of purines de novo in plastids of infected cells uses nitrogen from glutamine, glycine and aspartate to provide the four nitrogen atoms in the purine or ureide skeleton (Fig. 5). The five purine carbons come from glycine (2C), CO_2, and two C_1-THF derivatives.[128,139,140] The product of purine synthesis, inosinic acid (IMP), is oxidized by IMP dehydrogenase and xanthine dehydrogenase (XDH) to xanthosine monophosphate (XMP) and uric acid. The IMP or uric acid must migrate to uninfected cells, where uric acid is oxidized to allantoin by urease oxidase.[140] Most of the allantoin is converted to allantoic acid, the dominant transport ureide, by allantoinase which is associated with the endoplasmic reticulum in uninfected cells.[141-143]

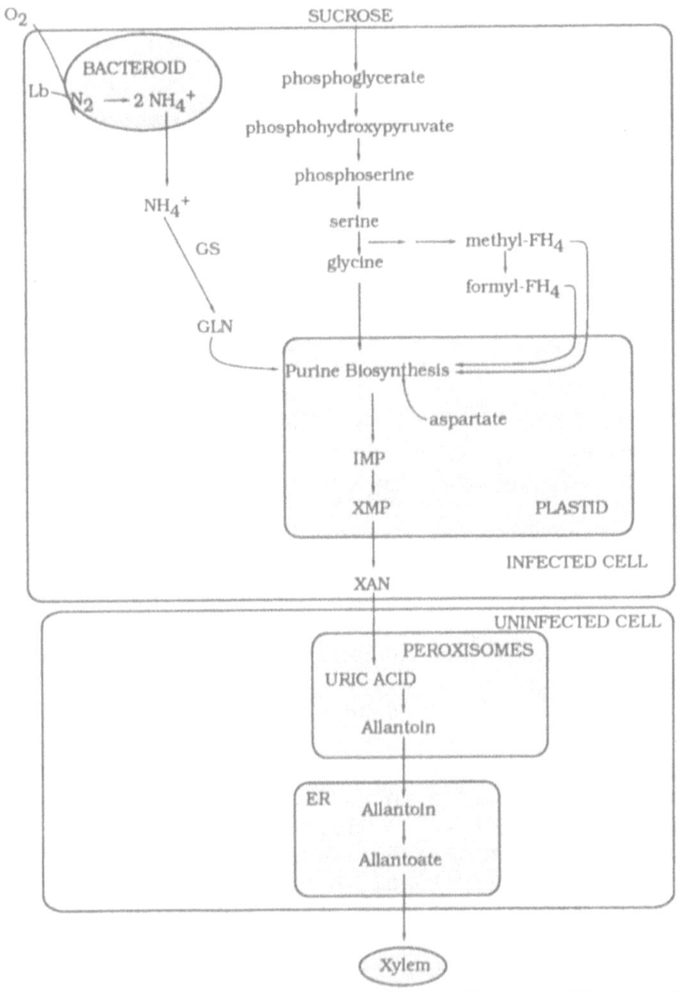

Fig. 5. Localization of the pathway for ureide synthesis in root nodules.

There has been some debate on the location of XDH, but apparently it is in the uninfected cells.[144] That urate oxidase is found in uninfected rather than infected cells of root nodules of ureide-transporting leguminous plants may be a consequence of the relatively higher levels of free O_2 in the uninfected cells (200 μM). Free

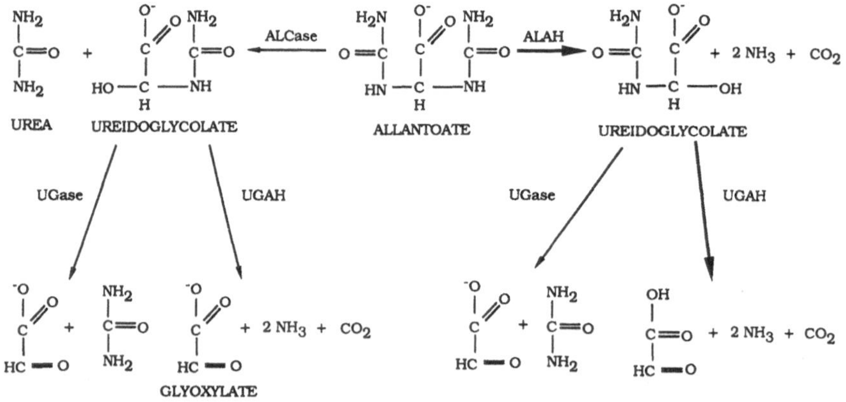

Fig. 6. Possible pathways for ureide catabolism in soybean leaves. The pathway depicted in bold lines is consistent with the most recent results.[150,151]

O_2 levels in infected cells are extremely low (10-200 nM) and the Km (O_2) for urate oxidase from soybean and cowpea nodules is around 30 μM.[145] Ureide synthesis in cell-free extracts of cowpea nodules was limited by O_2 supply[146] and this may also be the case in vivo.[140]

In soybeans, allantoinase is associated with the endoplasmic reticulum[147] in nodules and is membrane bound in leaf extracts.[148] Allantoinase is constitutive and found in nodules, leaves, developing fruit and mature seeds at levels 80 times those of allantoate degradation activities.[149 150] The nodule activity seems the most critical to ureide metabolism, since most analyses of xylem sap have found allantoate to be the predominant ureide.[128]

There are at least four possible combinations of enzymes that could be involved in allantoate catabolism[150] (Fig. 6). Most early reports indicated the involvement of allantoicase (allantoate amidinohydrolase). This may have been because allantoicase was the only allantoate degrading enzyme known in the early studies. Recent studies with intact leaf tissue from nodulated soybeans have indicated that allantoin is degraded to $4NH_4^+$, $2CO_2$ and glyoxylate without urea as an intermediate.[150,151]

$^{14}CO_2$ release from [2,7-^{14}C]allantoin was not affected by the irreversible urease inhibitor, phenylphosphordiamidate, at concentrations eliminating all urease activity. Neither ^{14}C-urea nor other potential intermediates could be detected in extracts of the leaf discs. $^{14}CO_2$, [^{14}C]glyoxylate, [^{14}C]glycine and [^{14}C]serine were released when leaf discs were incubated with [4,5-^{14}C] allantoin. These results indicated that allantoate amidohydrolase and a second amidohydrolase (ureidoglycolate amidohydrolase) were used, with subsequent metabolism of glyoxylate through the photorespiratory pathway. Cell free extracts from seed coats of developing soybeans degraded allantoate by a Mn^{2+}-dependent activity to CO_2 and glyoxylate without urea as an intermediate. That glyoxylate was released without urea indicated the two amidohydrolase reactions act to degrade allantoate to $4NH_4^+$, $2CO_3$ and glyoxylate. [Ureido-^{14}C]ureidoglycolate was degraded to $^{14}CO_2$; equimolar amounts of CO_2 and glyoxylate were released, while urease was inhibited. This is consistent with ureidoglycolate amidohydrolase involvement. The rate of production of $^{14}CO_2$ from [2,7-^{14}C]allantoate suggested that $^{14}CO_2$ production was not dependent upon the accumulation of free intermediates. $^{14}CO_2$ production from the same substrate was not proportionally diluted by unlabelled ureidoglycolate. Together these results are consistent with the idea that allantoate amidohydrolase and ureidoglycolate amidohydrolase are an enzyme complex.[151]

There is now very strong evidence that allantoate catabolism involves the enzymes, allantoate amidohydrolase and ureidoglycolate amidohydrolase, rather than enzymes which release urea as previously assumed (see bold arrows in Figure 6).

STORAGE PROTEINS

The major protein in leaves is Rubisco (ribulose bisphosphate carboxylase-oxygenase). Rubisco is so abundant in leaves that it could be thought of as a storage form of nitrogen especially for annual seed crops.[152] Millard lists three advantages of storing nitrogen as protein rather than as NO_3^-: 1) it maximizes the potential for carbon assimilation; 2) it avoids problems with leaf turgor; and 3) it allows the reduction of NO_3^-, an alternative storage form of nitrogen, to occur in the young, fully illuminated leaves.[152] In soybeans, Wittenbach[153,154] found that the loss of Rubisco following depodding was accompanied by the appearance of three major polypeptides in leaf tissue, two of which (27 and 29 KD) were glycosylated. Franceschi et al.[155] determined that these glycoproteins were exclusively localized in the vacuoles of paraveinal mesophyll cells. These specialized cells were shown to serve a unique role in the mobilization of leaf nitrogen reserves prior to and during pod fill.

Seeds contain many proteins including a few abundant ones called storage proteins. These proteins provide important nitrogen for amino acid synthesis during germination. Many seed storage proteins are rich in arginine.[156] Since arginine contains four nitrogen atoms, it is not surprising that as the nitrogen content of maize or lupine seeds increases, the percent arginine in seed protein dramatically increases.[157] This points out how versatile plants are in their ability to store nitrogen. When crop plants are low in sulfur, they increase the synthesis and storage of proteins that are low in the sulfur-amino acid, methionine.[158] When soybeans are fed methionine in cotyledon culture[152] or through stem infusions,[159,160] they increase storage of high methionine proteins. Several other plant nutrients and environmental conditions also affect the composition of seed storage proteins.[161,162]

During leaf senescence in the autumn, much of the leaf nitrogen in fruit trees is translocated to cells under the bark for storage during the winter.[164-167] This stored nitrogen is mobilized in the spring for the development of flower buds and leaves.[163,168,169] In apple trees there was a 30-fold increase in the total nitrogen content of the sap during the flowering period in early spring; this was followed by an abrupt decline in nitrogen as extension growth proceeded. Arginine was the predominent transport molecule in autumn and winter, while asparagine, aspartate and glutamine were major transport forms in summer.[166] Arginine was also the major amino acid found in the bark of dormant apple trees.[163]

The hydrolysis of arginine to ornithine and urea by arginase is the first step in the recovery of nitrogen and carbon from arginine. Most seeds contain 10-40% of the nitrogen as arginine.[156] When seeds germinate, storage proteins are hydrolyzed and amino acids or amides are translocated to growing points. During seed germination, arginase increases four- to ten-fold.[156] Arginase has a high pH optimum (pH - 9.5) and a requirement for Mn^{++}, similar to the pH requirement (8.75) and manganese requirement for allantoate amidohydrolase.[150,156] These properties are especially interesting when one considers the similar chemistry involved in the degradation of the ureido groups in arginine and allantoate.

Most seeds contain abundant, embryo-specific urease, which would degrade urea produced from arginine during germination. Another urease was identified in higher plants[170] and has been called the ubiquitous urease to contrast it with the abundant embryo-specific urease.[171] The ubiquitous urease has a low specific activity, but could be involved in turnover of leaf arginine. Despite the preceding statements, the role of urease remains unknown. Polacco and coworkers[172,173] have recovered a series of soybean mutants lacking urease that will help in defining the role of this enzyme in plant metabolism. These mutants lack one or both urease isozymes, i.e., the embryo-specific and the ubiquitous urease.

POLYAMINES

The polyamines putrescine, spermidine and spermine are apparently ubiquitous in living cells and have been implicated in a variety of regulatory processes, e.g., promotion of growth and cell division, pollination, sprouting of dormant buds, embryogenesis, phytochrome-induced morphogenesis, hormone-induced changes, normal and stress-induced senescence, ethylene production and free radical scavenging.[174,175] At physiological pH, polyamines are fully protonated and polycationic.[174] Physiologically they may affect various processes by associating with negative charges on nucleic acids and phospholipids. These associations stabilize chromosome and membrane functioning. A recent study[175] has shown that superoxide radical formation from xanthine oxidase, riboflavin or senescing microsomal membranes and superoxide-dependent conversion of 1-amino-cyclopropane-1-carboxylic acid (ACC) to ethylene were all inhibited by polyamines. Perhaps some of the physiological effects of polyamines, i.e., inhibition of lipid peroxidation on membranes and retardation of senescence, can be attributed to their radical-scavenging capabilities. In addition, putrescine is well-known for its accumulation in K^+-deficient plants. The correlation of putrescine accumulation with K^+ deficiency is so high, that putrescine concentration in leaf tissue has been proposed as an assay of the K^+ status of the tissue.[176]

REFERENCES

1. MARSCHNER, H. 1986. Mineral Nutrition in Higher Plants. Academic Press, New York, 674 pp.
2. SCHENK, M., J. WEHRMAN. 1979. The influence of ammonia in nutrient solution on growth and metabolism of cucumber plants. Plant Soil 52: 403-414.
3. BHAT, K.K.S. 1983. Nutrient in flows into apple roots. Plant Soil 71: 371-380.
4. KLEMM, K. 1966. Dereinfluss der N-form auf die ertragsbildung verschiedener kulturpflanzen. Bodenkultur 17: 265-284.
5. BUBAN, T., A. VARGA, J. TROMP, E. KNEGT, J. BRUINSMA. 1978. Effects of ammonium and nitrate nutrition on the level of zeatin and amino nitrogen in xylem sap of apple rootstocks. Z. Pflanzenphysiol. 89: 289-295.
6. MIFLIN, B.J., P.J. LEA. 1980. Ammonia assimilation. In The Biochemistry of Plants. (B.J. Miflin, ed.), Academic Press, New York, Vol. 5, pp. 169-202.
7. KEYS, A.J., I.F. BIRD, M.J. CORNELIUS, P.J. LEA, R.M. WALLSGROVE, B.J. MIFLIN. 1978. Photo-respiratory nitrogen cycle. Nature 275: 741-743.
8. SIECIECHOWICZ, K.A., K.W. JOY, R.J. IRELAND. 1988. The metabolism of asparagine in plants. Phyto-chemistry 27: 663-671.
9. SOMERVILLE, C.R., W.L. OGREN. 1980. Inhibition of photosynthesis in Arabidopsis mutants lacking leaf glutamate synthase activity. Nature 286: 257-259.
10. RAVEN, J.A., F.A. SMITH. 1976. Nitrogen assimila-tion and transport in vascular land plants in

relation to intracellular pH regulation. New Phytol. 76: 415-431.

11. DAVIES, D.D. 1973. Metabolic control in higher plants. In Biosynthesis and Its Control in Plants. (B.V. Millborrow, ed.), Academic Press, London, pp. 1-20.
12. BEN-ZIONI, A., Y. VAADIA, S.H. LIPS. 1970. Correlations between nitrate reduction, protein synthesis and malate accumulation. Physiol. Plant. 23: 1039-1047.
13. BLEVINS, D.G., N.M. BARNETT, W.B. FROST. 1978. Role of potassium and malate in nitrate uptake and translocation in wheat seedlings. Plant Physiol. 62: 784-788.
14. MULDER, E.G. 1948. Investigations on the nitrogen nutrition of pea plants. Plant Soil 1: 179-212.
15. KIRKBY, E.A., K. MENGEL. 1967. Ionic balance in different tissues of the tomato plant in relation to nitrate, urea, or ammonium nutrition. Plant Physiol. 42: 6-14.
16. EVANS, H.J., A. NASON. 1953. Pyridine nucleotide nitrate reductase from extracts of higher plants. Plant Physiol. 28: 233-254.
17. BEEVERS, L., D. FLESHER, R.H. HAGEMAN. 1964. Studies on the pyridine nucleotide specificity of nitrate reductase in higher plants and its relationship to sulfhydryl level. Biochim. Biophys. Acta 89: 453-464.
18. OAKS, A., B. HIREL. 1985. Nitrogen metabolism in roots. Annu. Rev. Plant Physiol. 36: 345-365.
19. GUERRERO, M.G., J.M. VEGA, M. LOSADO. 1981. The assimilatory nitrate-reducing system and its relation. Annu. Rev. Plant Physiol. 32: 169-204.
20. CAMPBELL, W.H., J. SMARRELLI, JR. 1978. Purification and kinetics of higher plant NADH-nitrate reductase. Plant Physiol. 61: 611-616.
21. CAMPBELL, W.H. 1985. The biochemistry of higher plant nitrate reductase. In Nitrogen Fixation and CO_2 Metabolism. (P.W. Ludden, J.E. Burris, eds.), Elsevier, Amsterdam, pp. 143-151.
22. KUO, M.G., D.A. SOMERS, A. KLEINHOFS, R.L. WARNER. 1982. NADH-nitrate reductase in barley leaves. Identification and amino acid composition of subunit protein. Biochim. Biophys. Acta 708: 75-81.

23. NAKAGAWA, H., Y. YONEMURA, H. YAMAMOTO, T. SATO, N. OGURA, R. SATO. 1985. Spinach nitrate reductase. Purification, molecular weight and subunit composition. Plant Physiol. 77: 124-128.
24. CAMPBELL, W.H., J.L. REMMLER. 1986. Regulation of corn leaf nitrate reductase. 1. Immunochemical methods for analysis of the enzymes protein component. Plant Physiol. 80: 435-441.
25. HARPER, J.E., R.S. NELSON, L. STREIT. 1985. Nitrate metabolism of soybean - physiology and genetics. In Proceedings, World Soybean Research Conference III. (R. Shibles, ed.), Westview Press, Boulder, Colorado, pp. 476-483.
26. HARKER, A.R., K.R. NARAYANAN, R.L. WARNER, A. KLEINHOFS. 1986. NAD(P)H bispecific nitrate reductase in barley leaves: partial purification and characterization. Phytochemistry 25: 1275-1279.
27. CHENIAE, G., H.J. EVANS. 1959. Properties of a particular nitrate reductase from the nodules of the soybean plant. Biochim. Biophys. Acta 35: 140-153.
28. SMITH, G.B., M.S. SMITH. 1986. Symbiotic and free-living denitrification by Bradyrhizobium japonicum. Soil Sci. Soc. Am. J. 50: 349-354.
29. HAGEMAN, R.H., D. FLESHER. 1960. Nitrate reductase activity in corn seedlings as affected by light and nitrate content of nutrient media. Plant Physiol. 35: 700-708.
30. SHANER, D.L., J.S. BOYER. 1976. Nitrate reductase activity in maize (Zea mays L.) leaves. I. Regulation by nitrate flux. Plant Physiol. 58: 499-504.
31. SOMERS, D.A., T.-M. KUO, A. KLEINHOFS, R.L. WARNER, A. OAKS. 1983. Synthesis and degradation of barley nitrate reductase. Plant Physiol. 72: 949-952.
32. REMMLER, J.L., W.H. CAMPBELL. 1986. Regulation of corn leaf nitrate reductase. II. Synthesis and turnover of the enzyme's activity and protein. Plant Physiol. 80: 442-447.
33. DALLING, M.J., N.E. TOLBERT, R.H. HAGEMAN. 1972. Intracellular location of nitrate reductase and nitrite reductase in spinach and tobacco leaves. Biochim. Biophys. Acta 283: 505-512.

34. GRANT, B.R., C.A. ATKINS, D.T. CANVIN. 1970. Intracellular location of nitrate reductase and nitrite reductase in spinach and sunflower leaves. Planta 94: 60-72.
35. VAUGHN, K.C., S.O. DUKE, E.A. FUNKHOUSER. 1984. Immunochemical characterization of nitrate reductase in nonflurazon-treated soybean cotyledons. Physiol. Plant 62: 481-484.
36. KAMACHI, K., Y. AMEMIYA, N. OGURA, H. NAKAGAWA. 1987. Immunogold localization of nitrate reductase in spinach (Spinacia oleracea) leaves. Plant Cell Physiol. 28: 333-338.
37. BEEVERS, L., R.H. HAGEMAN. 1980. Nitrate and nitrite reduction. Op. cit. Reference 6, pp. 115-168.
38. NINOMIYA, Y., S. SATO. 1984. A ferredoxin-like electron carrier from non-green cultured tobacco cells. Plant Cell Physiol. 25: 453-458.
39. JACOBS, T.W., T.T. EGELHOFF, S.R. LONG. 1985. Physical and genetic map of a Rhizobium meliloti nodulation gene region and nucleotide sequence of nod C. J. Bacteriol. 162: 469-476.
40. ROSSEN, L., A.W.B. JOHNSTON, J.A. DOWNIE. 1984. DNA sequence of the Rhizobium leguminosarum nodulation genes nod A, B and C required for root hair curling. Nucleic Acids Res. 12: 9497-9508.
41. INNES, R.W., P.L. KUEMPEL, J. PLAZINSKI, H. CANTER-CREMERS, B.G. ROLFE, M.A. DJORDJEVIC. 1985. Plant factors induce expression of nodulation and host-range genes in Rhizobium trifolii. Mol. Gen. Genet. 201: 426-432.
42. MULLIGAN, J.T., S.R. LONG. 1985. Regulation of nodulation genes in Rhizobium meliloti. Proc. Nat. Acad. Sci. USA 82: 6609-6613.
43. ROSSEN, L., C.A. SHEARMAN, A.W.B. JOHNSTON, J.A. DOWNIE. 1985. The nod D gene of Rhizobium leguminosarum is autoregulatory and in the presence of plant exudate induces the expression of nod ABC genes. EMBO J. 4: 3369-3373.
44. PETERS, N.K., J.W. FROST, S.R. LONG. 1986. A plant flavone, luteolin, induces expression of Rhizobium meliloti nodulation genes. Science 233: 977-980.

45. REDMOND, J.W., M. BATLEY, M.A. DJORDJEVIC, R.W. INNES, P.L. KUEMPEL, B.G. ROLFE. 1986. Flavones induce expression of nodulation genes in Rhizobium. Nature 323: 632-635.
46. DJORDJEVIC, M.A., D.W. GABRIEL, B.G. ROLFE. 1987. Rhizobium - the refined parasite of legumes. Annu. Rev. Phytopathol. 25: 145-168.
47. ROLFE, B.G., P.M. GRESSHOFF. 1988. Genetic analysis of legume nodule initiation. Annu. Rev. Plant Physiol. 39: 297-319.
48. KAPULNIK, Y., C.M. JOSEPH, D.A. PHILLIPS. 1987. Flavone limitations to root nodulation and symbiotic nitrogen fixation in alfalfa. Plant Physiol. 84: 1193-1196.
49. PANKHURST, C.E., A.S. CRAIG, W.T. JONES. 1979. Effectiveness of Lotus root nodules. I. Morphology and flavolan content of nodules formed on Lotus pedunculatus by fast-growing Lotus rhizobia. J. Exp. Bot. 30: 1085-1093.
50. PANKHURST, C.E., W.T. JONES. 1979. Effectiveness of Locus root nodules. II. Relationship between root nodule effectiveness and 'in vitro' sensitivity of fast-growing Lotus rhizobia to flavolans. J. Exp. Bot. 30: 1095-1107.
51. PANKHURST, C.E., W.T. JONES. 1979. Effectiveness of Lotus root nodules. III. Effect of combined nitrogen on nodule effectiveness and flavolan synthesis in plant roots. J. Exp. Bot. 30: 1109-1118.
52. FIRMIN, J.L., K.E. WILSON, L. ROSSEN, A.W.B. JOHNSTON. 1986. Flavonoid activation of nodulation genes in Rhizobium reversed by other compounds present in plants. Nature 324: 90-92.
53. BURN, J., L. ROSSEN, A.W.B. JOHNSTON. 1987. Four classes of mutations in the nod D gene of Rhizobium leguminosarum biovar. viciae that affect its ability to autoregulate and/or activate other nod genes in the presence of flavonoid inducers. Genes and Devel. 1: 456-464.
54. ROSSEN, L., E.O. DAVIS, A.W.B. JOHNSTON. 1987. Plant-induced expression of Rhizobium genes involved in host specificity and early stages of nodulation. Trends Biochem. Sci. 12: 430-433.
55. SADOWSKY, M.J., E.R. OLSON, V.E. FOSTER, R.M. KOSSLAK, D.P.S. VERMA. 1988. Two host-inducible genes of Rhizobium fredii and characterization

of the inducing compounds. J. Bacteriol. 170: 171-178.
56. SIMPSON, F.B. 1987. The hydrogen reactions of nitrogenase. Physiol. Plant. 69: 187-190.
57. WITTY, J.F., F.R. MINCHIN, L. SKØT, J.E. SHEEHY. 1986. Nitrogen fixation and oxygen in legume root nodules. Oxf. Surv. Plant Mol. Cell Biol. 3: 275-314.
58. BERGERSEN, F.J., G.L. TURNER. 1975. Leghaemoglobin and the supply of O_2 to nitrogen-fixing root nodule bacteroids: studies of an experimental system with no gas phase. J. Gen. Microbiol. 89: 31-47.
59. POSTGATE, J.R. 1982. The Fundamentals of Nitrogen Fixation. Cambridge University Press, Cambridge, 252 pp.
60. SHAK, V.K. 1980. Iron-molybdenum cofactor of nitrogenase. In Nitrogen Fixation. (W.E. Newton, W.H. Orme-Johnson, eds.), University Park Press, Baltimore, Maryland, Vol. 1, pp. 237-247.
61. SHAH, V.K., T.R. HOOVER, J. IMPERIAL, T.D. PAUSTIAN, G.P. ROBERTS, P.W. LUDDEN. 1988. Role of NIF gene products and homocitrate in the biosynthesis of iron-molybdenum cofactor. Abstr. 7th Intern'l. Congress on Nitrogen Fixation, Cologne, p. L-34.
62. LJONES, T., R.H. BURRIS. 1978. Evidence for one-electron transfer by the Fe protein of nitrogenase. Biochem. Biophys. Res. Commun. 80: 22-25.
63. TSO, M.-Y.W., R.H. BURRIS. 1973. The binding of ATP and ADP by nitrogenase components from Clostridium pasteurianum. Biochem. Biophys. Acta 39: 263-270.
64. EVANS, H.J., D.W. EMERICH, T. RUIZ-ARGUESO, S.H. ALBRECHT, R.J. MAIER, F. SIMPSON, S.A. RUSSELL. 1978. Hydrogen metabolism in legume nodules and rhizobia: some recent developments. In Hydrogenase: Their Catalytic Activity, Structure and Function. (H.G. Schlegel, K. Schneider, eds.), Erich Goltie KG, Göttingen, pp. 287-306.
65. LAMBERT, G.R., A.R. HARKER, M. ZUBER, D.A. DALTON, F.J. HANUS, S.A. RUSSELL, H.J. EVANS. 1985. Characterization, significance and transfer of hydrogen uptake genes from Rhizobium japonicum. In Nitrogen Fixation Research Progress. (H.J. Evans, P.J. Bottomley, W.E. Newton, eds.),

Martinus Nijhoff Publishers, Dordrecht, pp. 208-215.
66. MINAMISAWA, K. 1988. Hydrogenase phenotype, rhizobitoxine-producing ability and serinol content in soybean nodules formed with various strains of Bradyrhizobium japonicum. Op. cit. Reference 61, pp. 12-41.
67. LIGERO, F., C. LLUCH, J. OLIVARES, E.J. BEDMAR. 1987. Nitrate reductase activity in nodules of pea inoculated with hydrogenase positive and hydrogenase negative strains of Rhizobium leguminosarum. Physiol. Plant. 69: 313-316.
68. ARIMA, Y. 1981. Respiration and efficiency of N_2 fixation by the nodules formed with a H_2-uptake positive, strain of Rhizobium japonicum. Soil Sci. Plant Nutr. 27: 115-119.
69. MINAMISAWA, K., Y. ARIMA, K. KUMAZAWA. 1981. Rapid turnover of aspartic acid in soybean nodules formed with H_2-uptake positive Rhizobium japonicum strains. Soil Sci. Plant Nutr. 27: 387-391.
70. DEVINE, T.E., D.F. WEBER. 1977. Genetic specificity of nodulation. Euphytica 26: 527-535.
71. BERGERSEN, F. 1977. Physiological chemistry of dinitrogen fixation by legumes. In A Treatise on Dinitrogen Fixation, Section III. (R.W.F. Hardy, W.S. Silver, eds.), John Wiley & Sons, Inc., New York, pp. 519-555.
72. TRICHANT, J.C., J. RIGAUD. 1979. Sur les substrates energetiques utilises, lore de la reduction de C_2H_2, par les bacteroid extraits des nodosites de Phaseolus vulgaris L. Physiol. Veg. 17: 547-556.
73. REIBACH, P.H., J.G. STREETER. 1983. Metabolism of ^{14}C-labelled photosynthate and distribution of enzymes of glucose metabolism in soybean nodules. Plant Physiol. 72: 634-640.
74. KAHN, M.L., J. KRAUS, J.E. SOMERVILLE. 1985. A model of nutrient exchange in the Rhizobium-legume symbiosis. Op. cit. Reference 65, pp. 193-199.
75. KOUCHI, H., T. YONEYAMA. 1984. Dynamics of carbon photosynthetically assimilated in nodulated soya bean plants under steady-state conditions. 2. The incorporation of ^{13}C into carbohydrates, organic acids, amino acids and some storage compounds. Ann. Botany 53: 883-896.

76. KOUCHI, H., T. YONEYAMA. 1986. Metabolism of [^{13}C]-labelled photosynthate in plant cytosol and bacteroids of root nodules of *Glycine max*. Physiol. Plant. 68: 238-244.
77. KOUCHI, H., K. FUKAI, H. KATAGIRI, K. MINAMISAWA, S. TAJIMA. 1988. Isolation and enzymological characterization of infected and uninfected cell protoplasts from root nodules of *Glycine max*. Planta (in press).
78. SALMINEN, S.O., J.G. STREETER. 1987. Involvement of glutamate in the respiratory metabolism of *Bradyrhizobium japonicum* bacteroids. J. Bacteriol. 169: 495-499.
79. KOHN, D.H., K.R. SCHUBERT, M.B. CARTER, C.H. HAGEDORN, G. SHEARER. 1988. Proline metabolism in N_2-fixing root nodules: energy transfer and regulation of purine synthesis. Proc. Nat. Acad. Sci. USA 85: 2036-2040.
80. OHYAMA, T., K. KUMAZAWA. 1978. Incorporation of ^{15}N into various nitrogenase compounds in intact soybean nodules after exposure to ^{15}N gas. Soil Sci. Plant Nutr. 24: 525-533.
81. OHYAMA, T., K. KUMAZAWA. 1979. Assimilation and transport of nitrogenous compounds originated from $^{15}N_3$ absorption. Soil Sci. Plant Nutr. 25: 9-19.
82. OHYAMA, T., K. KUMAZAWA. 1980. Nitrogen assimilation in soybean nodules. I. The role of GS/GOGAT system in the assimilation of ammonia produced by N_2-fixation. Soil Sci. Plant Nutr. 26: 109-115.
83. FUJIHARA, S., M. YAMAGUCHI. 1981. Assimilation of $^{15}NH_3$ by root nodules detached from soybean plants. Plant Cell Physiol. 22: 797-806.
84. OHYAMA, T. 1984. Comparative studies on the distribution of nitrogen in soybean plants supplied with N_2 and NO_3^- at the pod filling stage. II. Assimilation and transport of nitrogenous constituents. Soil Sci. Plant Nutr. 30: 219-229.
85. DeVRIES, G.E., P. IN'T VELD, J.R. KIJNE. 1980. Production of organic acids in *Pisum sativum* root nodules as a result of oxygen stress. Plant Sci. Lett. 20: 115-123.
86. TAJIMA, S., T.A. LaRUE. 1982. Enzymes for acetaldehyde and ethanol formation in legume nodules. Plant Physiol. 70: 388-392.

87. SUGANUMA, N., Y. YAMAMOTO. 1987. Carbon metabolism related to nitrogen fixation in soybean root nodules. Soil Sci. Plant Nutr. 33: 79-91.
88. SUGANUMA, N., Y. YAMAMOTO. 1987. Respiratory metabolism of mitochondria in soybean root nodules. Soil Sci. Plant Nutr. 33: 93-101.
89. SUGANUMA, N., M. KITOU, Y. YAMAMOTO. 1987. Carbon metabolism in relation to cellular organization of soybean root nodules and respiration of mitochondria aided by leghemoglobin. Plant Cell Physiol. 28: 113-122.
90. CHRISTELLES, J.T., W.A. LAING, W.D. SUTTON. 1977. Carbon dioxide fixation by lupin root nodules. I. Characterization, association with phosphoenolpyruvate carboxylase and correlation with nitrogen fixation during nodule development. Plant Physiol. 60: 47-50.
91. PETERSON, J.B., H.J. EVANS. 1979. Phosphoenolpyruvate carboxylase from soybean nodule cytosol: evidence for isoenzymes and kinetics of the most active component. Biochim. Biophys. Acta 567: 445-452.
92. DAVIES, D.D. 1980. Anaerobic metabolism and the production of organic acids. In The Biochemistry of Plants. (D.D. Davies, ed.), Academic Press, New York, Vol. 2, pp. 581-611.
93. ANTOUN, H., L.M. BORDELEAU, R. SANVAGEAU. 1984. Utilization of the tricarboxylic acid cycle intermediates and symbiotic effectiveness in *Rhizobium meliloti*. Plant Soil 77: 29-38.
94. RONSON, C.W., P. LYTTLETON, J.G. ROBERTSON. 1981. C_4-decarboxylate transport mutants of *Rhizobium trifolii* form ineffective nodules on *Trifolium repens*. Proc. Natl. Acad. Sci. USA 78: 4284-4288.
95. FINAN, T.M., J.M. WOOD, D.C. JORDAN. 1981. Succinate transport in *Rhizobium leguminosarum*. J. Bacteriol. 148: 193-202.
96. GARDIOL, A., A. ARIAS, C. CERVENANSKY, G. MARTINEZ-DRETS. 1982. Succinate dehydrogenase mutant of *Rhizobium meliloti*. J. Bacteriol. 151: 1621-1623.
97. FINAN, T.M., J.M. WOOD, D.C. JORDAN. 1983. Symbiotic properties of C_4-dicarboxylic acid transport mutants of *Rhizobium leguminosarum*. J. Bacteriol. 154: 1403-1413.

98. ARWAS, R., A. McKAY, F.R. ROWNEY, M.J. DILWORTH, A.R. GLENN. 1985. Properties of organic acid utilization mutants of Rhizobium leguminosarium strain 300. J. Gen. Microbiol. 131: 2059-2066.

99. GARDIOL, A.E., G.L. TRUCHET, F.B. DAZZO. 1987. Requirement of succinate dehydrogenase activity for symbiotic bacteroid differentiation of Rhizobium meliloti in alfalfa nodules. Appl. Environ. Microbiol. 53: 1947-1950.

100. RAWSTHORNE, S., F.R. MINCHIN, R.J. SUMERFIELD, C. COOKSON, J. COOMBS. 1980. Carbon and nitrogen metabolism in legume root nodules. Phytochemistry 19: 341-355.

101. JOHNSON, G.V., H.J. EVANS, T.M. CHING. 1965. Enzymes of the glyoxylate cycle in rhizobia and nodules of legumes. Plant Physiol. 41: 1330-1336.

102. LAWRIE, A.C., C.T. WHEELER. 1975. Nitrogen fixation in root nodules of Vicia faba L. in relation to the assimilation of carbon. New Phytol. 74: 437-445.

103. VALLEAU, W.D., E.M. JOHNSON, S. DIACHUN. 1942. Association of tobacco leaf spot bacteria with roots of crop plants. Science 96: 164.

104. VALLEAU, W.D., E.M. JOHNSON, S. DIACHUN. 1944. Root infection of crop plants and weeds by tobacco leaf spot bacteria. Phytopathology 34: 163-174.

105. KNIGHT, T.J., R.D. DURBIN, P.J. LANGSTON-UNKEFER. 1986. Effects of tabtoxinine-β-lactam on nitrogen metabolism in Avena sativa L. roots. Plant Physiol. 82: 1045-1050.

106. KNIGHT, T.J., P.J. LANGSTON-UNKEFER. 1988. Enhancement of symbiotic dinitrogen-fixation by a toxin-releasing plant pathogen. Science, (in press).

107. PETERSON, J.B., H.J. EVANS. 1978. Properties of pyruvate kinase from soybean nodule cytosol. Plant Physiol. 61: 909-914.

108. TJEPKEMA, J.D., C.S. YOCUM. 1973. Respiration and oxygen transport in soybean nodules. Planta 115: 59-72.

109. TJEPKEMA, J.D., C.S. YOCUM. 1974. Measurement of oxygen partial pressure within soybean nodules by oxygen micro-electrodes. Planta 119: 351-360.

110. HARTWIG, U., B. BOLLER, J. NÖSBERGER. 1987. Oxygen supply limits nitrogenase activity of clover nodules after defoliation. Ann. Bot. 59: 285-291.

111. WITTY, J.F., L. SKØT, N.P. REVSBECH. 1987. Direct evidence for changes in the resistance of legume root nodules to O_2 diffusion. J. Exp. Bot. 38: 1129-1140.

112. HUNT, S., S.T. GAITO, O.B. LAYZELL. 1988. A model of gas exchange and diffusion in legume nodules. II. Characteristics of the diffusion barrier and estimation of the concentrations of CO_2, H_2 and N_2 in the infected cells. Planta 173: 128-141.

113. PANKHURST, C.E., J.I. SPRENT. 1975. Surface features of soybean nodules. Protoplasma 85: 85-98.

114. BOLAND, M.J., A.M. FORDYCE, R.M. GREENWOOD. 1978. Enzymes of nitrogen metabolism in legume nodules: A comparative study. Aust. J. Plant Physiol. 5: 553-559.

115. ROBERTSON, J.G., K.J.F. FARNDEN, M. WARBURTON, J.M. BANKS. 1975. Induction of glutamine synthetase during nodule development in lupin. Aust. J. Plant Physiol. 2: 265-272.

116. ROBERTSON, J.G., M. WARBURTON, K.J.F. FARNDEN. 1975. Induction of glutamate synthase during nodule development in lupin. FEBS Lett. 55: 33-37.

117. MEEKS, J.C., C.P. WOLK, N. SCHILLING, P.W. SHAFFER, Y. AVISSAR, W.S. CHIEN. 1978. Initial organic products of fixation of [^{13}N]dinitrogen by root nodules of soybean (<u>Glycine max</u>). Plant Physiol. 61: 980-983.

118. BOLAND, M.J., A.G. BENNY. 1977. Enzymes of nitrogen metabolism in legume nodules. Purification and properties of NADH-dependent glutamate synthase from lupin nodules. Eur. J. Biochem. 79: 355-362.

119. CULLIMORE, J.V., M. LARA, P.J. LEA, B.J. MIFLIN. 1983. Purification and properties of two forms of glutamine synthetase from the plant fraction of <u>Phaseolus</u> root nodules. Planta 157: 245-253.

120. GROAT, R.G., L.E. SCHRADER. 1982. Isolation and immunochemical characterization of plant glutamine synthetase in alfalfa (<u>Medicago sativa</u> L.) nodules. Plant Physiol. 70: 1759-1761.

121. McPARLAND, R.H., J.G. GUEVARA, R.R. BECKER, H.J. EVANS. 1976. The purification and properties of the glutamine synthetase from the cytosol of soybean root nodules. Biochem. J. 153: 597-606.

122. AWONAIKE, K.O., P.J. LEA, B.J. MIFLIN. 1981. The location of the enzymes of ammonia assimilation in root nodules of Phaseolus vulgaris L. Plant Sci. Lett. 23: 189-195.

123. BOLAND, M.J., J.F. HANKS, P.H.S. REYNOLDS, D.G. BLEVINS, N.E. TOLBERT, K.R. SCHUBERT. 1982. Subcellular organization of ureide biogenesis from glycolytic intermediates and ammonium in nitrogen-fixing soybean nodules. Planta 155: 45-51.

124. SHELP, B.J., C.A. ATKINS. 1984. Subcellular location of enzymes of ammonia assimilation and asparagine synthesis in root nodules of Lupinus alba L. Plant Sci. Lett. 36: 225-230.

125. SHELP, B.J., C.A. ATKINS, P.J. STORER, D.T. CANVIN. 1983. Cellular and subcellular organization of pathway of ammonia assimilation and ureide synthesis in nodules of cowpea (vigna unguiculata L. Walp.). Arch. Biochem. Biophys. 224: 429-441.

126. ROBERTSON, J.G., K.J.F. FARNDEN. 1980. Ultrastructure and metabolism of the developing legume root nodule. Op. cit. Reference 6, pp. 65-113.

127. NEWELL, F.B., R.R. SCHMIDT. 1987. Purification and partial kinetic and physical characterization of two chloroplast-localized NADP-specific glutamate dehydrogenase isoenzymes and their preferential accumulation in Chlorella sorokiniano cells cultured at low or high ammonium levels. Plant Physiol. 83: 75-84.

128. SCHUBERT, K.R. 1986. Products of biological nitrogen fixation in higher plants. Annu. Rev. Plant Physiol. 37: 539-574.

129. MATSUMOTO, T., M. YATAZAWA, Y. YAMAMOTO. 1977. Effects of exogenous nitrogen compounds on the concentrations of allantoin and various constituents in several organs of soybean plants. Plant Cell Physiol. 18: 613-624.

130. McCLURE, P.R., D.W. ISRAEL. 1979. Transport of nitrogen in the xylem of soybean plants.

Plant Physiol. 64: 411-416.
131. FUJIHARA, S., M. YAMAGUCHI. 1980. Asparagine formation in soybean nodules. Plant Physiol. 66: 139-141.
132. COKER, G.T. III, K.R. SCHUBERT. 1981. Carbon dioxide fixation in soybean roots and nodules. I. Characterization and comparison with N_2 fixation and composition of xylem exudate during early nodule development. Plant Physiol. 67: 691-696.
133. VANCE, C.P., S. STADE, C.A. MAXWELL. 1983. Alfalfa root nodule carbon dioxide fixation. I. Association with nitrogen fixation and incorporation into amino acids. Plant Physiol. 72: 469-473.
134. VANCE, C.P., S. STADE. 1984. Alfalfa root nodule carbon dioxide fixation. II. Partial purification and characterization of root nodule phosphoenolpyruvate carboxylase. Plant Physiol. 75: 261-164.
135. HUBER, T.A., J.G. STREETER. 1984. Asparagine biosynthesis in soybean nodules. Plant Physiol. 74: 605-610.
136. REYNOLDS, P.H., K.J. FARNDEN. 1979. The involvement of aspartate aminotransferases in ammonium assimilation in lupin nodules. Phytochemistry 18: 1625-1630.
137. REYNOLDS, P.H.S., M.J. BOLAND, K.J.F. FARNDEN. 1981. Enzymes of nitrogen metabolism in legume nodules. Partial purification and properties of the aspartate aminotransferase from lupine nodules. Arch. Biochem. Biophys. 209: 524-533.
138. RYAN, E., F. BODLEY, P.F. FOTRELL. 1972. Purification and characterization of aspartate aminotransferases from soybean root nodules and Rhizobium japonicum. Phytochemistry 11: 957-963.
139. REYNOLDS, P.H.S., M.J. BOLAND, D.G. BLEVINS, D.D. RANDALL, K.R. SCHUBERT. 1982. Ureide biogenesis in leguminous plants. Trends Biochem. Sci. 7: 366-368.
140. ATKINS, C.A. 1987. Metabolism and translocation of fixed nitrogen in the nodulated legume. Plant Soil 100: 157-169.
141. VAUGHN, K.C., S.O. DUKE, C.A. HENSON. 1982. Ultrastructural localization of urate oxidase in nodules of Sesbania exaltata, Glycine max, and

Medicago sativa. Histochemistry 74: 309-318.
142. VandenBOSCH, K.A., E.H. NEWCOMB. 1986. Immunogold localization of nodule-specific urease in developing soybean root nodules. Planta 167: 425-436.
143. NEWCOMB, E.H., SH. R. TANDON, R.R. KOWAL. 1985. Ultrastructural specialization for ureide production in uninfected cells of soybean root nodules. Protoplasma 125: 1-12.
144. NGUYEN, J., L. MACHAL, J. VIDAL, C. PERROT-RECHENMANN, P. GADAL. 1986. Immunochemical studies on xanthine dehydrogenase of soybean root nodules. Planta 167: 190-195.
145. RAINBIRD, R.M., C.A. ATKINS. 1981. Purification and some properties of urate oxidase from nitrogen-fixing nodules of cowpea (_Vigna unguiculata_ L. Walp.). Biochim. Biophys. Acta 659: 132-140.
146. WOO, K.C., C.A. ATKINS, J.S. PATE. 1981. Ureide synthesis in a cell-free system from cowpea (_Vigna unguiculata_ [L.] Walp.) studies with oxygen, pH and purine metabolism. Plant Physiol. 67: 1156-1160.
147. THOMAS, R.J., S.P. MEYERS, L.E. SCHRADER. 1983. Allantoinase from shoot tissues of soybeans. Phytochemistry 22: 1117-1120.
148. HANKS, J.F., N.E. TOLBERT, K.R. SCHUBERT. 1981. Localization of enzymes of ureide biosynthesis in peroxisomes and nuciosomes of nodules. Plant Physiol. 68: 65-69.
149. THOMAS, R.J., L.E. SCHRADER. 1981. The assimilation of ureides in shoot tissues of soybeans. I. Changes in allantoinase activity and ureide contents of leaves and fruits. Plant Physiol. 67: 973-976.
150. WINKLER, R.G., D.G. BLEVINS, J.C. POLACCO, D.D. RANDALL. 1988. Ureide catabolism in nitrogen-fixing legumes. Trends Biochem. Sci. 13: 97-100.
151. WINKLER, R.G., D.G. BLEVINS, D.D. RANDALL. 1988. Ureide catabolism in soybeans. III. Ureidoglycolate amidohydrolase and allantoate amidohydrolase are activities of an allantoate degrading enzyme complex. Plant Physiol. 86: 1084-1088.
152. MILLAD, P. 1988. The accumulation and storage of nitrogen by herbaceous plants. Plant Cell Environ. 11: 1-8.

153. WITTENBACH, V.A. 1982. The effects of pod removal on leaf senescence in soybeans. Plant Physiol. 70: 1544-1548.
154. WITTENBACH, V.A. 1983. Purification and characterization of a soybean leaf storage glycoprotein. Plant Physiol. 73: 125-129.
155. FRANCESCHI, V.R., V.A. WITTENBACH, R.T. GIAQUINTA. 1983. Paroveinal mesophyll of soybean leaves in relation to assimilate transfer and compartmentation. III. Immunohistochemical localization of specific glycopeptides in the vacuole after depodding. Plant Physiol. 72: 586-589.
156. THOMPSON, J.F. 1980. Arginine synthesis, proline synthesis and related processes. Op. cit. Reference 6, pp. 375-402.
157. MOSSE, J., J.-C. HUET, J. BAUDET. 1987. Relationships between nitrogen, amino acids and storage proteins in Lupinus albus seeds. Phytochemistry 26: 2454-2458.
158. CASTLE, S.L., P.J. RANDALL. 1987. Effects of sulfur deficiency on the synthesis and accumulation of proteins in the developing wheat seed. Aust. J. Plant Physiol. 14: 503-516.
159. THOMPSON, J.F., J.T. MADISON, M.A. WATERMAN, A.M.E. MUENSTER. 1981. Effect of methionine on growth and protein composition of cultured soybean cotyledons. Phytochemistry 20: 941-945.
160. GRABAU, L.J., D.G. BLEVINS, H.C. MINOR. 1986. Stem infusions enhanced methionine content of soybean storage protein. Plant Physiol. 82: 1013-1018.
161. RANDALL, P.J., J.A. THOMSON, H.E. SCHROEDER. 1978. Cotyledonary storage proteins in Pisum sativum. IV. Effects of sulfur, phosphorus potassium and magnesium deficiencies. Aust. J. Plant Physiol. 5: 519-534.
162. GAYLER, K.R., G.E. SYKES. 1985. Effects of nutritional stress on the storage proteins of soybeans. Plant Physiol. 78: 582-585.
163. O'KENNEDY, B.T., J.S. TITUS. 1979. Isolation and mobilization of storage proteins from apple shoot bark. Physiol. Plant 45: 419-424.
164. OLAND, K. 1963. Changes in the content of dry matter and major nutrient elements of apple foliage during senescence and abscission. Physiol. Plant. 16: 682-684.

165. SPENCER, P.W., J.S. TITUS. 1972. Biochemical and enzymatic changes in apple leaf during autumnal senescence. Plant Physiol. 49: 746-750.
166. O'KENNEDY, B.T., J.S. TITUS. 1975. Changes in nitrogen reserves of apple shoots during the dormant season. J. Hort. Sci. 50: 321-329.
167. KANG, S.M., J.S. TITUS. 1980. Qualitative and quantitative changes in nitrogenous compounds in senescing leaf and bark tissues of the apple. Physiol. Plant. 50: 285-290.
168. TROMP, J., J.C. OVAA. 1971. Spring mobilization of storage nitrogen in isolated shoot sections of apple. Physiol. Plant. 25: 16-22.
169. TROMP, J., J.C. OVAA. 1973. Spring mobilization of storage nitrogen in apple. Physiol. Plant. 29: 1-5.
170. KERR, P.S., D.G. BLEVINS, B. RAPP, D.D. RANDALL. 1983. Soybean leaf urease: comparison with seed urease. Physiol. Plant. 57: 339-345.
171. POLACCO, J.C., R.W. KRUGER, R.G. WINKLER. 1985. Structure and possible ureide degrading function of the ubiquitous urease of soybean. Plant Physiol. 79: 794-800.
172. MEYER_BOTHLING, L.E., J.C. POLACCO, S.R. CIANZIO. 1987. Pleiotropic soybean mutants defective in both urease isozymes. Mol. Gen. Genet. 209: 432-438.
173. MEYER-BOTHLING, L.E., J.C. POLACCO. 1987. Mutational analysis of the embryo-specific urease locus of soybean. Mol. Gen. Genet. 209: 439-444.
174. GALSTON, A.W. 1983. Polyamines as modulators of plant development. BioScience 33: 382-388.
175. DROLET, G., E.B. DUMBROFF, R.L. LEGGE, J.E. THOMPSON. 1986. Radical scavenging properties of polyamines. Phytochemistry 25: 367-371.
176. SMITH, G.S., D.R. LAUREN, I.S. CORNFORTH, M.P. AGNEW. 1982. Evaluation of Putrescine as a biochemical indicator of the potassium requirements of lucerne. New Phytol. 91: 419-428.

Chapter Two

PLANT GENES INVOLVED IN CARBON AND NITROGEN ASSIMILATION IN ROOT NODULES

DESH PAL S. VERMA

Biotechnology Center and
Department of Molecular Genetics
The Ohio State University
Columbus, Ohio 43210

INTRODUCTION

The availability of reduced nitrogen is the most limiting factor in plant growth. In legume plants, nitrogen is provided by symbiotic bacteria (Rhizobium spp.) in exchange for carbon supplied by the host as photosynthates. A significant portion (25 to 30%) of the carbon received by nodules from photosynthates is returned to the shoot as nitrogenous solutes. Thus, not only are large amounts of energy required to support nitrogen fixation, but assimilation of that nitrogen is also energy intensive. It has been suggested that CO_2 fixation in the dark, catalyzed by phosphoenolpyruvate (PEP) carboxylase, can function as an anapleurotic pathway to supply part of the carbon skeleton needed for nitrogen assimilation.[1] Legumes of temperate regions generally assimilate and export fixed nitrogen as amides, while the tropical legumes, e.g., soybean, synthesize ureides (allantoin and allantoic acid) that account for > 90% of the nitrogen in their xylem sap.[2] Extensive studies have been carried out to calculate the carbon/nitrogen budget of several legumes, but our knowledge of the regulatory mechanisms controlling the expression of genes involved in carbon and nitrogen pathways in nodules is very limited. Moreover, we do not know how the regulation of these genes is affected by the supply of carbon and nitrogen and different levels of oxygen present in infected and uninfected cells of the nodules.

PRIMARY ASSIMILATION OF REDUCED NITROGEN IN NODULES

Immediately following reduction of dinitrogen by bacteroids, the ammonia is transported to the host cell cytoplasm where it is converted into amides and in some cases further metabolized into ureides before being translocated to the shoot. Between the amides and ureides, the ureides appear to be the more efficient form of transport since they use less metabolic energy by loading more nitrogen per carbon skeleton. The energy cost has been calculated to be 5 to 6 times higher for amides as compared to ureides in terms of ATP equivalents.[1,2] Ureides can be formed by 1) condensation of urea and glyoxylate or glycine, or 2) oxidative catabolism of purines. The latter route is now firmly established in nodules, which requires _de novo_ synthesis of purines.[1,3,4] High levels of activity of xanthine dehydrogenase and uricase are present in ureide-producing nodules while they are low in amide-producing nodules,[5-8] and cell-free synthesis of ureides has been shown from labelled hypoxanthine.[7] Treatment with allopurinol, an irreversible inhibitor of xanthine dehydrogenase, reduces the level of ureides and concomitantly increases the xanthine pool in the nodules.[9] While nitrogen fixation is not affected, the glutamine produced as a result of assimilation of ammonia is apparently not accessible for transport to the plant and the plant shows symptoms of nitrogen starvation. At the same time, externally supplied ammonia or glutamine is assimilated, apparently by the rest of the root tissue. This lack of utilization of glutamine following a block in ureide production suggests the presence of a one way "channel" for the assimilation of reduced nitrogen in nodules of ureide-producing legumes.

Nitrogen metabolism is tightly compartmentalized in ureide-producing nodules both within the infected cells and between uninfected and infected cells. Figure 1 shows a general outline of the proposed compartmentalization of various nitrogen metabolites in ureide-producing nodules. The metabolic requirements and constraints of the infected nodule cell are very different from those of the uninfected cell. Since the oxygen level is not uniform in nodule tissue and several enzyme systems are sensitive to oxygen, the oxygen-requiring and oxygen-sensitive processes need to be compartmentalized.[10] However, there is no clear understanding of the role of oxygen in controlling carbon and nitrogen metabolism in nodules. Several enzymes of the purine and ureide biosynthetic pathways in fungi are known to be controlled at the transcriptional level by metabolites of these pathways, but the inducers are not known in legume plants.

Glutamine Synthetase (EC 6.3.1.2)

Ammonia produced by nitrogen fixation is assimilated by the sequential activities of glutamine synthetase (GS) and glutamate synthase. Since GS plays a central role in the nitrogen metabolism of the plant, the biochemistry and the genetic organization and regulation of this enzyme have been extensively studied in a number of systems. The initial pathway of ammonia assimilation in the two types of legumes, i.e., amide and ureide producers, is apparently the same. Two forms of glutamine synthetase, cytosolic and chloroplastic, exist in plants.[11] The root is the primary site of nitrogen assimilation, and cytosolic GS in roots controls the entry of ammonia into

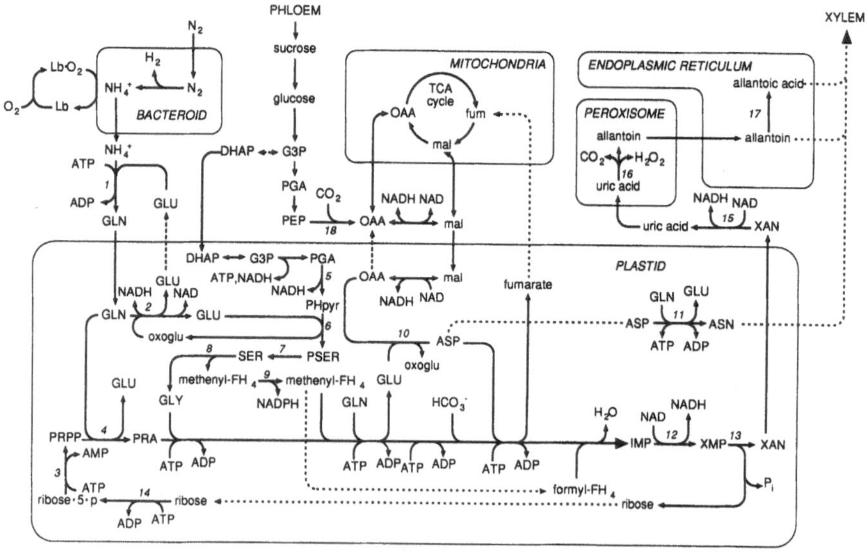

Fig. 1. Proposed pathways and subcellular organization of the reactions involved in amide synthesis and ureide biogenesis in ureide-exporting legume nodules. Numbered reactions: 1) glutamine synthetase; 2) glutamate synthase; 3) PRPP synthetase; 4) PRPP amidotransferase; 5) PGA dehydrogenase; 6) phosphoserine aminotransferase; 7) phosphoserine phosphatase; 8) serine hydroxymethylase; 9) methylene FH_4 dehydrogenase; 10) aspartate aminotransferase; 11) asparagine synthetase; 12) IMP dehydrogenase; 13) nucleotidase nucleosidase; 14) ribokinase; 15) xanthine dehydrogenase; 16) uricase; 17) allantoinase; 18) PEP carboxylase. (From Schubert,[1] reprinted with permission).

the nitrogen metabolism of the whole plant. Nodules in some legumes (e.g., Phaseolus and alfalfa) contain a unique form of glutamine synthetase which appears to be a nodule-specific protein.[12-14] Glutamine synthetase is encoded by a small family of genes, the members of which are expressed in different organs of the plant. In pea and Phaseolus, several GS genes have been shown to be differentially expressed in root, leaf and nodule tissues.[14,15] It has been shown that the cytosolic GS in Phaseolus vulgaris consists of 4 different subunits, one of which is exclusively made in nodules. Quantitative nuclease protection experiments showed that mRNAs for γ and β subunits are most abundant in nodules, and mRNA for α and δ are present at a much lower level.[16]

The sequence for a nodule-specific GS has also been reported in soybean.[17] However, our initial studies on the soybean glutamine synthetase gene have indicated that the availability of ammonia may be directly responsible for the induction of the GS gene in soybean.[18] We have localized this enzyme in nodules at the subcellular level using antibodies and protein-A gold labelling[19] and the enzyme was found in the cytoplasm of the infected cells. Two cDNA clones were isolated from a fraction of the nodule cDNA library that contained sequences common to the root and the nodules.[20] Both clones are expressed in a ratio of 1:10 in root tissue. Following exposure of roots to ammonium sulfate their expression in the roots was increased to a level equal to that in nodules but the ratio of two sequences remained the same.[18] We have isolated four GS genes from a genomic library of soybean and have made transcriptional fusions with the β-glucuronidase (GUS) reporter gene using a putative promoter fragment from one of these genes (Fig. 2). This construct was found to be expressed effectively in a root-specific manner when introduced into tobacco using Agrobacterium tumefaciens but does not appear to be stimulated by ammonia. The other GS genes have not yet been characterized and it is possible that one of them is nodule-specific or ammonia-stimulated.

Recently, it has been demonstrated that a toxin, tabtoxinine-β-lactam, produced by a pathogen (Pseudomonas syringae pv. tabaci) known to inhibit glutamine synthetase activity[21] differentially affects nodule and root forms of GS enzymes.[22] The nodule form of GS is much less sensitive to this toxin. Co-inoculation of Rhizobium with the Pseudomonas that produce the toxin leads to an apparent increase in nitrogen fixation and, consequently, in growth yield. This effect is more pronounced in alfalfa than in soybean. The fact that the nodule form of GS is not inhibited by tabtoxinine-β-lactam, suggests major alterations in the catalytic site of the nodule GS and the mode of regulation of the nodule-specific GS gene. If this observation is correct, then it may be possible to directly control the level of GS activity in legumes

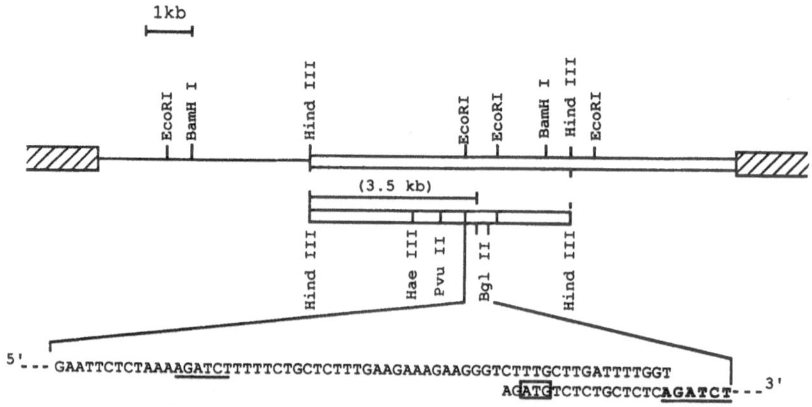

Fig. 2. Restriction map of the genomic clone of soybean glutamine synthetase genes and the sequence of the region where the transcription fusion was made with the - glucuronidase (GUS) gene. Two Bgl II sites in the sequence are underlined and the initiation codon boxed. Fusion was made with the 2nd upstream Bgl II site. The hatched areas represent EMBL 3 vector arms.

to enhance nitrogen fixation. This can be achieved by stable transformation of legume plants with an antisense GS sequence using a nodule-specific gene promoter like that of leghemoglobin (Lb). We have successfully demonstrated[23] such an approach to control expression of a targeted gene and this approach could be applied to control the expression of any nodulin genes.

Aspartate Aminotransferase (L-Aspartate:2-Oxoglutarate Aminotransferase, EC 2.6.1.1)

Some root nodules, e.g., alfalfa (*Medicago sativa* L.), export fixed nitrogen primarily in the form of asparagine and aspartate.[24] Conversion of $[^{14}C]$asparatate and $[^{14}C]$ CO_2 to $[^{14}C]$asparagine has been demonstrated and it was found that the flow of carbon between the organic acid and amino acid pools is regulated by aspartate aminotransferase (AAT). Multiple forms of AAT have been reported in soybean, lupin (*Lupinus angustifolius* L.),[25] and cowpea (*Vigna unguiculata* L.).[26] In soybean root nodules, 87% of the total AAT activity is apparently confined to bacteroids. However, Boland et al.[4] reported that soybean nodules contained two forms of AAT, one in the plant cytoplasm and the other located in plastids. In cowpea nodules AAT activity was localized to both infected and uninfected cells and is suggested to be primarily of host origin.[26] The bulk of the activity (> 90%) in alfalfa nodules is in the cytoplasm.[27] Thus, this enzyme in most legume nodules appears to be of host origin. It has been purified from soybean and alfalfa nodules; however, it has not yet been demonstrated that it is a product of a nodule-specific gene.

REGULATION OF GENES INVOLVED IN UREIDE BIOSYNTHESIS

Ureides are produced by purine biosynthesis de novo in nodules. Synthesis of allantoin and allantoic acid from xanthine and hypoxanthine has been demonstrated in cell-free extracts of nodules from tropical legumes.[7,28] Two key enzymes of the ureide pathway, xanthine dehydrogenase (EC 1.2.1.37) and uricase (EC 1.7.3.3), have been purified from soybean and their inter- and intra-cellular distribution has been determined. Uricase, a tetrameric enzyme with a subunit of 35 kDa (nodulin-35) is restricted to the peroxisomes of uninfected cells of nodules,[29,30] while xanthine dehydrogenase appears to be a cytosolic enzyme in the infected cells.[31]

The activities of several other enzymes of the purine biosynthetic pathway are found to be increased in soybean nodules, but their precise subcellular location is not fully established. Four enzymes, phosphoribosylpyrophosphate amidotransferase (EC 2.4.2.14), phosphoglycerate dehydrogenase (EC 1.1.1.95), serine hydroxymethylase (EC 2.1.2.1), and methylenetetrahydrofolate dehydrogenase (EC 1.5.1.5) are found in the plastid fraction of the nodules.[4] However, the location of the genes encoding these enzymes (nuclear or organellar) is not known.

Attempts have been made in a number of laboratories to isolate genes encoding various enzymes of the ureide biosynthetic pathway in legumes. However, except for the uricase (nodulin-35) gene that we isolated using standard molecular techniques employing immunoprecipitation of polysomes,[29] no other plant gene of this pathway has yet been isolated. Moreover, most of the nodulin clones isolated to date, which may encode some of these enzymes, have not been characterized with respect to their functions in nodules. This is primarily due to the fact that none of the enzymes concerned have been purified to a stage that specific antisera could be raised and also the transcripts of the genes involved appear to be low in abundance in nodules.[32] From the data on the putative size of polypeptides of some of the enzymes of purine and ureide pathways and the size of the hybrid-released translation products of known nodulin sequences,[33] it appears that none of the uncharacterized soybean nodulins encode enzymes of these pathways. Thus, the respective clones need to be isolated using approaches different from those applied to date.

Purine biosynthesis is similar in bacteria, plants, animals and fungi. The pathway is well characterized in a number of bacteria and yeast (*Schizosaccharomyces pombe* and *Saccharomyces cerevisiae*), and many mutants exist in this pathway affecting both structural and regulatory genes.[34-36] Since these pathways are highly conserved, it is possible to obtain functional complementation of some of the mutant yeast and *Escherichia coli* genes of these pathways by soybean cDNA sequences driven by appropriate promoters active in yeast and *E. coli*. Functional complementation by higher eucaryotic genes of mutation in *E. coli* and *S. pombe* have been demonstrated. A plant glutamine synthetase gene has been shown to complement a *glnA* mutation in *E. coli*,[37] and a human homologue of a yeast cell cycle control gene *cdc2* has been obtained by complementation in *S. pombe*.[38] Thus, it may be possible to isolate soybean genes encoding enzymes of the purine and ureide pathways by the functional complementation approach using appropriate bacterial and yeast mutants.

Phosphoribosylpyrophosphate amidotransferase (EC 2.4.2.14)

This enzyme, referred to as PRPP-AT, catalyzes the following reaction in purine biosynthesis:

5-phosphoribosyl-α-1-pyrophosphate + glutamine = phosphoribosylamine + glutamate + PPi.

This reaction, which is the committed step in the pathway, has been shown to be tightly regulated by feedback control with purine nucleotide monophosphates in bacteria, yeast and animal cells.[39,40] The enzyme has been purified from bacteria,[41] yeast and animals.[42,43] The gene has been cloned from E. coli, Bacillus subtilis, and S. cerevisiae, and is shown to encode a conserved protein with 47% sequence homology between the bacteria and yeast.[44] This gene has not yet been isolated from any higher eucaryotes. Studies on the yeast gene confirm that it is regulated at the transcriptional level and the level of mRNA is reduced 5-fold in the presence of excess adenine. It has been reported that in avian liver this enzyme is associated with the other enzymes of the purine biosynthetic pathway; the same may be true for the plant enzyme which has been partially purified.[45] It is competitively inhibited by IMP, and has been suggested to be a key regulatory point in purine biosynthesis leading to ureide production in nodules.

Uricase II (Urate:Oxygen Oxido-Reductase, EC 1.7.3.3)

The second most abundant nodulin in soybean nodules is an approximately 125 kDa protein comprising four 35 kDa subunits identified as nodulin-35.[30,32] Using an immunofluorescent technique, nodulin-35 was shown to be present in the uninfected cells of soybean nodules[30] and the enzyme was subsequently localized to the peroxisomes of these cells using immunogold labelling.[29,46] Nodulin-35 was shown to exhibit uricase activity.[30] An enzyme that catabolizes uric acid is present in many organisms. In soybean, this activity (detectable in uninfected root tissue and leaves) has been designated uricase I.[5,30] The antibodies against the nodule-specific enzyme, uricase II, showed no immunological cross-reactivity with uricase I nor with any other proteins in uninfected roots and leaves. Confirming the nodule-specificity of uricase II, a nodulin-35 cDNA clone hybridized only to nodule RNA and not to RNA from roots or leaves.[29]

The gene encoding soybean nodulin-35 has been isolated and was found to comprise almost 5 kb of DNA with 7 introns.[29] The processed mRNA encodes a 35,100 M_r polypeptide. The mRNA is translated on free polysomes and the protein lacks a signal peptide, suggesting that it is post-translationally directed to the peroxisomes. The mechanism by which nodulin-35 is targeted to the peroxisomes has not been elucidated. Based on recent studies on other peroxisomal proteins,[47] the targeting sequence may be encoded in the carboxy terminus of nodulin-35.

Xanthine Dehydrogenase (EC 12.1.37)

The ureides in nodules are produced by the oxidative catabolism of purines and, accordingly, the enzymes of purine catabolism occur at high levels of nodules or ureide producers compared to the nodules of amide producers.[1] Similar to nodulin-35 (uricase II), it might be expected that other enzymes involved in ureide biosynthesis are induced specifically in the nodule. One of these enzymes, xanthine dehydrogenase (XDH), catalyses the oxidation of hypoxanthine to xanthine and oxidation of xanthine to uric acid under microaerophilic conditions. XDH has recently been shown to be present in high concentrations in the infected cells of soybean nodules.[48] Soybean XDH showed antigenic cross-reactivity with nodule extracts from two ureide producers, cowpea and lima bean, but not from the amide producers alfalfa and lupin. Strictly speaking, however, XDH may not be nodule-specific since low levels of the enzyme were immunologically detectable in soybean leaves, roots, stems and pods.[48] The cloning of a legume XDH gene has not yet been reported.

Purine Nucleosidase (EC 3.2.2.1)

This enzyme is involved in the catabolism of the purine nucleosides, inosine and xanthosine, to hypoxanthine and xanthine respectively.[1] Purine nucleosidase activity is present in soybean roots and leaves but is elevated in nodules about 15-fold over root levels. Larsen and Jochimsen[49] have purified this enzyme from soybean leaves, roots and nodules and their data indicate that leaves and roots contain one isozyme of purine nucleosidase (form I), whereas nodules contain, in addition to increased amounts of form I, approximately equal levels of a novel isoform of the enzyme (designated form II). Purine nucleosidase II is apparently a nodule-specific enzyme, though the application of more sensitive techniques is required to confirm that this isoform is indeed absent from other plant organs and that it is a nodulin.

REGULATION OF GENES OF CARBON METABOLISM IN NODULES

The establishment of an effective symbiosis requires that an abundant supply of carbon substrates be delivered to the nodule. In addition to serving as the primary nutrient for both plant-derived nodule tissue and the bacteroids, carbohydrates are utilized as substrates in the assimilation of fixed nitrogen. In fact, the nodule constitutes one of the strongest sinks for carbon substrates, tapping up to 30% of the net photosynthate of the plant.[50] Thus, a number of enzymes involved in carbon metabolism (e.g., invertase, PEP carboxylase, sucrose synthase) show elevated activities in nodules[51,52] and it is conceivable that some of these enzymes occur in nodule-specific forms.

Sucrose Synthase (UDP-Glucose:Fructose 2-α-D Glucosyltransferase, EC 2.4.1.13)

Sucrose is the main carbohydrate transported from the leaves to the nodules.[53] There is evidence that sucrose synthase activity is high in tissues concerned with sucrose utilization (e.g., developing endosperm) and low in sucrose-synthesizing tissues such as leaves.[54] The enzyme catalyzes the following reversible reaction:

$$\text{Sucrose} + \text{UDP} \rightleftharpoons \text{UDP-Glucose} + \text{D-Fructose}.$$

Thus, this enzyme appears to be involved in the cleavage of sucrose to support the carbon requirements of the nodules.[52] It is a tetrameric enzyme with a 90,000 M_r subunit and a large quantity of this enzyme is produced in nodules. Sucrose has several important functions in nodules: 1) it serves as a primary nutrient to the root tissue and bacteroids; 2) it provides the skeleton for the development of the cellulosic structure of the cell wall; and 3) it provides carbon intermediates for the assimilation of fixed nitrogen. Excess sucrose is converted into starch and thus serves as an energy reserve in the nodule. Very little is known about the regulation of sucrose synthase and the control of the activity of the gene encoding this enzyme in nodules. Two non-allelic sucrose synthase genes have recently been isolated from maize.[55] Both genes are found to be expressed in various parts of the plant and both homotetrameric and heterotetrameric forms of this enzyme occur.

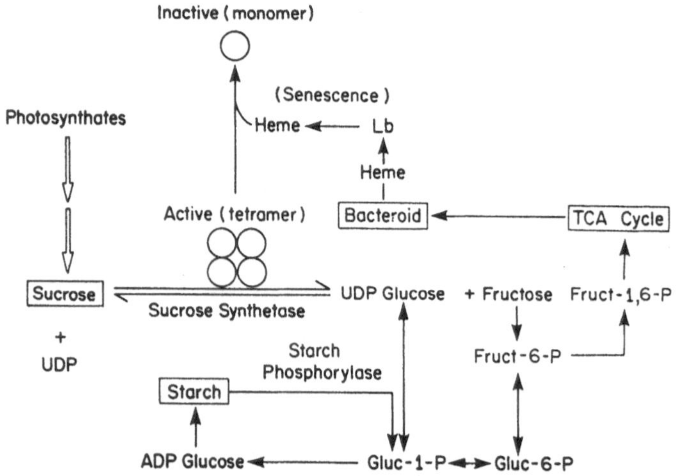

Fig. 3. Role of sucrose synthase in utilization of sucrose transported to root nodules and the putative inhibition/suppression of sucrose synthase activity by free heme released during nodule senescence. Starch accumulates in ineffective nodules when UDP-glucose is not rapidly utilized via the tricarboxylic acid cycle (TCA cycle) for the supply of carbon to bacteroids.

Recently, we have found[56] that soybean nodulin-100 is a subunit of this enzyme. The nodule sucrose synthase was found to be dissociated into monomers in the presence of heme, suggesting that the availability of free heme may regulate this enzyme. Most of the heme in nodules is bound to apo-leghemoglobin and there is no free heme available in the active phase of symbiosis. However, as senescence commences and degradation of leghemoglobin occurs, a concomitant decrease in the activity of this enzyme was observed. Heme added externally to the purified enzyme also renders it inactive. We have proposed a model (Fig. 3) which suggests that this enzyme plays a key role in the carbon economy of the nodule. We have isolated two genes from soybean encoding sucrose synthase (unpublished data). By fusion to a reporter gene and transfer to a heterologous legume we may be able to study more precisely the temporal and spatial regulation of expression of these genes during root nodule development and under ineffective states.

Starch Phosphorylase

Starch is a major energy reserve in nodules. It is stored in amyloplasts which are very abundant in the uninfected cells of the nodules.[46,57] During early development both cell types, infected and uninfected, contain significant amounts of starch, but its concentration rapidly declines in the infected cells. In ineffective nodules, the concentration of starch builds up to 6-fold the amount in uninfected cells of the effective nodules.[58] This is particularly evident in nodules formed by a mutant of *Bradyrhizobium japonicum* that failed to release from the infection thread.[57] Thus, it appears that the two cell types assume their respective roles in nodule development prior to the actual release of bacteria from the infection thread. These studies also suggest that the uninfected cells play an important role, not only in nitrogen assimilation, but also in carbon metabolism. Degradation of starch occurs via starch phosphorylase and amylase. Activities of these enzymes may be differentially distributed between infected and uninfected cells of the nodules. Moreover, the starch phosphorylase gene may be subject to control by the level of oxygen [as is sucrose synthase (B. Jochimson, personal communication)] which is different in the two cell types. Since the soybean starch phosphorylase sequence cross hybridizes with that from potato,[59] it may allow isolation of this gene from soybean.

PROLINE METABOLISM AND ENERGY TRANSFER

In tropical legumes synthesizing large amounts of purines for ureide production, an alternative route for energy transfer has recently been proposed. The proposal is based upon observations in animal systems indicating that purine biosynthesis may be regulated by the levels of ribose 5-phosphate produced by the oxidative limb of the pentose phosphate pathway. It has been proposed[60] that proline biosynthesis by pyrroline-5-carboxylate reductase (P5CR) is used to produce the NADP required for the synthesis of the purine precursor (ribose 5-phosphate). The activity of P5CR is found to be very high in the nodule cytoplasmic fraction and is higher in ureide-producing nodules as compared to that in amide-producing nodules.[60] These authors have proposed

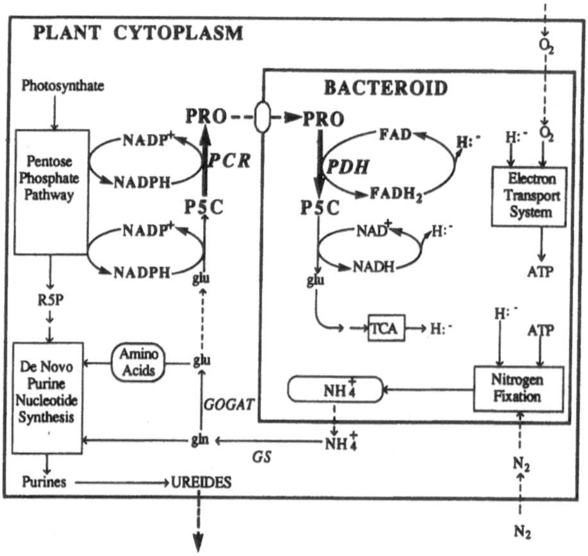

Fig. 4. The proposed role of proline in N_2-fixing, ureide-exporting nodules. Broken lines represent transport, and solid lines show individual reactions. A putative, membrane-bound proline-transport protein is represented on the bacteroid membrane. Reducing equivalents produced in the oxidation of proline, P5C, and glutamate are represented by H^-. Glutamine synthetase and glutamate synthase are represented by GS and GOGAT respectively. P5C, proline-5-carboxylate; P5CR, pyrroline-5-carboxylate reductase; PDH proline dehydrogenase; PRO, proline (see Reference 60 for details). Reprinted with permission from Kohl et al.[60]

a shuttle system (Fig. 4) where some of the proline synthesized in nodule cell cytoplasm by P5CR is catabolized in the bacteroids by proline dehydrogenase. This novel mechanism can indeed save a significant amount of energy for purine biosynthesis. In this regard, the endosymbiont behaves like a mitochondrion. This also suggests that a specific transport system for proline must be present in the peribacteroid membrane. This membrane contains several nodulins,[10] some of which may be involved in such a transport mechanism.

PERSPECTIVES

The nodule tissue constitutes a unique environment where many metabolic reactions require low oxygen while others are highly energy demanding. This paradoxical situation is dealt with by tight compartmentation in nodules, and leghemoglobin plays a major role in balancing the oxygen requirements of this tissue. It has been demonstrated that leghemoglobin not only facilitates diffusion of oxygen for bacteroid respiration, but also helps host mitochondrial respiration.[61] Several enzymes of the carbon and nitrogen metabolic pathways in nodules have evolved into nodule-specific forms which may help improve the efficiency of these enzymes. The genes encoding these enzymes are apparently regulated by different metabolites and _trans_-acting factors that are produced following infection of the plant by _Rhizobium_. Understanding these regulatory circuits would greatly facilitate eventual manipulation of the genes involved. Since the symbiotic state is largely dependent upon the proper balance of carbon and nitrogen in this tissue, an understanding of the regulation of the genes involved in these pathways is of paramount importance in order to improve the process of symbiotic nitrogen fixation.

ACKNOWLEDGMENTS

I wish to thank A. Delauney and S. Barsel for their suggestions and comments on this manuscript and Ms. P. Robinson and P. Snyder for their help in preparation of this manuscript. This study was supported by NSERC Canada at the McGill University and the College of Biological Sciences, the College of Agriculture and the Faculty of Graduate Studies and Research, The Ohio State University.

REFERENCES

1. SCHUBERT, K.R. 1986. Products of biological nitrogen fixation in higher plants:synthesis, transport and metabolism. Annu Rev. Plant Physiol. 37: 539-574.
2. McCLURE, P.E., D.W. ISRAEL. 1979. Transport of nitrogen in the xylem of soybean plants. Plant Physiol. 64: 411-416.

3. ATKINS, C.A., R.M. RAINBIRD, J.S. PATE. 1980. Evidence for a purine pathway of ureide synthesis in nitrogen-fixing nodules of cowpea (Vigna unguiculata L. Walp). Z. Pflanzenphysiol. 97: 249-260.
4. BOLAND, M.J., J.F. HANKS, P.H.S. REYNOLDS, D.G. BLEVINS, N.E. TOLBERT, K.R. SCHUBERT. 1982. Subcellular organization of ureide biogenesis from glycolytic intermediates and ammonium in nitrogen-fixing soybean nodules. Planta 155: 45-51.
5. TAJIMA, S., Y. YAMAMOTO. 1975. Enzymes of purine catabolism in soybean plants. Plant Cell Physiol. 16: 271-282.
6. SCHUBERT, K.R. 1981. Enzymes of purine biosynthesis and catabolism in Glycine max L. Plant Physiol. 68: 1115-1122.
7. ATKINS, C.A. 1981. Metabolism of purine nucleotides to form ureides in nitrogen-fixing nodules of cowpea (Vigna unguiculata L. Walp). FEBS Lett. 125: 89-93.
8. REYNOLDS, P.H.S., M.J. BOLAND, D.G. BLEVINS, K.R. SCHUBERT, D.D. RANDALL. 1982. Enzymes of amide and ureide biogenesis in developing soybean nodules. Plant Physiol. 69: 1334-1338.
9. ATKINS, C.A., P.J. STORER, J.S. PATE. 1988. Pathways of nitrogen assimilation in cowpea nodules studied using $^{15}N_2$ and allopurinol. Plant Physiol. 86: 204-207.
10. VERMA, D.P.S., M.G. FORTIN. 1988. Nodule development and formation of the endosymbiotic compartment. In The Molecular Biology of Nuclear Genes. (I.K. Vasil, ed.), Academic Press, New York, in press.
11. OAKS, A., B. HIREL. 1985. Nitrogen metabolism in roots. Annu. Rev. Plant Physiol. 36: 345-365.
12. CULLIMORE, J.V., M. LARA, P.J. LEA, B.J. MIFLIN. 1983. Purification and properties of two forms of glutamine synthetase from the plant fraction of Phaseolus root nodules. Planta 147: 245-253.
13. DUNN, K., R. DICKSTEIN, R. FEINBAUM, B.K. BURNETT, T.K. PETERMAN, G. THOIDIS, H.M. GOODMAN, F.M. AUSUBEL. 1988. Developmental regulation of nodule-specific genes in alfalfa root nodules. Mol. Plant Microbe Interactions 1: 66-75.

14. GEBHARDT, C., J.E. OLIVER, B.G. FORDE, R. SAARELAINEN, B.J. MIFLIN. 1986. Primary structure and differential expression of glutamine synthetase genes in nodules, roots and leaves of Phaseolus vulgaris. EMBO J. 5: 1429-1435.
15. TINGEY, S.V., E.L. WALKER, G.M. CORUZZI. 1987. Glutamine synthetase genes of pea encode distinct polypeptides which are differentially expressed in leaves, roots, and nodules. EMBO J. 6: 1-9.
16. CULLIMORE, J.V., D.A. LIGHTFOOT, M.J. BENNETT, F.L. CHEM, C. GEBHARDT, J. OLIVER, B.G. FORDE. 1988. Expression of glutamine synthetase and glutamate synthase in root nodules of Phaseolus vulgaris L. In Molecular Genetics of Plant-Microbe Interactions --1988. APS Press, St. Paul, Minnesota, pp. 340-345.
17. SENGUPTA-GOPALAN, C., J.W. PITAS. 1986. Expression of nodule-specific glutamine synthetase genes during nodule development in soybeans. Plant Mol. Biol. 7: 189-199.
18. HIREL, B., C. BOUET, B. KING, D. LAYZELL, F. JACOBS, D.P.S. VERMA. 1987. Glutamine synthetase genes are regulated by ammonia provided externally or by symbiotic nitrogen fixation. EMBO J. 6: 1167-1171.
19. VERMA, D.P.S., M.G. FORTIN, J. STANLEY, V.P. MAURO, S. PUROHIT, N. MORRISON. 1986. Nodulins and nodulin genes. Plant Mol. Biol. 7: 51-61.
20. FULLER, F., P.W. KUNSTNER, R. NGUYEN, D.P.S. VERMA. 1983. Soybean nodulin genes: analysis of cDNA clones reveals several major tissue-specific sequences in nitrogen-fixing root nodules. Proc. Nat. Acad. Sci. USA 80: 2594-2598.
21. LANGSTON-UNKEFER, P.J., A.C. ROBINSON, T.J. KNIGHT, R.D. DURBIN. 1987. Inactivation of pea seed glutamine synthetase by the toxin, tabtoxinine-β-lactam. J. Biol. Chem. 262: 1608-1613.
22. KNIGHT, T.J., R. DICKSTEIN, C. SENGUPTA-COPALAN, P.J. UNKEFER. 1988. Enhancement of symbiotic N_2 fixation in alfalfa and soybean. Fourth International Symposium on Molecular Genetics of Plant-Microbe Interactions, Acapulco, Mexico.
23. DELAUNEY, A.J., Z. TABAEIZADEH, D.P.S. VERMA. 1988. A stable bifunctional antisense transcript inhibiting gene expression in transgenic plants. Proc. Nat. Acad. Sci. USA 85: 4300-4304.

24. SNAPP, S.S., C.P. VANCE. 1986. Asparagine biosynthesis in alfalfa (*Medicago sativa* L.) root nodules. Plant Physiol. 82: 390-395.

25. REYNOLDS, P.M., K.J.F. FARNDEN. 1979. The involvement of aspartate aminotransferase in ammonia assimilation in *Lupin* nodules. Phytochemistry 18: 1625-1630.

26. SHELP, B.J., C.A. ATKINS, P.J. STOVER, D.T. CAANVIN. 1983. Cellular and subcellular organization of patterns of ammonia assimilation and ureides synthesis in nodules of cow pea (*Vigna unguiculata* L. Walp). Arch. Biochem. Biophys. 224: 429-441.

27. HENSON, C.A., M. COLINS, S.M. DUKE. 1982. Subcellular locationalization of enzymes of carbon and nitrogen metabolism in noduler of *Medicago sativa*. Plant Cell Physiol. 23: 227-235.

28. WOO, K.C. 1981. Ureide synthesis in a cell-free system from cowpea (*Vigna unguiculata* L. Walp). Plant Physiol. 37: 1156-1160.

29. NGUYEN, T., M. ZELECHOWSKA, V. FOSTER, H. BERGMANN, D.P.S. VERMA. 1985. Primary structure of the soybean nodulin-35 gene encoding uricase II localized in the peroxisomes of uninfected cells of nodules. Proc. Nat. Acad. Sci. USA 82: 5040-5044.

30. BERGMANN, H., E. PREDDIE, D.P.S. VERMA. 1983. Nodulin-35: a subunit of specific uricase (uricase II) induced and localized in the uninfected cells of soybean nodules. EMBO J. 2: 2333-2339.

31. NGUYEN, J., L. MACHAL, H. VIDEL, C. PERROT-RICHENMANN, P. GADAL. 1986. Immunochemical studies on xanthine dehydrogenase of soybean root nodule. Planta 167: 190-195.

32. LEGOCKI, R., D.P.S. VERMA. 1979. A nodule-specific plant protein (Nodulin-35) from soybean. Science 205: 190-193.

33. VERMA, D.P.S., A.J. DELAUNEY. 1988. Root nodule symbiosis: nodulins and nodulin genes. *In* Temporal and Spatial Regulation of Plant Genes. (D.P.S. Verma, R. Goldber, eds.), Springer Verlag, New York, pp. 169-199.

34. GUTZ, H., H. HESLOT, V. LEUPOLEL. 1974. *Schizosaccharomyces pombe* 2. *In* Handbook of

Genetics. (R.C. King, ed.), Plenum Press, New York, First Edition, pp. 395-446.
35. NIETO, D., R.A. WOODS. 1983. Studies on mutants affecting amidopyrophosphoribosyltransferase activity in Saccharomyces cerevisiae. Can. J. Microbiol. 29: 681-688.
36. FLURI, R., J.R. KINGHORN. 1985. The all2 gene is required for the induction of the purine deamination pathway in Schizosaccharomyces pombe. J. Gen. Microbiol. 131: 527-532.
37. DasSARMA, S., E. TISCHER, H.M. GOODMAN. 1986. Plant glutamine synthetase complements a glnA mutation in Escherichia coli. Science 232: 1242-1244.
38. LEE, M.G., P. NURSE. 1987. Complementation used to clone a human homologue of the fission yeast cell cycle control gene cdc2. Nature 327: 31-35.
39. NIERLICH, D.P., B. MAGASANIK. 1965. Regulation of purine nucleotide synthesis by end-product inhibition. The effect of adenine and guanine ribonucleotides on the 5'-phosphoribosylpyrophosphate amidotransferase of Aerobacter aerogenes. J. Biol. Chem. 240: 358-365.
40. MARTIN, D.W., N.T. OWEN. 1972. Repression and derepression of purine biosynthesis in mammalian hepatoma cells in culture. J. Biol. Chem. 247: 5477-5488.
41. MESSENGER, L.J., H. ZALKIN. 1979. Glutamine phosphoribosylpyrophosphate amidotransferase from Escherichia coli. J. Biol. Chem. 254: 3382-3389.
42. NAGY, M. 1970. Regulation of the biosynthesis of purine nucleotides in Schizosaccharomyces pombe. I. Properties of the phosphoribosylpyrophosphate: glutamine aminotransferase of the wild strain and of a mutant desensitized towards feedback modifiers. Biochim. Biophys. Acta 198: 471-481.
43. TSUDA, M., N. KATUNUMA, G. WEBER. 1979. Rat liver glutamine 5-phosphoribosyl-1-pyrophosphate amidotransferase (EC 2.4.2.14). Purification and properties. J. Biochem. 85: 1347-1354.
44. MANTSALA, P., H. ZALKIN. 1984. Glutamine nucleotide sequence of Saccharomyces cerevisiae Ade4 encoding phosphoribosylpyrophosphate amidotransferase. J. Biol. Chem. 259: 8478-8484.

45. REYNOLDS, P.H.S., D.G. BLEVINS, D.D. RANDALL. 1984. 5-Phosphoribosylpyrophosphate amidotransferase from soybean root nodules: kinetics and regulatory properties. Arch. Biochem. Biophys. 229: 623-631.
46. KANIKO, Y., E.H. NEWCOMB. 1987. Cytochemical localization of uricase and catalase in developing root nodules of soybean. Protoplasma 140: 1-12.
47. GOULD, S.J., G.A. KELLER, S. SUBRAMANI. 1987. Identification of a peroxisomal targetting signal at the carboxy terminus of firefly luciferase. J. Cell Biol. 105: 2923-2931.
48. TRIPLETT, E.W. 1985. Intracellular nodule localization in nodule specificty of xanthine dehydrogenase in soybean. Plant Physiol. 77: 1004-1009.
49. LARSEN, K., B.U. JOCHIMSEN. 1987. Appearance of purine-catabolizing enzymes in fix-plus and fix-minus root nodules on soybean and effect of oxygen on the expression of the enzymes in callus tissue. Plant Physiol. 85: 452-456.
50. MINCHIN, F.R., R.J. SUMMERFIELD, P. HADLEY, E.H. ROBERTS, S. RAWSTHORNE. 1981. Carbon and nitrogen nutrition of nodulated roots of grain legumes. Plant Cell Environ. 4: 5-26.
51. VERMA, D.P.S., K. NADLER. 1984. Legume-rhizobium symbiosis: host's point of view. In Genes Involved in Microbe-Plant Interactions. (D.P.S. Verma, T. Hohn, eds.), Springer-Verlag, Wien, New York, pp. 58-93.
52. MORELL, M., L. COPELAND. 1985. Hexose kinases from the plant cytosolic fraction of soybean nodules. Plant Physiol. 79: 114-117.
53. REIBACH, P.H., J.G. STREETER. 1983. Metabolism of ^{14}C-labeled photosynthate and distribution of enzymes of glucose metabolism in soybean nodules. Plant Physiol. 72: 634-640.
54. DELMAR, D.P. 1972. The purification and properties of sucrose synthetase from etiolated *Phaseolus aureus* seedlings. J. Biol. Chem. 247: 3822-3828.
55. GUPTA, M., P.S. CHOUREY, B. BURR, P.E. STILL. 1988. cDNAs of two non-allelic sucrose synthase genes in maize: cloning, expression, characterization and molecular mapping of the sucrose synthase-2 gene. Plant Mol. Biol. 10: 215-224.
56. THUMMLER, F., D.P.S. VERMA. 1987. Nodulin 100 of soybean is the subunit of sucrose synthase

regulated by the availability of free heme in nodules. J. Biol. Chem. 262: 14730-14736.

57. MORRISON, N., D.P.S. VERMA. 1987. A block in the endocytosis of Rhizobium allows cellular differentiation in nodules but affects the expression of some peribacteroid membrane nodulins. Plant Mol. Biol. 9: 185-196.

58. FORREST, S. 1987. The metabolism of starch in effective and ineffective nodules of soybean. M.Sc. Thesis, McGill University, Canada.

59. BRISSON, N., H. GIROUX, M. ZOLLINGEN, A. CAMIRAND, C. SIMERD. 1988. Maturation and subcellular compartmentation of potato starch phosphorylase. J. Cell Biol., submitted.

60. KOHL, D.H., K.R. SCHUBERT, M.B. CARTER, C.H. HAGEDORN, G. SHEARER. 1988. Proline metabolism in N2-fixing root nodules: energy transfer and regulation of purine synthesis. Proc. Nat. Acac. Sci. USA 85: 2036-2040.

61. SUGANUMA, H., M. KITOU, Y. YAMAMOTO. 1987. Carbon metabolism in relation to cellular organization of soybean root nodules and respiration of mitochondria aided by leghemoglobin. Plant Cell Physiol. 28: 113-122.

Chapter Three

SYNTHESIS, TRANSPORT, AND UTILIZATION OF PRODUCTS OF SYMBIOTIC NITROGEN FIXATION

JOHN S. PATE

Department of Botany
University of Western Australia
Nedlands
Western Australia 6009

INTRODUCTION

Nitrogen fixation by symbiotic associations between bacteria, cyanobacteria or actinomycetes and certain taxa of vascular plants provides ground for study of a number of aspects of plant nitrogen metabolism. Interest for biochemists derives from both the commonality and diversity displayed by these associations in metabolic pathways for nitrogen, starting with synthesis of NH_4^+ as first stable product of fixation by the microsymbiont, through primary and secondary products of NH_4^+ assimilation to final incorporation of the nitrogen into protein and solute reserves of host plant tissues. For plant physiologists, principal concerns are the mechanisms and routings by which fixation products are exported to and thence partitioned within the host, the interaction of nitrogen fixation and photosynthetic fixation of carbon, and the relationships between nitrogen fixation and plant growth and reproduction. Agronomists and ecologists are especially interested in the performance of N_2-fixing associations in replenishing the nitrogen capital of ecosystems, leading to consideration of the cost effectiveness of nitrogen fixation versus nitrogen fertilization in agricultural systems, and the competitiveness of nitrogen fixing associations in stress- and predator-prone natural environments.

This chapter highlights the cardinal features of the nitrogen metabolism of a diverse range of N_2-fixing associations, wherever possible in relation to the issues raised above. Great uneveness in knowledge of different associations leads to heavy bias towards a few well investigated herbaceous legumes, and it must be stressed that the picture presented here for such species is not necessarily indicative of how other associations might perform. Space limitation prevents an exhaustive review of the topic, so the bibliography is restricted to key reviews and especially relevant experimentally-based investigations.

AMMONIUM ASSIMILATION FOLLOWING NITROGEN FIXATION, AND SUBSEQUENT TRANSFORMATIONS IN SYMBIOTIC ORGANS

There is now undisputed evidence from studies using ^{15}N- or ^{13}N-labelled nitrogen gas that nodules of a number of legumes genera and the Azolla:Anabaena association generate free ammonia as the first fixation product of the microsymbiont.[1-4] Although some NH_4^+ may be utilized directly, most is immediately excreted to the host, either apoplastically via the leaf cavity in the case of Azolla, or from the bacteroid across bacterial and peribacteroid membranes to the host cytoplasm in the case of the legume nodule.[5,6] Involvement of NH_4^+ in the initial transfer of nitrogen to host has been less convincingly demonstrated for symbioses involving cycads and the cyanobacterium Nostoc,[7] or between various angiosperm genera and the Actinomycete Frankia.[8]

Primary Assimilation Products

Support for the existence of separate NH_4^+-assimilating systems in host and microsymbiont comes from short term $^{15}N_2$ labelling studies on legume nodules.[6,8] Soybean bacteroids have been shown to become heavily labelled in glutamic acid, followed by alanine and glutamine, whereas ^{15}N in the host cytosol is recovered principally in glutamine and to a lesser extent in glutamate, alanine and the ureide alantoin.[9,10] Similarly, ^{15}N labelling of alfalfa bacteroids is in glutamate, aspartate and asparagine, while that of the host tissues is mainly in asparagine.[11] Labelling studies with $^{15}N_2$ on the Azolla:Anabaena association also suggest glutamine, glutamic acid and alanine to be primary acceptors of fixed nitrogen.[2,12]

Where the ^{15}N labelling of the two nitrogen atoms of glutamine has been examined, consistently greater enrichment of the amide nitrogen than the amino nitrogen atoms has been demonstrated. This applies to $^{15}N_2$ feeding experiments on legume nodules,[1] to $^{13}NH_4^+$ feeding of nodules of soybean and Alnus,[4,8] and to $^{15}N_2$ feeding of Nostoc-containing coralloid roots of the cycad Macrozamia riedlei.[7] It is also generally found that ^{13}N- or ^{15}N-enrichment of glutamine and glutamate precedes that of other products, whether these be asparagine or the ureides citrulline, allantoin and allantoic acid.[4,9,13,14]

It is now widely accepted that the glutamine synthetase:glutamate synthase (GS:GOGAT) system, which operates widely in NH_4^+ assimilation of a wide variety of plants heterotrophic for nitrogen,[15,16] fulfills a similar function in legume nodules fixing nitrogen and, in certain instances, is operative in both bacteroid and host cytosol.[4,5,17] Glutamate dehydrogenase (GDH) may also be active, though probably not as a major contributor to ammonia assimilation. These enzymes of ammonia assimilation are present in ineffective nodules or in potentially effective nodules not yet fixing nitrogen.[4,18] Three isozymes of GS have been demonstrated in nodule extracts of Phaseolus, one specific to nodules that increases with increased nitrogen fixation.[19]

Convincing support for involvement of the GS:GOGAT system in soybean nodules comes from the demonstration that if $^{15}N_2$ feeding of attached nodules is accompanied by injections of methionine sulfoximine (MSX), an inhibitor of GS but not GDH, unassimilated $^{15}NH_4^+$ builds up to levels six times those in non-injected nodules. Similar treatment with azaserine, an analogue of glutamine which inhibits all glutamine-mediated amide transfer reactions, leads to lower ^{15}N-incorporation into the amino nitrogen atom of glutamine than in control nodules treated only with $^{15}N_2$.[9,10]

The situation is less clear regarding ammonia assimilation in non-leguminous symbioses. Results of $^{13}NH_4^+$ feeding of Alnus nodules indicate a major involvement of GS but not GOGAT, consistent with demonstrations of GS and GDH but not appreciable GOGAT activity in host cytosol of nodules of this association.[4,8,20] Greater GDH activity than GS activity has been recorded in Azolla:Anabaena associations.[2,12] Some 75% of the GS:GOGAT activity in this association is in host tissue where most assimilation of ammonia is presumed to occur.[12] In cycad:Nostoc symbioses, immuno-gold labelling techniques have demonstrated GS in both heterocysts and vegetative cells of the cyanobacterium in coralloid roots of Cycas revoluta and Zamia skinneri, suggesting that at least some assimilation of ammonia takes place in the microsymbiont, albeit possibly only for its own growth.[21,22] $^{15}N_2$ labelling of coralloid roots of C. revoluta leads to almost equal labelling of the two nitrogen atoms of the glutamine pool of the whole coralloid root with ^{15}N, and 95% of the nitrogen of the ethanol-soluble fraction of the coralloid root remains attached to this amide 24 hours after feeding $^{15}N_2$.[7]

Secondary Assimilation Products

Assuming a major role for GS, GDH and/or the GS:GOGAT system in assimilation of ammonia generated in nitrogen fixation, the resulting glutamyl nitrogen will obviously serve as a donor of nitrogen to synthesis of a number of additional fixation products. The compounds formed, and the proportion which each represents of the total pool of fixed nitrogen, vary widely between symbioses, and, as will be seen, these differences are ultimately reflected in the spectrum of products passed from symbiotic organ to host. Four compounds are principally involved, viz. asparagine, citrulline, allantoin and allantoic acid. Mechanisms of synthesis of each is now considered in turn, using Figures 1 and 2 to detail the respective synthetic pathways and suggest possible relationships with ammonia assimilation and dark fixation of CO_2.

Asparagine synthesis by legume nodules is generally held to be mediated by asparagine synthetase, a glutamine-specific transamidase converting aspartate to asparagine (Fig. 1).[4,11,23,24] This enzyme has been purified from the cytosol of lupin nodules and also shown to be active in nodules of soybean and alfalfa.[23,25-27] It and aspartate aminotransferase increase in activity with the onset of fixation, but rates of activity are difficult to measure in crude nodule extracts because of the rapid metabolism of aspartate and asparagine,[4] the latter caused largely by the active asparaginase system of bacteroids.[28] However, the dark $^{14}CO_2$ fixation activity shown by intact nodules, attributable mostly to their extremely active phosphoenolpyruvate (PEP) carboxylase system,[29-31] has been used as an indirect assay of asparaginase synthesis, by measuring rates of incorporation of ^{14}C into the C1-carboxyl group of the amide.[4] This anapleurotic pathway has been estimated to supply up to 25% of the carbon returned from nodule to host plant in asparagine-forming symbioses such as lupin and alfalfa.[4,27,31,32] Its significance in the overall economy of carbon in legume nodules will be considered later.

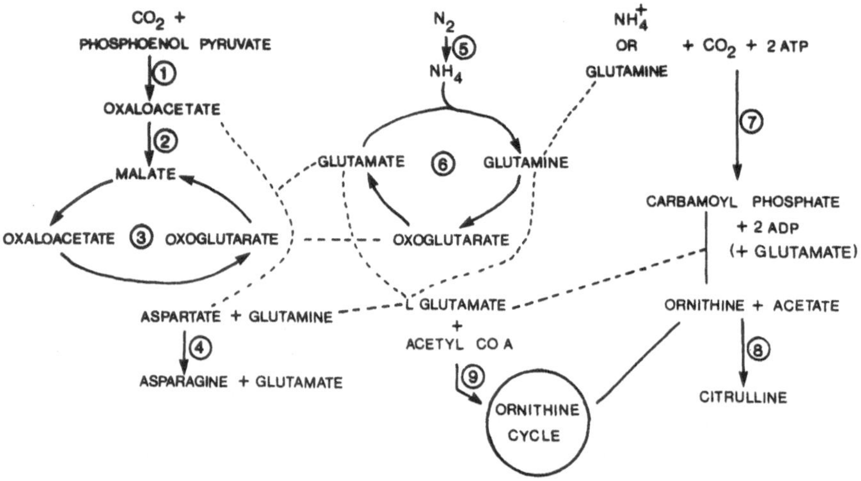

Fig. 1. Metabolic relationships between nitrogen fixation, dark CO_2 fixation and amide and citrulline synthesis in symbiotic associations. Enzymes in the various pathways as follows: (1) PEP carboxylase; (2) malate dehydrogenase; (3) enzymes of the TCA cycle; (4) asparagine synthetase; (5) nitrogenase; (6) GS:GOGAT system; (7) carbamoylphosphate synthetase; (8) ornithine carbamoyltransferase; (9) acetyl CoA:glutamate N-acetyltransferase. See text for key references supporting the scheme depicted.

More direct evidence of the involvement of asparagine synthetase comes from experiments in which $^{15}NH_4^+$ and ^{15}N (amide label)-glutamine were fed to soybean nodules and asparagine labelled in both nitrogen atoms was subsequently recovered from the nodules. However, inhibitor studies using MSX and azaserine on alfalfa nodules have disputed any absolute requirement for glutamine in asparagine synthesis.[75] Not all legume nodules are active in asparagine synthesis, for example negligible incorporation of ^{15}N from $^{15}N_2$ occurs into the amide in cowpea nodules, despite heavy labelling of ureide and glutamine.[33-35] Mechanisms of asparagine synthesis are not clear for non-leguminous symbioses, although these associations may show highly active PEP carboxylase sponsored-$^{14}CO_2$ fixation.[7,36,37]

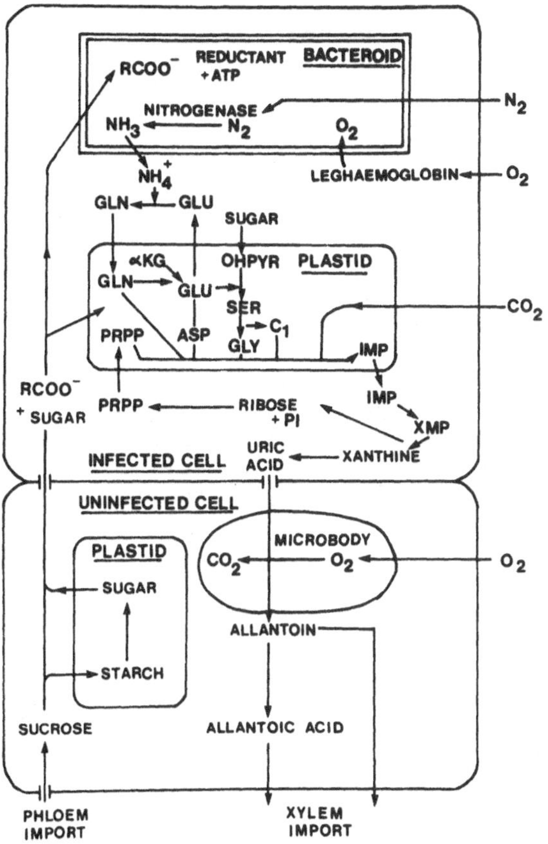

Fig. 2. Principal metabolic events and intercellular and subcellular locations in the synthesis of the ureides allantoin and allantoic acid in the nitrogen-fixing legume nodule. See text for further description of the pathway and the significance of its compartmentation within the bacterial tissue of the nodule. The author is greatly indebted to C.A. Atkins for providing this figure as a synopsis of current progress in elucidating the probable pathways involved.

Studies of the synthesis of citrulline by symbiotic organs have been confined almost entirely to the *Alnus*: *Frankia* association. Synthesis of the ureide has been suggested to occur in mitochondria and to involve the condensation of ornithine and carbamoyl phosphate by the enzyme ornithine carbamoyltransferase (Fig. 1).[4,38] The ornithine involved is likely to be formed via the N-acetylornithine cycle, with acetyl-CoA, glutamate and ATP as substrates, while the required carbamoyl phosphate probably results from activity of carbamoyl phosphate synthetase.[36] The latter reaction would require inputs of CO_2, ATP and NH_4^+ or the amide nitrogen of glutamine. $^{14}CO_2$ feeding of *Alnus* labels the carbamoyl carbon of citrulline, consistent with the operation of carbamoyl phosphate synthetase.[36] A second anapleurotic input into the Cl position of citrulline might also result from PEP carboxylase activity.[4,36,37] As indicated in Figure 1, this would involve malate, formed by PEP carboxylase activity, transferring fixed carbon to oxoglutarate via the tricarboxylic acid cycle, and thence to glutamate and finally to the ornithine moiety of citrulline.

In agreement with the above, $^{15}N_2$ labelling of *Alnus* nodules[4] and coralloid roots of the citrulline-forming cycad *M. riedlei*[7] show heavy incorporation initially into the amide group of glutamine and thence into the carbamoyl nitrogen of citrulline. $^{14}CO_2$ feeding of coralloid roots of *M. riedlei*[7] shows labelling patterns similar to those indicated above for *Alnus*. The extent of coupling, if any, between GS:GOGAT, GS and GDH activity and citrulline synthesis has not been properly explored in either the cycad or the *Alnus* systems.

Detailed discussion of the pathway of ureide biogenesis in legume nodules has been well reviewed recently[4,39-42] and will not be repeated here. In brief, allantoin and allantoic acid are formed by oxidative catabolism of purines, the latter having been synthesized from fixed nitrogen and host-derived carbon.[43] The scheme shown in Figure 2 summarizes the principal events, and the likely subcellular and intercellular location of these events within the bacterial tissues of a ureide-producing species such as cowpea or soybean. Synthesis of purines *de novo* apparently occurs in the plastids of the infected cells, and requires provision of glutamine, aspartate, and glycine derived directly or indirectly from the ammonia-assimilating system of the host cytosol.[40] Glycine provides 2 carbon atoms as well as one nitrogen atom to the purine ring; glutamine and aspartate 2 carbon and 1 nitrogen atoms respectively.[4,39,41] One of the three remaining carbon atoms is derived anapleurotically from CO_2, the other two from methenyl tetrahydrofolic acid and formyl tetrahydrofolic acid.[4] The compound 5-phosphoribosyl-1-pyrophosphate (PRPP), derived from ATP and ribose-5-phosphate, is identified as a purine precursor in that inosine 5-monophosphate (IMP) is the final product exported from the plastid.[4,44] In the host cytosol, IMP is converted to xanthine monophosphate (XMP) by IMP dehydrogenase;[44] the XMP is then converted to xanthine, and the latter to uric acid by xanthine dehydrogenase.[34,45-47] At this point uric acid is suggested to move symplastically to the better-oxygenated uninfected cells of the bacterial tissue (Fig. 2) where it is oxidized to allantoin by uricase,[48] located in the prominent microbodies within these cells.[49,50] Finally, allantoin is partly converted to allantoic acid by allantoinase, and this product and unmetabolized allantoin then become available for xylem export from the nodule.[4]

General evidence in support of de novo purine synthesis comes from the demonstration of high levels of enzymes involved in synthesis of the purine precursor and the various purines in nodules which engage in substantial net synthesis of ureides.[4,35] These enzymes are induced at the onset of nitrogen fixation and decline in activity when that fixation is temporarily inhibited.[51-53] Evidence for oxidative catabolism of purines is similarly based, levels of xanthine dehydrogenase, uricase and allantoinase being high in ureide producers but not in systems forming amides.[4,35] The localization of uricase in microbodies of uninfected cells is suggested by both immuno-gold labelling techniques[50,54-56] and enzymatic analyses following ultracentrifugation of organelles on density gradients.[4,21] The presence of populations of uninfected cells, strategically dispersed among the bacterial zone of all legume nodules so far known to specialize in synthesizing allantoin and allantoic acid,[39,54,56,57] attests to a probable universality in the type of compartmentalization illustrated in Figure 2.

IDENTITY OF SOLUTES CARRYING FIXED NITROGEN FROM SYMBIOTIC ORGAN TO HOST

Having ascertained the major enzymatic pathways for the assimilation of ammonia in symbiotic organs, the next obvious step is to determine what types and amounts of nitrogen compounds are transferred to the host. This has been studied mainly by examining the composition of xylem sap exuding under root pressure from decapitated, symbiotically-dependent plants. This of course assumes that fixed nitrogen does indeed move to the host exclusively through the xylem elements. However, significant proportions of the nitrogen ascending in the xylem stream clearly can result from root catabolism, or may be in the process of circulating between shoot and root. Therefore, more definitive procedures are required to distinguish geniune fixation products.[58,59] Thus, if freshly detached symbiotic organs bleed spontaneously from xylem, these exudates can be examined directly without any possible contamination from the products of root xylem transport.[58,60] Alternatively, if xylem exudates can be obtained from root systems cut above or below the zone of symbiotic organs, comparisons of the classes and relative amounts of solutes present in the respective samples will provide an indirect measure of which types and amounts of solutes are likely to have originated directly from recent nitrogen fixation.[58] More definitively, ^{15}N-labelled solutes in xylem exudates can be identified following exposure of detached symbiotic organs or nodulated root systems to $^{15}N_2$, but even this may provide misleading information unless time courses of ^{15}N-exchange with all components of the system are followed.[7]

Several authors have compiled data on nitrogenous solutes in the xylem sap of nitrogen-fixing plants.[4,58,60,61] These and other unpublished data from the author's laboratory are summarized in Table 1, from which the following conclusions emerge:

1. Cycads studied thus far comprise genera whose cyano-bacterial coralloid roots export principally glutamine and some glutamic acid (Bowenia and Cycas), and genera exporting the ureide citrulline and, in some cases, also some glutamine (Macrozamia, Dioon, Certozamia, Stangeria, Encephalartos and Lepidozamia).[7]

This distinction concurs with the above mentioned data for the $^{15}N_2$ fixation products identified for M. riedlei and C. revoluta.[7]

2. Actinorhizal plants, involving Frankia and any one of a taxonomically diverse range of non-leguminous angiosperm genera, provide examples of associations in which asparagine is the principal export,[62] sometimes accompanied by relatively large amounts of glutamine. In other taxa (e.g., Alnus spp.) citrulline predominates.[4]

3. Symbioses involving leguminous plants of the family Fabaceae conform to the general rule that members of tribes of temperate origin (e.g., Trifolieae, Genisteae and Vicieae) export the amides asparagine and glutamine, whereas species of certain tribes of tropical origin (e.g., Phaseoleae) transport relatively large amounts of the ureides allantoin and allantoic acid.[59,63] In certain cases the ureide is accompanied by appreciable amounts of amide.[14,59] It is generally true that the amide component of xylem sap of ureide producers is richer in glutamine than asparagine, whereas that of non-ureide producing species shows higher levels of asparagine than glutamine, often accompanied by appreciable quantities of aspartic acid but not of glutamic acid.[4,64]

4. All Fabaceae which export mainly ureides when fixing nitrogen transport principally asparagine, a lesser amount of glutamine, and very much reduced amounts of allantoin and allantoic acid when growing non-symbiotically on NH_4^+ or NO_3^-.[61,65] Accordingly, ratios of ureide nitrogen:non-ureide amino nitrogen in xylem sap have been utilized as a basis for field assays of plant dependence on nodule fixation as opposed to soil mineral nitrogen.[64,65] The ratios of glutamine to asparagine and of glutamine + ureide to asparagine + NO_3^- also decline predictably with decreasing dependence on symbiosis.[64] However, in asparagine-dominated species, this amide remains as the major organic solute of xylem regardless of whether plants are using nitrogen, NH_4^+ or NO_3^-.[59] In such plants the ratio of asparagine:nitrate in sap may be useful in assessing symbiotic dependency,

Table 1. Major nitrogenous solutes exported in xylem of symbiotic N_2-fixing associations.[1]

	Cycads	Actinorhizal Plants	Legumes (Fabaceae and Mimosaceae)
Asparagine	–	Ceanothus, Comptonia, Eleagnus, Gale, Myrica, Parasponia	Trifolium, Vicia faba[2]
Asparagine and Glutamine	–	Datisca, Myrica cerifera	Acacia,* Cicer, Genista,* Lathyrus, Lens, Lupinus, Pisum, Spartium, Vicia spp.
Glutamine	Cycas, Bowenia	–	–
Citrulline	Stangeria,* Dioon*	Alnus	–
Citrulline and Glutamine	Macrozamia, Encephalartos Lepidozamia Ceratozamia*	–	–
Allantoin and Allantoic Acid (+ some amide)	–	–	Cyamopsis, Glycine, Macrotyloma, Phaseolus, Psophocarpus,[2] Vigna[2]
Asparagine and Glutamine (+ some citrulline)	–	Casuarina, Coriaria	Albizia

Asparagine and Glutamine (+ some allantoin and allantoic acid)	*Erythrine*,[2*] *Hardenbergia*,[*] *Kennedia*,[*] *Pisum*, *Bossiaea*,[*] *Cajanus*

[1]Category rated on the basis of the major designated compounds accounting for 70% or more of xylem sap nitrogen. Major compounds listed in order of abundance.

[2]From assays of bleeding sap of detached nodules, otherwise root bleeding sap of symbiotically-dependent plants.

Data in Table derived from references cited in text, except for asterisked entries which represent unpublished data from the author's laboratory.

but only if nitrate is the major source of nitrogen available to the rooting medium, and if the species in question reduces nitrate principally in its shoot system.[66-70]

5. A number of Fabaceae, (e.g., Pisum, Hardenbergia, Erythrina, Bossiaea, Kennedia) show detectable levels of allantoin and allantoic acid in their xylem in addition to the dominant compounds asparagine and glutamine, but the relationship of this ureide component to nitrogen fixation has yet to be evaluated.[59]

6. Species of mimosoid legumes (Mimosaceae) have been poorly studies with regard to xylem sap composition, but both asparagine exporters (e.g., Acacia spp.) and amide ureide exporters (e.g., Albizia) have been identified.[4,59] Information on translocated products in caesalpinioid legumes (Caesalpiniaceae) is lacking, largely because of the paucity of nodulated species in this family.

7. In addition to the major amide and/or ureide component mentioned above, xylem sap of leguminous, actinorhizal and cycad symbioses normally contains a broad spectrum of minor constituents, including most or all protein amino acids and γ-amino butyric acid.[4,7,59,71-73] In certain instances unusual non-protein amino acids may be encountered [e.g., homoserine in Vicieae,[59,74,75] djenkolic acid in Acacia spp.[76] canavanine in Canavalia spp[77-79] and γ-methylene glutamine and γ-methylene glutamic acid in peanut (Arachis hypogaea)].[67,80] The significance of these is discussed later.

Evidence presently available for legumes points to a strong correlation between ureide (allantoin and allantoic acid) production and the possession of determinate nodules, in which the vascular network consists of a number of closed looped strands joining at either end to the vascular connection with xylem of the host root.[59,63] In such a system it is possible for fixation products to be flushed out of the nodule vasculature by the host plant transpiration stream. Indeed, the high water fluxes likely within such a system may be critically important in view of the six to eight fold lower solubility of ureides than amides.[63] Xylem exudates from detached nodules of Psophocarpus tetragonolobus, Vigna unguiculata and Erythrina crista galli show total ureide levels (allantoin + allantoic acid) as high as 40-60 mM, causing these compounds to precipitate when sap is cooled.[59] However, studies on cold tolerant cultivars of the ureide-producing bean Phaseolus vulgaris indicate no tendency for plants to switch to an amide-based metabolism when grown at low temperature,[81] nor for their ability to fix and export nitrogen to be reduced under those conditions. This would suggest that the low solubility of ureide is not a real impediment to its transport from the presumably well-flushed xylem of nodules of intact plants of ureide forming species.

In direct contrast to the above, virtually all temperature herbaceous legumes, and a large number of tropical and subtropical ones with an asparagine-dominated metabolism, possess indeterminate nodules with apical or peripheral meristems.[63] Here the vascular system consists of a number of branched strands emanating from the main vascular tissue entering the nodule from the host root and ending blindly in proximal parts of the nodule close behind the nodule meristem(s).[60,63] Certain species in this category possess specialized pericycle transfer cells which apparently secrete amides and other fixation products to the xylem elements.[60] Water is then attracted osmotically into the solute-enriched apoplast of the vascular strand across the well-developed endodermis of the nodule vascular strands, so that when a nodule is detached and placed with its meristem downwards on moist filter paper, fixation products bleed out of the cut nodule xylem under positive pressure.[60] Osmotically-operated export systems of this nature are not restricted to legume nodules bearing transfer cells, since determinate nodules of a number of ureide-producing legumes, and the coralloid roots of cycads, bleed profusely from xylem when detached from their parent root.[7,59,60,67] Incidentally, citrulline has a solubility several times that of most amino acids and almost 20 times that of allantoin and allantoic acid, and may therefore be an ideal solute for nitrogen transport in open unflushed xylem strands.

Fluxes of amino compounds across the plasma membranes of the pericycle transfer cells of vascular strands of rapidly fixing nodules of *Pisum sativum* have been estimated at 24 p moles·cm^{-2}·s^{-1}. This is a relatively high flux value, even in comparison with secretory structures such as nectaries and salt glands.[82] High throughputs of nitrogen must certainly be required when one considers that nodules, collectively comprising only 6 to 10% of the total mass of the plant, supply the total requirement of the host for nitrogen, yet do so through a vascular network which occupies only a few percent of the total volume of the nodule.[60]

CATABOLISM OF PRODUCTS OF NITROGEN FIXATION IN HOST TISSUES AND ORGANS

Two possible mechanisms of asparagine catabolism have been suggested[4,83] by nitrogen fixing associations (Fig. 3A). One involves the direct breakdown of the amide to aspartate and ammonium by asparaginase, with the possible reassimilation of ammonia by the GS:GOGAT system.[34] The other consists of the transfer of the amino nitrogen atom of asparagine by asparagine aminotransferase to amino acids such as alanine, glycine or homoserine, with the attendant production of 2-oxosuccinamate.[83] The latter product may then be reduced to 2-hydroxysuccinamate, followed by utilization of the amide N through the activity of appropriate amidohydrolases.[83,84] Present evidence suggests that asparagine aminotransferase is generally important in the metabolism of leaves of asparagine-forming legumes such as lupin[85] and pea,[83] while asparagine is the predominant catabolic system of developing seeds[86] and nodules in these species and in soybean.[23,87-89] Using the deamidase inhibitor 5-diazo, 4-oxo-L-norvaline and the transaminase inhibitor aminooxyacetate in conjunction with xylem feeding of ^{15}N-(amide)labeled asparagine, it has been concluded that utilization of the amide nitrogen atom of asparagine in pea leaves involves some direct deamidation of asparagine and some deamidation of its transmination product 2-hydroxysuccinamate.[84] In control plants not fed inhibitors, the ^{15}N-labelled amide group of both molecules accumulates predominantly in the amide group of glutamine, but also in other amino acids.[84,90] This transfer is effectively blocked by MSX, an inhibitor of GS.[84]

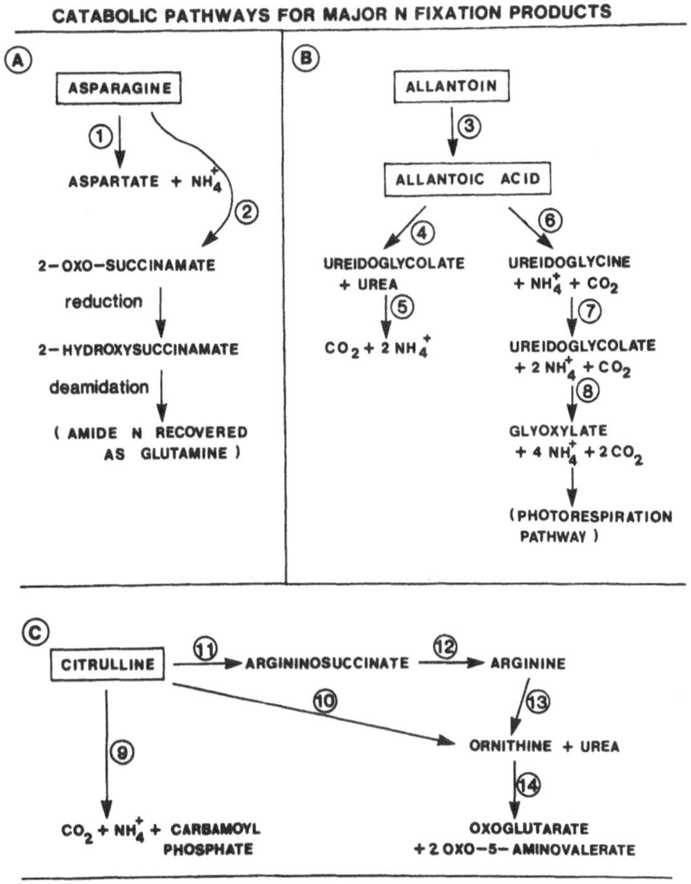

Fig. 3. Suggested pathways of breakdown of nitrogen fixation products in host plants of symbiotic associations fixing atmospheric nitrogen. Two alternative pathways of asparagine catabolism are indicated, one involving asparaginase (Reaction 1) and the other asparagine aminotransferase (2). Allantoin catabolism is first to allantoic acid via allantoinase (3) and thence either to carbon dioxide and ammonia by allantoicase (4) and urease (5), or to glyoxylate, carbon dioxide and ammonia by the successive action of allantoate amidohydrolase (6), ureidoglycine aminohydrolase (7) and ureidoglycolate amidohydrolase (8). Citrulline degradation is pictured as reversal of the citrulline synthesis pathway (9), or

(continued next page)

Asparaginase plays a key role in the nitrogen nutrition of the seed storage proteins of legumes, as indicated by feeding experiments using ^{14}C, ^{13}N, ^{15}N-asparagine coupled to cross polarization NMR in soybean,[91] or ^{14}C, ^{15}N-(amide)labelled asparagine in lupin.[92] However, direct incorporation of unmetabolized asparagine accounts for half of the asparagine residues in storage protein of these species.[91 92] The asparaginase of legume seeds is K^+ dependent,[93] and peaks first in the seed coat and then in the developing embryo.[92,93]

Bacteroids of legume nodules contain an especially active asparaginase,[23,28] and since early growth of bacterial tissue is dependent on amide nitrogen released from cotyledon reserves, this capacity for amide catabolism may be crucial for nodule development.[33,58] Empirically based models of carbon and nitrogen exchanges between nitrogen-fixing nodules and parent plants of cowpea and lupin (see Fig. 4) indicate that the growth of nodules remains partly dependent on phloem-borne asparagine from the host,[32] so the bacteroid asparaginase system may continue to play a key role in nodule metabolism. This would be all the more important were the microsymbiont inactive in synthesis of the amide, despite proven ability to incorporate some of its own fixed NH_4^+ into bacteroid protein.[33]

The mechanism for degradation of citrulline in nitrogen fixing associations is still not properly understood. Drawing from the general literature on plant, animal and bacterial systems, three possible catabolic pathways have been suggested[4] (see Fig. 3B). One would involve argininosuccinate and arginine with subsequent degradation of the arginine by arginase to

---

(Fig. 3 continued)

by formation of ornithine and urea by citrulline hydrolase (10) or by the sequential action of arginosuccinate synthetase (11), arginosuccinate lyase (12) and arginase (13) with subsequent catabolism of ornithine by 2-oxoglutarate aminotransferase (14). See text for key references on which the schemes are based.

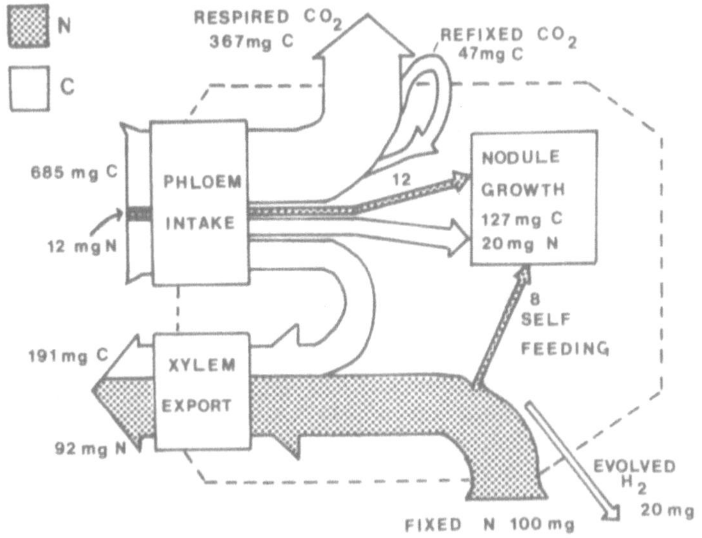

Fig. 4. Exchanges of C and N between nodule and parent plant of white lupin (Lupinus albus) in relation to nitrogen fixation, incorporation of C and N into nodule dry matter, and respiratory loss and dark fixation gains of CO_2 by the nodule. The model depicts the relative extents to which the nitrogen requirements of nodule growth are met by phloem import or direct incorporation of fixed nitrogen. Amounts of H_2 evolved by the nodule are indicated. Data refer to a 10-day period of growth immediately preceding flowering, and items of the nodule budget are given relative to fixation of 100 mg N by the nodule. Data from Reference 32.

yield urea and ornithine. A second would be by reversal of the already mentioned synthetic pathway for citrulline synthesis. The third would require direct hydrolysis of citrulline by citrulline hydrolase to form ornithine. With the last mentioned, ornithine catabolism would be effected by ornithine:2-oxoglutarate aminotransferase yielding 2-oxo-5-aminovaleric acid and glutamate.[4]

The enzymology of breakdown of allantoin and allantoic acid in ureide-forming symbioses is also not fully elucidated, despite the importance of ureide-bound nitrogen in the nitrogen nutrition of such plants. Assuming allantoin to be the major starting compound, the first step would be hydrolysis to allantoic acid by allantoinase.[4,94] This enzyme occurs widely in the tissues and organs of seedling, vegetative and reproductive stages of a number of legume species. It has been isolated and purified from leaves and fruits of soybean, and even shown to cause hydrolysis of exogenously supplied allantoin in water cultures of soybean.[95] Particularly active preparations of allantoinase have been obtained from leaves where much of the ureide arriving from the root is likely to be metabolized.[42,96,97] Non-photosynthetic paraveinal mesophyll of soybean shows activity, just as does mesophyll tissue.[98] The unusually high levels of allantoic acid in petiole, stem and fruit stalks of soybean and cowpea suggest that degradation of allantoic acid may be rate limiting in the utilization of ureide.[42,96,99]

Allantoicase, the enzyme degrading allantoic acid to urea, $2NH_4^+$ and ureidoglycolate (Fig. 3B) has been demonstrated for cotyledons of peanut,[4] but as yet not conclusively for other legume tissues. The other possible candidate is allantoate amidohydrolase, which degrades allantoic into CO_2, NH_4^+ and ureidoglycine (but not urea). This enzyme has been demonstrated in the seed coat of maturing seeds of soybean.[4] Evidence suggesting a major involvement of allantoicase in soybean leaves comes from the finding that acetohydroxamate, an inhibitor or urease, greatly reduces the production of NH_4^+ and $^{14}CO_2$ from 2-^{14}C-allantoin by tissue slices, a result consistent with the requirement for urea catabolism in the pathway via allantoicase but not via amidohydrolase.[100] However, completely opposite findings have been recently reported for intact soybean leaf tissue, based on the finding that phenyl phosphordiamidate, a potent inhibitor of urease, does not inhibit the release of $^{14}CO_2$ from 2,7-^{14}C-allantoin, but does inhibit strongly the release of $^{14}CO_2$ from ^{14}C-urea. In addition the feeding of allantoin ^{14}C-labelled in the 4- and 5-positions yields $^{14}CO_2$ and ^{14}C-labelled glyoxylate, glycine and serine.[101] This would suggest that urea and urease[102,103] are not involved in allantoate metabolism and that carbons 4 and 5 of the ureide may enter the photorespiratory pathway, possibly through amidohydrolases located in plastids.[101,104]

QUANTITATIVELY-BASED CASE STUDY OF THE PARTITIONING AND UTILIZATION OF NITROGENOUS SOLUTES IN THE HOST PLANT OF AN AMIDE-PRODUCING SYMBIOSIS

The species selected was white lupin (Lupinus albus). This plant bleeds profusely from phloem when shallow incisions are made in old or young stem tissue, leaf petioles, inflorescence stalks and distal tips of fruits, thereby enabling virtually all major translocation systems to be assayed for solutes.[75,105,106] The xylem stream of N_2-fixing plants carried over half of its nitrogen as asparagine, a further 15-25% as glutamine, and most of the remaining nitrogen as aspartic acid and lesser amounts of serine, valine and lysine.[107] Comparisons of the composition and total concentrations of nitrogen in samples of xylem sap collected above and below the zone of nodules suggest that some 90% of the nitrogen carried to a shoot is newly fixed nitrogen exported from nodules.[58] This component is more heavily biased towards asparagine, glutamine and aspartic acid than is that originating in roots distal to nodules.[58,71] Phloem sap from different regions of the shoot shows remarkable constancy in the relative proportions of nitrogenous solutes,[71] though differing widely from site to site on a shoot in balance between sugar and amino compounds, and hence in the C:N ratio of the translocation streams.[85] The spectrum of amino compounds in phloem sap is essentially that found in xylem, but concentrations of total nitrogen are 10-15 times greater, and amides and aspartic acid comprise a smaller, though still dominant, fraction of sap nitrogen.[71,85,107]

Partitioning of Nitrogen in Vegetative Stages of Growth

Studies involving the supplying of ^{14}C- and ^{15}N-labelled xylem amino acids to detached vegetative shoots of white lupin have indicated that each amino compound is partitioned and metabolized in a highly characteristic fashion. Short feeding times (20 min) followed by a 10-min chase in unlabelled xylem sap have been used to simulate primary distribution of each compound in the transpiration stream.[108] Autoradiographs of such shoots indicate that the ^{14}C of basic xylem amino compounds such as arginine is abstracted mainly by stem and leaf vascular tissue, so that hardly any anionic label breaks through to the mesophyll.[85,108] In complete contrast the ^{14}C of the dicarboxylic amino acids aspartate and glutamate accumulates intensely in leaf mesophyll and not appreciably in vascular tissue.[108] Valine, glutamine and asparagine yield an intermediate type of labelling pattern in which the ^{14}C is fairly evenly shared between stem, petiole and leaflets.[108]

In longer labelling studies, the fate of both the carbon and nitrogen of an amino compound has been followed within the shoot, especially in relation to the subsequent appearance in phloem sap of ^{14}C or ^{15}N still attached to the compound fed or transferred to other amino compounds or non-amino components.[59,105] By considering this information alongside that on relative proportions of amino compounds in xylem and phloem streams, a quantitatively based picture has been obtained of the principal routings for carbon and nitrogen during transfer from the xylem to the phloem within the leaf.[85,107] Transfer of the carbon and nitrogen of the amides, as well as valine, serine and threonine, is both rapid and effective with little metabolism prior to transfer. Conversely, aspartate and glutamate are transferred much less readily as such, and their carbon and nitrogen accordingly appears mostly in other phloem solutes, apparently following breakdown in interveinal regions of the leaf.[85,108]

An alternative approach has been to construct net balances for amino compounds based on ontogenetic changes in their respective pool sizes in the soluble and insoluble nitrogen fractions of the leaf, and on measured rates of import and export by the leaf, as determined from xylem and phloem sap composition and empirical models of leaf nitrogen balance. Import of asparagine and aspartic acid at all times exceeds the present requirement for these compounds in the leaf, while glutamine, the third major nitrogen solute in the xylem is supplied in excess only until the leaf becomes fully expanded.[85] Net synthesis and export of glutamine then occurs, especially when the leaf is losing large amounts of protein prior to its senescence.[85] The balance sheets indicate that synthesis of some 16 amino compounds supplied inadequately through xylem is accomplished from nitrogen released in catabolism of asparagine, aspartate and glutamine.[85] Asparagine provides up to 68% of this nitrogen, and in early development asparaginase, and later, asparagine: pyruvate amino transferase, show *in vitro* activities fully accounting for currently observed rates of utilization of the amide. Similarly activity of the leaf's GS: GOGAT system in vitro is compatible with estimated rates of NH_4^+ generation in amide catabolism and photorespiration.[85]

When ^{14}C-amino compounds are fed to cut shoots, autoradiographic evidence of intense labelling of stem tissue indicates a capacity to abstract certain solutes laterally from the xylem, and, for such activity to be concentrated, the surrounding departing leaf traces in mature regions of the stem.[108] The functional importance of this activity becomes apparent when empirical models of total nitrogen flow are constructed for white lupin.[85] These indicate that nitrogen absorbed laterally from vascular tissue serving mature leaves is reintroduced into xylem streams passing further up the shoot. This progressively enriches the xylem stream available to young expanded leaves as well as weakly transpiring structures such as shoot apices, inflorescences and fruits.[85] Apically-sited organs thus obtain more than their expected shares of nitrogen in terms of transpirational activity, and do so at the expense of lower, older leaves. Direct evidence of enrichment of stem xylem with nitrogen comes from the finding that vacuum-extracted tracheal (xylem) fluid recovered from terminal parts of a lupin shoot may be 3 or 4 times more concentrated in nitrogen than that leaving the root system.[85] Despite this concentrating effect, the balance of amino compounds in the xylem stream does not change appreciably from top to base of stem, so xylem to xylem transfer is apparently not accompanied by significant metabolism of the participating solutes.[58]

The models of nitrogen flow in lupin so far constructed all suggest that a second class of exchange of amino compounds occurs in the uppermost parts of lupin stems, and that this involves direct transfer of amino compounds from the already enriched xylem stream to the phloem stream transporting assimilates from upper leaves to terminal parts of the shoot. The supporting evidence for this is that phloem sap collected just below the shoot apex is several times more concentrated in amino compounds than is the sugar-rich phloem stream flowing from upper leaves to upper parts of the shoot.[85]

As a result of these differential partitioning processes in stem and leaf, the distribution of nitrogen in phloem becomes inherently biased towards apical regions of the shoot, and to this extent does not conform to patterns of translocation of photosynthetically-produced sucrose.[85] For example, a nodulated root of a white lupin plant starting to flower consumes over half of the shoot's current net photosynthate, and in so doing monopolizes the translocate of the lower 12 of 16 main stem leaves. Yet below ground parts attract only one-third of the plant's current nitrogen resource, and most of this is nitrogen mobilized from lower senescing leaves.[58,85]

A scheme summarizing the principal pathways for nitrogen flow during the late vegetative growth of a white lupin plant is presented in Figure 5. In addition to highlighting the stem- and leaf-based exchanges of nitrogen alluded to above, it provides information on the extent of self feeding of nodules with their own products of nitrogen fixation, and on the amount of nitrogen surplus to root requirements which cycles back to the shoot following phloem to xylem transfer in organs below ground. Net gains or losses of nitrogen by plant parts during the study interval are specified, and numbers appended in brackets beside each transport activity indicate the percentages of the nitrogen of that stream which are comprised of asparagine (upper number) and glutamine (lower number).[71] These percentage values attest to the dominant role of these two compounds in lupin metabolism and the aptness of the term "amide-plant" when referring to lupin and similarly disposed legumes.

Transport of Nitrogen to and Utilization by Fruit and Seed

As with other parts of the plant, xylem and phloem of the fruit stalk carry broad spectra of amino compounds, but especially large amounts of asparagine and glutamine.[92] Estimates of the relative extents to which xylem and phloem contribute to the nitrogen nutrition of fruits have been derived using data on the rates of consumption of carbon and nitrogen in fruit dry matter, carbon exchanges of the fruit with the atmosphere as CO_2, and the C:N weight ratios of xylem and phloem streams entering the fruit.[92] The data indicate that close to 90% of the

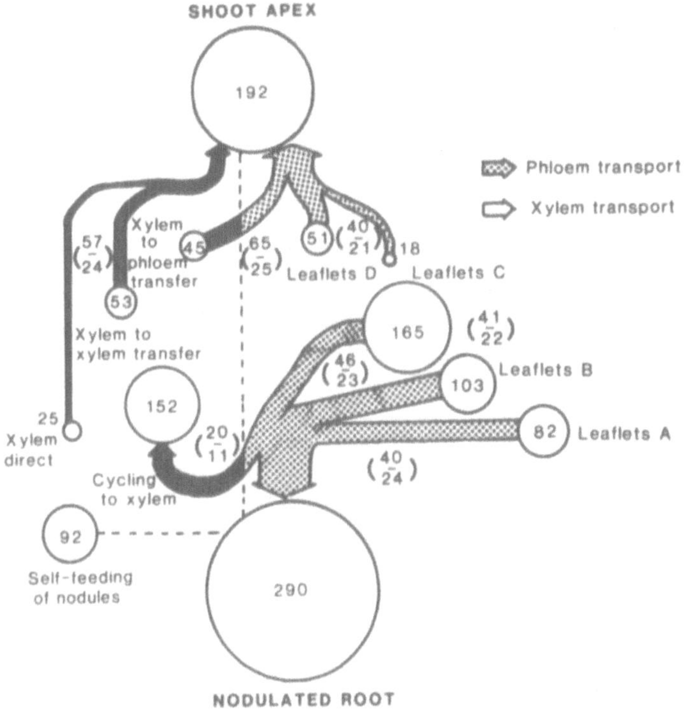

Fig. 5. Diagrammatic representation of the source:sink relationships for nitrogen within symbiotically-dependent plants of white lupin (Lupinus albus) during the stage 51-58 days after sowing. Amounts (mg) of nitrogen supplied from each source agency to the shoot apex or nodulated root during the study interval are indicated, and the involvement of phloem and xylem transport and xylem to phloem and xylem to xylem interchanges are quantified by the number in circles. Self feeding of nodules refers to their capacity to incorporate fixed nitrogen directly into their dry matter (see also Fig. 4). The numbers appended in brackets beside each transport activity indicate percentages of nitrogen in that stream which are comprised of asparagine (upper number) and glutamine (lower number). Data from References 58, 71, 85.

nitrogen acquired by the fruit during its growth enters through phloem, and that this reflects the 10 to 20 times higher concentration of nitrogen in phloem sap than in xylem sap, and the relatively poor transpiration loss of the fruit relative to mass flow intake of phloem-borne photosynthate.[88] However, xylem does deliver a transiently high proportion of nitrogen very early in fruit growth, when pools of soluble nitrogen are being established, yet before seed growth starts to create a large demand for phloem solutes.[88,92]

The fruits of grain legumes generally exhibit a progressive increase in requirements for nitrogen relative to those for carbon, as they switch from laying down carbon-rich fibrous tissues in their pod walls to building up protein reserves in their seeds. Increasing demand for nitrogen by the lupin fruit is matched by a progressive decrease from 25:1 to 12:1 in the C:N ratio of phloem translocate entering from the parent plant, this major change in composition being mediated by accelerating mobilization of nitrogen from rapidly senescing vegetative parts compounded with declining production of net photosynthate by leaves.[88] However, white lupin exhibit high rates of nitrogen fixation well into fruiting and early formed fruits clearly benefit from this.

Since the nutritional demands of the maturing seeds of grain legumes are met almost entirely if not exclusively by import through phloem, the principal metabolic changes for nitrogen required by the maturing seed can be conveniently deduced by comparing the amino acid composition of the entering phloem translocate with that of the cotyledon storage protein. Applying such an analysis to white lupin, asparagine and glutamine turn out to be the principal providers of amino nitrogen for synthesis within the seed of a number of amino compounds inadequately supplied in phloem. Arginine is prominent among these, comprising up to 15% of the nitrogen in the seed's storage protein yet present in only trace amounts in the entering phloem stream.[88,92]

Studies in which ^{14}C,^{15}N (amide-N)-labelled asparagine was fed to fruiting shoots through the transpiration stream indicate that over 90% of the compound arrives in the fruit in an unmetabolized state.[92] During early growth of the seed greater asparaginase activity is recorded from extracts of the seed coat than from the juvenile embryo. Consistent with this, the ^{14}C of the fed asparagine is recovered from the endospermic fluid mostly attached to non-amino compounds, whereas the ^{15}N of its amide group is traced to free NH_4^+, alanine and glutamine in endospermic fluid, and to a wide range of protein amino acids in the embryo.[92] The older embryo absorbs asparagine and other amino compounds directly from the phloem, and, consistent with this, the asparaginase activity of the embryo increases sharply while that of the seed coat declines.[92]

A QUANTITATIVELY-BASED CASE STUDY OF PARTITIONING AND UTILIZATION OF NITROGENOUS SOLUTES IN THE HOST PLANT OF A UREIDE-FORMING SYMBIOSIS

Cowpea (*Vigna unguiculata*) is selected for this presentation because it is the only ureide-producing species known to bleed spontaneously when its phloem is cut.[109] However, this capacity is apparently restricted to vascular tissue of mature fruits and their stalks, so direct information on phloem transport is restricted to the mobilization of nitrogen from vegetative parts to fruit and seed.[88 109-111]

The ureides allantoin and allantoic acid regularly comprise up to three quarters of the nitrogen supplied to shoots through the xylem.[109] Asparagine, aspartic acid, valine, threonine and serine are the only other major constituents.[109] The proportion of total nitrogen in an ethanol soluble form varies considerably with the age and type of plant part, namely 20 to 70% for stem plus petioles, 18 to 36% for nodulated roots, and within the range of 4 to 18% for fruits and leaflets of varying age.[112] Ureides comprise 10 to 30% of the total soluble nitrogen of leaflets and nodulated roots and young fruits, approximately half of that of mature fruits, and up to 80% of that of stem plus petioles.[112] Allantoin predominates in all plant parts except stem plus petioles, in which it is present at approximately the same level as allantoic acid.[112] The two ureides are also present in roughly equal proportions in xylem sap.[112]

Transformations of Nitrogen in Vegetative Organs of Plants Fed N_2 and NO_3^-

Attempts to follow the initial distribution and eventual utilization of ureide in vegetative organs of cowpea have involved xylem feeding of 2-^{14}C,1,3-^{15}N-allantoin followed by a study of the fate of the two isotopes in plant parts, or xylem feeding of 2-^{14}C-allantoin to aphid-infested plants and subsequent assay of the aphids for labelled ureide and its breakdown products.[96] Ureide catabolism has also been investigated by assaying the metabolism of 2-^{14}C-allantoin in fresh tissue slices, or by measuring the ureolytic activity of tissue homogenates *in vitro* with ^{14}C-urea as substrate.[96]

All xylem feeding experiments show extensive metabolism of the ^{15}N-^{14}C-labelled ureide in the shoot, with leaflets serving as the principal organs of assimilation.[96] Approximately one-third of its ^{14}C is lost as respired CO_2, and a much greater proportion of the carbon label remains in the soluble fractions of the plant than in the case of ^{15}N.[96] Similarly, most amino compounds of the ethanol soluble fraction of the shoot show much lower ^{14}C:^{15}N ratios than the supplied allantoin, consistent with a greater involvement of the ureide-N than the urea-C of the molecule in amino acid synthesis.[96] Principal recipients of ^{14}C and ^{15}N in extracts of stem plus petiole and leaflet are asparagine, glutamine, threonine, serine and glutamic acid. Substantial proportions (14 to 40%) of the ^{14}C recovered in ethanol extracts of the aphids of plants fed 2-^{14}C allantoin are still present as unmetabolized ureide, indicating that the translocation stream from leaflets to fruits of cowpea does carry some ureide.[96]

Phloem bleeding can be induced in certain cultivars of cowpea by inserting a fine needle previously cooled in liquid nitrogen into the vascular traces of the placental suture of fully-grown fruits.[109] This so-called 'cryopuncture' phloem exudate contains 0.4 to 0.8 M sugar, 98% of this sucrose, and 2 to 3 g/liter of nitrogen as nitrogenous solutes.[109] The amino-acid N:ureide-N ratio of fruit phloem sap of symbiotic plants is heavily biased towards amino acids (89:11), compared with a ratio of 23:77 in root xylem sap from the same plants.[88,109] This directly confirms labelling and enzymatic data which imply intense metabolic transfer of ureide-N to amide and amino acids in vegetative organs of ureide-based legumes such as cowpea. Interestingly, the cryopuncture sap of nitrate-fed, non-nodulated plants, shows an amino acid: ureide:nitrate ratio (based on nitrogen) of 90:10:01, i.e., closely resembling that of symbiotic plants.[109,111] However, the xylem sap of these NO_3^--fed plants is ureide-poor and nitrate-rich,[109] having a corresponding ratio of 20:2:78. This suggests that nitrate reduction in the shoot leads to phloem loading of a large amount of reduced nitrogen as the amide but not as ureide. Despite large differences in the ureide:amino acid balance of the xylem and phloem of symbiotic plants, the amino acid composition of the two conducting channels is virtually identical.[109-111] Amides comprise over half of this fraction on a molar basis, with glutamine consistently exceeding asparagine.[109] Of the non-amide acids, histidine, serine, valine and leucine(s) are relatively major fractions. Arginine is at a high level in the xylem but not in the phloem.[106] The root xylem sap of NO_3^--fed plants is also amide-rich but asparagine is present at 1.5 times the level of glutamine.[111] Cryopuncture phloem sap of the same NO_3^--fed plants carries slightly more asparagine than glutamine.[111]

A series of labelling studies has provided evidence supporting the above conclusions. Thus, when $^{15}NO_3^-$ is fed for 24 h to roots of fruiting plants, 42% of the ^{15}N of fruit cryopuncture phloem sap is attached to asparagine, 33% to glutamine and only 0.9% to ureide.[109] Conversely, when ^{14}C-asparagine is fed to cut fruiting shoots of symbiotically-dependent plants through the xylem, over 80% of the ^{14}C of fruit phloem sap is still present as asparagine, whereas similar feeding of 2-^{14}C-allantoin results in over half of the phloem-borne ^{14}C being attached to compounds other than allantoin.[109]

Metabolic Interconversions of Nitrogenous Solutes in Developing Fruit and Seed

Using a modelling procedure identical to that mentioned earlier for white lupin fruits, the nutritional economy of the developing fruit of the "Vita 3" cultivar of cowpea has been assessed using data on CO_2 and H_2O exchanges of the fruit with the atmosphere, carbon and nitrogen gains in pod and seed dry matter, and C:N weight ratios in the xylem and phloem streams entering the fruit.[88] Phloem supplies 97% of the fruit's carbon and 72% of its nitrogen, and the small contribution from the xylem is made mainly in early growth.[88] During later growth water is supplied through the phloem during the day at rates faster than are lost in transpiration, and the fruit is presumed to return this excess water to the parent plant through the xylem.[113] At night, however, transpiration loss exceeds phloem import of water, so some intake through the xylem is possible.[113] This is likely to bring in critically important amounts of minieral elements mobile in the xylem but sparingly mobile in the phloem.[88]

Except in their very early development, seeds are supplied with a considerable surplus of water through their phloem import, and therefore record a continuous efflux of water through the xylem.[113] This back flow from seeds in the xylem may possibly mobilize senescence-inducing factors from fruit to the adjacent foliage of "vita 3" cowpea,[88] as has been suggested for certain cultivars of soybean.[114] In absence of xylem intake, assays of cryopuncture phloem sap obtained from the placental suture of the pod are likely to provide an accurate picture of what solutes are being supplied to the seed from pod and parent plant.

Extracts of pod, seed coats and embryos of seeds of cowpea all exhibit measurable activities of the ureide-catabolizing enzymes allantoinase and urease, the asparagine-utilizing enzyme asparaginase, the ammonia-assimilating GS:GOGAT system, and the transaminases, aspartate amino transferase and alanine aminotransferase.[110] Asparagine:pyruvate amino transferase is a major enzyme of pods but not of seeds.[110]

Balance sheets for the net intake and utilization of ureide and amino compounds have been constructed for specific intervals in fruit growth by comparing the net input of each compound through the phloem and xylem against net gains of the same compounds in tissue soluble pools and protein.[110] Over the first half of fruit growth, when seeds are still a relatively small component of fruit mass, a number of nitrogenous solutes are supplied in excess, namely ureide, glutamine, asparagine, γ-amino butyric acid, proline, valine, cysteine and methionine.[110]

However, only the first three of these furnish appreciable fractions of the relatively large amounts of nitrogen necessary for synthesis of the dozen or so other amino compounds supplied inadequately from the parent plant. Over two-thirds of the requirement for some of these (glycine, aspartate, glutamate, serine, alanine) is met by synthesis within the fruit.[88,110]

Study of embryo nutrition in maturing seeds of a number of legumes including cowpea has been greatly aided by the use of a seed coat cup leakage technique carried out in vivo.[115-119] This involves cutting a small window in the lateral wall of the fruit to allow removal of the distal halves of the coats and the whole enclosed embryos of one or more seeds.[120] Solutes subsequently leaking into the still attached half-seed coat cups are then collected, and the composition of the resulting fluid then measured to reflect the mixture of compounds which normally traverses the apoplastic boundary between seed coat and embryo.[121] When applied to cowpea fruits in mid-pod fill, rates of seed coat cup leakage of glutamine and asparagine are highest followed by those of histidine, arginine, serine and alanine and of a group of other, minor components.[110] Significantly, ureides comprise less than 2% of the total nitrogen released into the seed coat cup, although ureide nitrogen in phloem sap supplying the fruit is found to account for 10 to 15% of the translocated nitrogen.[110] Armed with this information, and information on soluble pool composition and rates of incorporation of individual amino acids into the proteins of seed coat and embryo, quantitatively-based pictures have been constructed depicting the principal flow pathways and metabolic routings for nitrogenous solutes during seed development of cowpea (see Fig. 6).

The flow model for early seed development (Fig. 6A) shows that the seed coat features prominently in the processing of incoming solutes, and that its main metabolic activity is the breakdown of the substantial amounts of ureide, glutamine and asparagine supplied in excess of the seed's immediate requirements for these compounds. The nitrogen released from these solutes is pictured as being used for the synthesis of amino compounds such as histidine, glycine, aspartate, glutamate and alanine. These, together with unmetabolized glutamine, asparagine and a small amount of remaining ureide, are

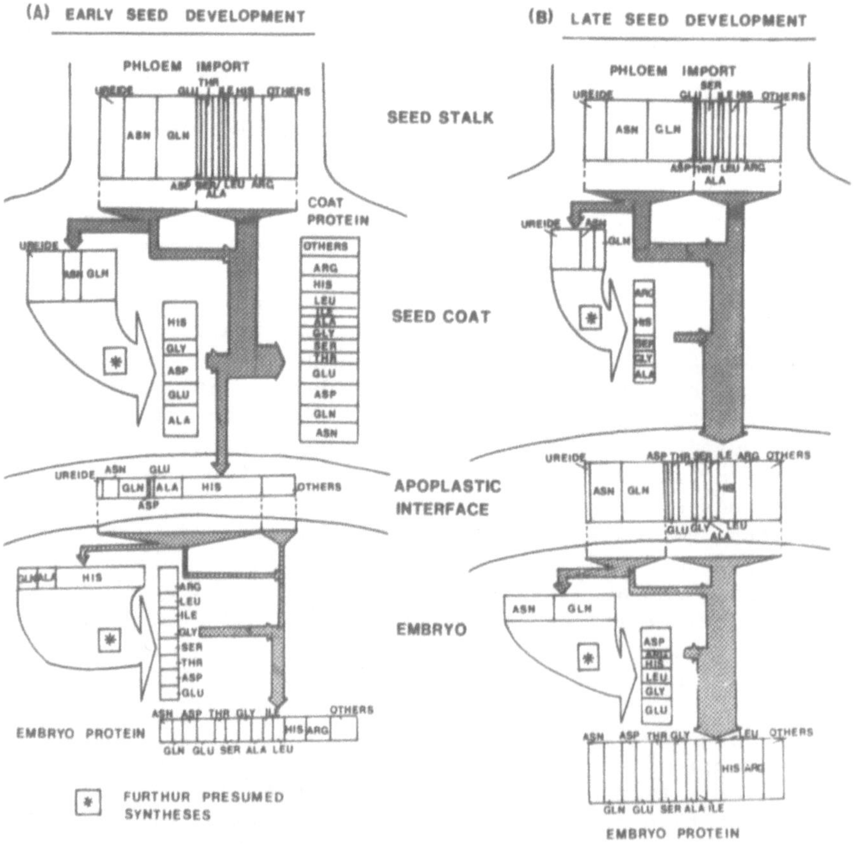

Fig. 6. Flow pathways and suggested metabolic transformations of phloem-delivered nitrogenous solutes during two stages (A, 4 to 9 d; B, 12 to 18 d after anthesis) in the development of seeds of symbiotically-dependent cowpea (*Vigna unguiculata* cv. Vita 3). Incorporated are data on the composition of phloem cryopuncture sap (Stages A and B), embryo sac fluid (Stage A) or seed coat leakage products (Stage B), and composition of solute and protein of seed coat (Stage A) and embryo (Stages A and B). Arrow breadths for total nitrogen flow and areas of rectangles depicting nitrogenous solutes are drawn proportional to net amounts of nitrogen transported or consumed. See text and Reference 110 for further details.

then passed on to the embryo via the embryo sac fluid.[110] Further metabolic rearrangement within the embryo involves the net production of a range of protein amino acids, in this case mainly at the expense of the entering histidine, alanine and glutamine.[110] When a similar nitrogen balance is constructed for the main cotyledon filling stage (Fig. 6B), a somewhat similar picture emerges, but net protein synthesis at this stage is occurring only in the embryo, and an accordingly greater proportion of the incoming nitrogen is metabolized by the embryo than the seed coat. However, catabolism of incoming ureide is still conducted principally in the seed coat, and most nitrogen passes directly through to the embryo in the form of asparagine and glutamine.[110]

COST EFFECTIVENESS AND COMPETITIVENESS OF ASSOCIATIONS FIXING NITROGEN

Nitrogen-fixing associations are conspicuous components of a broad range of natural ecosystems world-wide, whether as long-lived perennials contributing to the maintenance of nitrogen capital within climax stages of ecosystems, or as short-lived pioneers injecting nitrogen into habitats ravaged by fire, flood, glacial activity or other environmental disturbances. In the latter cases nitrogen may be the key limiting revegetation, enabling nitrogen-fixing associations to dominate transiently over species unable to utilize atmospheric nitrogen. Here the competitiveness of the nitrogen fixer will be obviously maximized, even if the association concerned is less cost effective in its metabolism and resource utilization than companion non-fixing species.

Where nitrogen is not the limiting nutrient element, two basic reasons may be suggested as to why nitrogen-fixers might be disadvantaged when competing with species heterotrophic for nitrogen. Firstly, there may be greater energetic costs to the nitrogen fixer through the establishment, maintenance and functioning of its symbiotic apparatus, compared with species employing root-mediated uptake of soil nitrogen. Secondly, it generally holds that nitrogen-fixing components of ecosystems are richer in nitrogen,[122,123] and hence less nitrogen efficient than non-symbiotic counterparts. Flowing from this are cost penalties in the assimilation of this extra large load of nitrogen, and an attendant allocation of resources for chemical defence of nitrogen-rich foliage, seeds or seedlings against predators attracted especially to the species because of its high protein content.

A general indication of the likely costs of construction of symbiotic organs can be obtained simply by determining the proportion of seedling or adult plant dry matter which consists of nodule, coralloid root or actinorhizal root.[4 123] In legumes, for example, nodule biomass comprises from 5 to 15% of plant dry weight, by no means an insignificant expenditure when balanced against the extra root extension and improved efficiency in acquisition of water and nutrients which would have been possible were symbiotic organs not to have been formed.[123] Particularly crucial, is the large initial investment of nitrogen in symbiotic organs before they commence to fix nitrogen, especially in seedlings where a limited seed reserve of nitrogen is being deployed for this purpose.[123]

Additional extra costs are envisaged where the symbiotic organs require appreciable amounts of certain elements (e.g., Mo, Co, and possibly P) specifically in connection with their nitrogen-fixing activity.[122] This might assume unusual importance in soils deficient in these elements, since there will be selection pressure for symbiotic associations to invest more than non-fixers in mechanisms for enhancing nutrient uptake, e.g., mycorrhizae or specialized cluster roots.[122] Formation and functioning of these is known to be costly.[122]

It should be remembered that any cost penalty, measurable in additional photosynthate, will incur extra requirements for water, making nitrogen fixing associations automatically less competitive in low rainfall environments. Shaded environments would similarly disadvantage a species requiring extra photosynthate through its symbiosis.

A large number of investigations have compared the energetics of nitrogen fixation with other forms of nitrogen nutrition.[4,124-127] and it is generally agreed that symbiotic fixation is the most expensive in terms of photosynthate utilization per unit of nitrogen assimilated. Depending upon whether the association evolves H_2 when fixing nitrogen,[128] and whether some or all of this H_2 is reassimilated by uptake hydrogenase,[129] nitrogenase functioning per se has been estimated to cost of the order of 13 to 26 mols of ATP per mol N_2 assimilated,[125] compared with a suggested maximum cost for respiration-based nitrate reduction of 12 mol ATP per mol NO_3^-.[130] Photosynthetically-linked nitrate reduction would of course be less costly, depending on the extent to which reductant and ATP were being generated surplus to the requirements of CO_2 assimilation. Direct uptake of NH_4^+ from the environment would be at essentially zero cost compared to that of N_2 or NO_3^-.[122,123]

Regardless of the form of nitrogen utilized, further costs occur when converting NH_4^+ into organic solutes of nitrogen. Metabolic inputs will vary with the compounds being formed, especially in relation to the number of nitrogen atoms in the assimilation product.[123] For instance, as indicated in Figure 7A, the monoamino compounds asparate and glutamate are considerably more costly to produce than their corresponding amides, and the nitrogen-rich ureides. However, respiratory losses of CO_2 accompanying synthesis of these compounds may not reflect consumption of translocated sucrose, since in certain cases (see Fig. 7B), anapleurotic inputs of carbon through dark CO_2-fixation are likely to be high; in other compounds those inputs may be negligible or non-existent.[123]

Against the above costs of NH_4^+ assimilation into specific compounds there must be matched the metabolic implications of utilizing these same compounds as sources of carbon and nitrogen at some other time and place within the plant. In this case (Fig. 7C), an energy input (3ATP equivalents per NH_4^+ released) is required when allantoin/allantoate are metabolized, while net energetic gains are to be expected when amino acids and amides are broken down.[123]

By comparing these theoretically based costings of synthesis and metabolism, one reaches the conclusion (see values listed at the bottom of Fig. 7) that nitrogen utilization via allantoin/allantoic acid is almost twice as costly as the least expensive pathway listed via asparagine. However, cost disadvantages in ureide catabolism might be partly negated were their catabolism

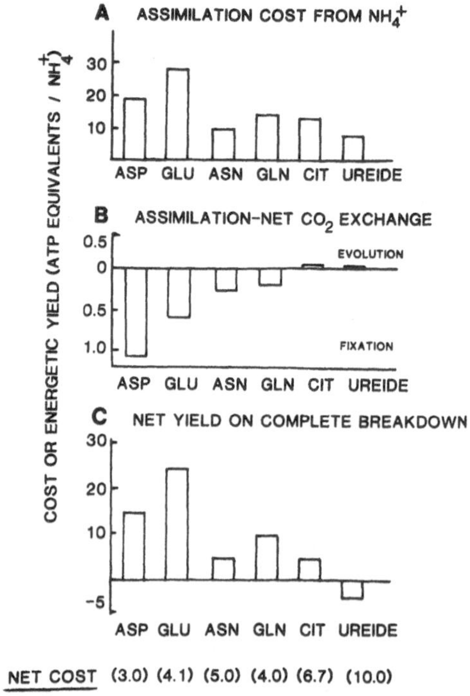

Fig. 7. Theoretically-based costs or energetic yields estimated to occur in assimilation (A) and complete metabolism (C) of each of a range of nitrogenous solutes which feature prominently in the nitrogen metabolism of various nitrogen-fixing association. All items are expressed in terms of molar equivalents of ATP consumed or produced in assimilation or catabolic release of 1 mol of NH_4^+. Net CO_2 exchanges (B) associated with the assimilation of each compound are based on the assumption of maximum anapleurotic inputs of CO_2 during assimilation. Relative net costs of utilizing each compound in plant nitrogen metabolism are listed at the bottom of the figure by subtracting values in C from those in A. See text and reference[123] for further details.

to occur photosynthetically in leaves, where supplies of carbon and reductant might be non-limiting. Conversely, at the site of synthesis, the cost advantage in utilizing ureide as opposed to amide as recipient of fixed nitrogen is well apparent, as shown in a comparison of the carbon and nitrogen economies of nodules of cowpea and lupin.[32] Nodules of the former, utilizing ureide as the principal fixation product require import of 5.4 mg C as phloem translocate per mg of nitrogen fixed versus 6.9 (same units) for the asparagine-forming nodules of lupin (see Fig. 4).[32]

Based on the above, comparisons on the cost effectiveness of fixing and non-fixing species will obviously be invalid if the species in question utilize different classes of compounds in their nitrogen metabolism. One should therefore compare species of similar metabolic characteristics, or, better still, assess the performance of the same spcies when fixing N_2 or assimilating NH_4^+ or NO_3^-. The latter type of comparison provides overwhelming evidence of substantially greater costs associated with nitrogen fixation than NO_3^- assimilation, and the implications of this to the efficiency of agricultural systems relying on inputs of molecular nitrogen have been assessed.[4,58,124,125,130]

Turning to possible costs to nitrogen-fixing associations of producing nitrogenous compounds in chemical defense, one again meets difficulties when attempting to compare broadly the investments made by such systems against those by non-fixing species. However, it is abundantly clear that nitrogen-fixers, particularly legumes, collectively display an especially diverse range of protective compounds.[78,122,131-133] prominent amongst which are cyanogens, seleno-amino compounds, trypsin inhibitors, alkaloids, glucosinolates, non-protein amino acids and amines, and lectin-type phytohaemogglutinins. Some act as outright poisons; some act by interfering with protein synthesis; some are antinutritional factors, and some are carcinogens. Detailed accounts of the metabolism of these unusual metabolites cannot be attempted here, but of particular relevance are the likely proportions of a plant's total nitrogen resource likely to be involved directly in protection, and the attendant specificities in deployment in time and space expressed by the species at tissue and organ level.

Proportions of plant nitrogen devoted to defense will vary widely with the compounds concerned and with their location. However, it has been suggested[132] that of the order of 1 to 4% of the total nitrogen of vegetative parts of a mature plant may be devoted to glycosides, 1 to 8% to non-protein amino acids such as homoserine[74,135] and djenkolic acid,[76] 0.5% to glycosinates, yet only 0.08 to 0.6% to bitter principles such as alkaloids.[132] However, seedling stages may be inordinately high in certain non-protein amino compounds [e.g., 56% of the seedling nitrogen as homoserine in Pisum,[134,135] 30% as γ-methylene glutamine and γ-methylene glutamate in germinating peanut (Arachis)],[80] and seeds may be specifically concentrated in trypsin inhibitor (0.06% of total nitrogen), glycoproteins (2 to 10%), and neurotoxins or neurotoxin-generating principles (e.g., canavanine and 3-hydroxyphenylalanine at concentrations up to 2% of seed nitrogen).[123,131,132] Assuming that costs of these unusual metabolites are not radically different from those suggested above for common metabolites of nitrogen, one might conclude that anywhere from 10 to 50% of a legume's current budget for nitrogen metabolism might be devoted to defense, with attendant penalties accruing from the progressive diversion of photosynthate and nitrogen away from protein synthesis and growth.

The ultimate cost of any investment in defense will of course be minimized were the nitrogen involved to become available as a general nitrogen source at some later stage, as is believed to occur for example with the nitrogen of canavanine[131] and of certain non-protein amino acids in legume seeds and seedlings.[123,134] Economies will also be effected if synthesis of a particular compound is restricted to those stages of the growth cycle in which its presence carries greatest impact, and if the compound in question accumulates only in those plant parts most likely to suffer predation. There is abundant evidence that these principles apply to many species, fixers and non-fixers. Amongst legumes, for example, one may quote the restriction of synthesis of certain protein amino acids to early seedling development,[134] the marked concentration of alkaloids, the depilating agent minosine,[72] protective compounds such as homoarginine, pipecolic acid, 4-hydroxy pipecolic acid and cyanogenic principles in grazing susceptible-foliage as opposed to other plant parts,[131,132] and the high levels of a whole range of protective materials in seeds, presumably for defense against bruchids and other seed predating agencies.[70,133]

CONCLUSIONS

It is apparent from this review that much research is still required on the most basic biochemical events of nitrogen assimilation and metabolism even in those few species which have been investigated in depth over recent years. Whole gaps exist at all levels in our knowledge of symbiotic systems other than legumes, and until the nitrogen metabolism of these is known we will remain ignorant of how the biochemical and physiological attributes of nitrogen fixers are likely to have evolved, and in which respects their functional attributes differ from those of species heterotrophic for nitrogen. Despite encouraging advances in understanding of the biochemical pathways of nitrogen metabolism in symbiotic organs of a number of nitrogen-fixing species, and the demonstration that the enzymatic systems involved and the products formed are not basically different from those of root-based nitrogen metabolism of other species, there remain enormous gaps in our understanding of how the process of nitrogen fixation and subsequent nitrogen transformations within symbiotic organs are regulated and modulated with respect to demands of the host plant for nitrogen and provisions of carbon. One would expect to find in nitrogen fixing associations much more tightly coupled systems for carbon and nitrogen than in species autotrophic only for carbon. If so, future effort might be sensibly directed at attempting to unravel the biochemical and physiological principles shaping the coordination of the symbiotic performance of below ground structures with that of shoot growth and functioning.

REFERENCES

1. KENNEDY, I.R. 1966. Primary products of symbiotic nitrogen fixation. I. Short-term exposures of serradella nodules to $^{15}N_2$. Biochim. Biophys. Acta 130: 285-294.

2. MEEKS, J.C., N.S. STEINBERG, C.S. ENDERLIN, C.M. JOSEPH, G.A. PETERS. 1987. Azolla-Anabaena relationship. Plant Physiol. 84: 883-886.
3. PETERS, G.A., R.E. TOIA, H.E. CALVERT, R.H. MARSH. 1986. Lichens to Gunnera with emphasis on Azolla. Plant Soil 90: 17-34.
4. SCHUBERT, K.R. 1986. Products of biological nitrogen fixation in higher plants: synthesis, transport and metabolism. Annu. Rev. Plant Physiol. 37: 339-374.
5. ROBERTSON, J.G., K.J.F. FARNDEN, M. WARBURTON, J.M. BANKS. 1975. Induction of glutamine synthetase during nodule development in lupin. Aust. J. Plant Physiol. 2: 265-272.
6. VAIRINHOS, F., B. BHANDARI, D.J.D. NICHOLAS. 1983. Glutamine synthetase, glutamate synthase and glutamate dehydrogenase in Rhizobium japonicum strains grown in cultures and in bacteroids from root nodules of Glycine max. Planta 159: 207-215.
7. PATE, J.S., P. LINDBLAD, C.A. ATKINS. 1989. Pathways of assimilation and transfer of fixed nitrogen in coralloid roots of Cycad-Nostoc symbioses. Planta 176: 461-471.
8. SCHUBERT, K.R., G.T. COKER, III. 1982. Studies of nitrogen and carbon assimilation in N_2-fixing plants: short-term studies using [^{13}N] and [^{11}C]. In Recent Developments in Biological and Chemical Research with Short-lived Radioisotopes. (J.W. Root, K.A. Krohn, eds.), American Chemical Society, Washington, D.C., Advances in Chemistry Series, Volume 197, pp. 317-339.
9. OHYAMA, T., K. KUMAZAWA. 1978. Incorporation of ^{15}N into various nitrogenous compounds in intact soybean nodules after exposure to $^{15}N_2$ gas. Soil Sci. Plant Nutr., 24: 525-533.
10. OHYAMA, T., K. KUMAZAWA. 1980. Nitrogen assimilation in soybean nodules. Soil Sci. Plant Nutr. 26: 109-115.
11. TA, T.-C., M.A. FARIS, F.D.H. MacDOWALL. 1986. Pathways of nitrogen metabolism in nodules of alfalfa (Medicago sativa L.). Plant Physiol. 80: 1002-1005.
12. RAY, T.B., B.C. MAYNE, R.E. TOIA, G.A. PETERS. 1979. Azolla-Anabaena relationship. Plant Physiol. 64: 791-795.

13. FUJIHARA, S., M. YAMAGUCHI. 1981. Assimilation of $^{15}NH_3$ by root nodules detached from soybean plants. Plant Cell Physiol. 22: 797-806.
14. MEEKS, J.C., C.P. WOLK, N. SCHILLING, P.W. SHAFFER, Y. AVISSAR, W.S. CHIEN. 1978. Initial organic products of fixation of ^{13}N dinitrogen by root nodules of soybean (*Glycine max*). Plant Physiol. 61: 980-983.
15. MIFLIN, B.J. 1980. Nitrogen metabolism and amino acid biosynthesis in crop plants. *In* The Biology of Crop Productivity. (P.S. Carlson, ed.), Academic Press, New York, pp. 255-296.
16. MIFLIN, B.J., P.J. LEA. 1980. Ammonia assimilation. *In* The Biochemistry of Plants. (B.J. Miflin, ed.), Academic Press, New York, Vol. 5, pp. 169-202.
17. SEN, D., H.M. SCHULMAN. 1980. Enzymes of ammonia assimilation in the cytosol of developing soybean root nodules. New Phytol. 85: 243-250.
18. GROAT, R.G., C.P. VANCE. 1981. Root nodule enzymes of ammonia assimilation in alfalfa (*Medicago sativa* L.). Plant Physiol. 67: 1198-1203.
19. PADILLA, J.E., F. CAMPOS, V. CONDE. M. LARA, F. SANCHEZ. 1987. Nodule-specific glutamine synthetase is expressed before the onset of nitrogen fixation in *Phaseolus vulgaris* L. Plant Mol. Biol. 9: 65-74.
20. HIREL, B., C. PERROT-RECHENMANN, B. MAUDINAS, P. GADAL. 1982. Glutamine synthetase in alder (*Alnus glutinosa*) root nodules. Purification, properties and cytoimmunochemical localization. Physiol. Plant. 55: 197-203.
21. BERGMAN, B., P. LINDBLAD, A. PETTERSSON, E. RENSTROM, E. TIBERG. 1985. Immuno-gold localization of glutamine synthetase in a nitrogen-fixing cyanobacterium (*Anabaena cylindrica*). Planta 166: 329-334.
22. LINDBLAD, P., B. BERGMAN. 1986. Glutamine synthetase: activity and localization in cyanobacteria of the cycads *Cycas revoluta* and *Zamia skinneri*. Planta 169: 1-7.
23. STREETER, J.G. 1977. Asparaginase and asparagine transaminase in soybean leaves and root nodules. Plant Physiol. 60: 235-239.
24. ROGNES, S.E. 1975. Glutamine-dependent asparagine synthetase from *Lupinus luteus*. Phytochemistry 14: 1975-1982.

25. FUJIHARA, S., M. YAMAGUCHI. 1980. Asparagine formation in soybean nodules. Plant Physiol. 66: 139-141.
26. HUBER, T.A., J.G. STREETER. 1984. Asparagine biosynthesis in soybean nodules. Plant Physiol. 74: 605-610.
27. SNAPP, S.S., C.P. VANCE. 1986. Asparagine biosynthesis in alfalfa (*Medicago sativa* L.). Plant Physiol. 82: 390-395.
28. LEES, E.M., A.B. BLAKENEY. 1970. The distribution of asparaginase activity in legumes. Biochim. Biophys. Acta 215: 145-151.
29. CHRISTELLER, J.T., W.A. LAING, W.D. SUTTON. 1977. Carbon dioxide fixation by lupin root nodules. I. Characterization, association with phosphoenolpyruvate carboxylase, and correlation with nitrogen fixation during nodule development. Plant Physiol. 60: 47-50.
30. COKER, G.T., K.R. SCHUBERT. 1981. Carbon dioxide fixation in soybean roots and nodules. I. Characterization and comparison with N_2 fixation and composition of xylem exudate during early nodule development. Plant Physiol. 67: 691-696.
31. VANCE, C.P., S. STADE, C.A. MAXWELL. 1983. Alfalfa root nodule carbon dioxide fixation. I. Association with nitrogen fixation and incorporation into amino acids. Plant Physiol. 72: 469-473.
32. LAYZELL, D.B., R.M. RAINBIRD, C.A. ATKINS, J.S. PATE. 1979. Economy of photosynthate use in N-fixing legume nodules: observations on two contrasting symbioses. Plant Physiol. 64: 888-891.
33. ATKINS, C.A., J.S. PATE, B.J. SHELP. 1984. Effects of short-term N_2 deficiency on N metabolism in legume nodules. Plant Physiol. 76: 705-710.
34. ATKINS, C.A., A. RITCHIE, P.B. ROWE, E. McCAIRNS, D. SAUER. 1982. *De novo* purine synthesis in nitrogen-fixing nodules of cowpea (*Vigna unguiculata* L. Walp.). Plant Physiol. 70: 55-60.
35. WOO, K.C., C.A. ATKINS, J.S. PATE. 1981. Ureide synthesis in a cell-free system from cowpea (*Vigna unguiculata* [L.] Walp.) nodules. Plant

Physiol. 67: 1156-1160.
36. McCLURE, P.R., G.T. COKER, K.R. SCHUBERT. 1983. Carbon dioxide fixation in roots and nodules of Alnus glutinosa. 1. Role of phosphoenolpyruvate carboxylase and carbomyl phosphate synthetase in dark CO_2 fixation, citrulline synthesis, and N_2 fixation. Plant Physiol. 71: 652-657.
37. PERROT-RECHENMANN, C., J. VIDAL, B. MAUDINAS, P. GADAL. 1981. Immunocytochemical study of phosphoenolpyruvate carboxylase in nodulated Alnus glutinosa. Planta 153: 14-17.
38. MARTIN, F., B. HIREL, P. GADAL. 1983. Purification and properties of ornithine carbamoyl transferase 1 from Alnus glutinosa root nodules. Z. Pflanzenphysiol. 111: 413-422.
39. ATKINS, C.A. 1982. Ureide metabolism and the significance of ureides in legumes. In Advances in Agricultural Microbiology. (N.S. Subba Rao, ed.), Oxford and IBH Publishing Co., New Delhi, pp. 53-88.
40. REYNOLDS, P.H., M.J. BOLAND, D.G. BLEVINS, D.G. RANDALL, K.R. SCHUBERT. 1982. Ureide biogenesis in leguminous plants. Trends Biochem. Sci. 10: 366-368.
41. REYNOLDS, P.H., M.J. BOLAND, D.G. BLEVINS, K.R. SCHUBERT, D.D. RANDALL. 1982. Enzymes of amide and ureide biogenesis in developing soybean nodules. Plant Physiol. 69: 1334-1338.
42. THOMAS, R.J., L.E. SCHRADER. 1981. Ureide metabolism in higher plants. Phytochemistry 20: 361-371.
43. MATSUMOTO, T., M. YATAZAWA, Y. YAMAMOTO. 1977. Incorporation of ^{15}N into allantoin in nodulated soybean plants supplied with $^{15}N_2$. Plant Cell Physiol. 18: 459-462.
44. SHELP, B.J., C.A. ATKINS. 1983. Role of inosine monophosphate oxidoreductase in the formation of ureides in nitrogen-fixing nodules of cowpea (Vigna unguiculata L. Walp.). Plant Physiol. 72: 1029-1034.
45. NGUYEN, J., L. MACHAL, J. VIDAL, C. PERROT-RECHENMANN, P. GADAL. 1986. Immunochemical studies on xanthine dehydrogenase of soybean root nodules. Planta 167: 190-195.
46. TRIPLETT, E.W. 1985. Intercellular nodule localization and nodule specificity of xanthine dehydrogenase in soybean. Plant Physiol. 77:

1004-1009.

47. WOO, K.C., C.A. ATKINS, J.S. PATE. 1980. Biosynthesis of ureides from purines in a cell-free system from nodule extracts of cowpea Vigna unguiculata (l) Walp. Plant Physiol. 66: 735-739.

48. LUCAS, K., M.J. BOLAND, K.R. SCHUBERT. 1983. Uricase from soybean root nodules: purification, properties, and comparison with the enzyme from cowpea. Arch. Biochem. Biophys. 226: 190-197.

49. HANKS, J.F., N.E. TOLBERT, K.R. SCHUBERT. 1981. Localization of enzymes of ureide biosynthesis in peroxisomes and microsomes of nodules. Plant Physiol. 68: 65-69.

50. NEWCOMB, E.H., S.K. TANDON. 1981. Uninfected cells of soybean root nodules: ultrastructure suggests key role in ureide production. Science 212: 1394-1396.

51. MATSUMOTO, T., M. YATAZAWA, Y. YAMAMOTO. 1977. Distribution and change in the contents of allantoin and allantoic acid in developing nodulating and non-nodulating soybean plants. Plant Cell Physiol. 18: 353-359.

52. SCHUBERT, K.R. 1981. Enzymes of purine biosynthesis and catabolism in Glycine max: comparison of activities with N_2 fixation and composition of xylem exudate during nodule development. Plant Physiol. 68: 1115-1122.

53. SCHUBERT, K.R., M.J. BOLAND. 1984. The cellular and intracellular organization of the reactions of ureide biogenesis in nodules of tropical legumes. In Advances in Nitrogen Fixation Research. (C. Veeger, W.E. Newton, eds.), Nijoff/Junk, The Hague, pp. 445-451.

54. KANEKO, Y., E.H. NEWCOMB. 1987. Cytochemical localization of uricase and catalase in developing root nodules in soybean. Protoplasma 140: 1-12.

55. NEWCOMB, E.H., S.K. TANDON, R.R. KOWAL. 1985. Ultrastructural specialization for ureide production in uninfected cells of soybean root nodules. Protoplasma 125: 1-12.

56. VAN DEN BOSCH, K.A., E.H. NEWCOMB. 1986. Immunogold localization of nodule-specific uricase in developing soybean root nodules. Planta 167: 425-436.

57. SELKER, J.M.L., E.H. NEWCOMB. 1985. Spatial relationships between uninfected and infected cells in root nodules of soybean. Planta 165: 446-454.
58. PATE, J.S. 1980. Transport and partitioning of nitrogenous solutes. Annu. Rev. Plant Physiol. 31: 313-340.
59. PATE, J.S., C.A. ATKINS. 1983. Nitrogen uptake, transport, and utilization. In Ecology of Nitrogen Fixation. (W.J. Broughton, ed.), Oxford University Press, Oxford, Vol. 3, pp. 245-298.
60. PATE, J.S., B.E.S. GUNNING, L.G. BRIARTY. 1969. Ultrastructure and functioning of the transport system of the leguminous root nodule. Planta 85: 11-34.
61. ISRAEL, D.W., P.R. McCLURE. 1980. Nitrogen translocation in the xylem of soybeans. In World Soybean Research Conference II: Proceedings. (F.T. Corbin, ed.), Boulder, Westview, pp. 111-127.
62. WALSH, K.B., B.H. NG, G.E. CHANDLER. 1984. Effects of nitrogen nutrition on xylem sap composition of Casuarinaceae. Plant Soil 81: 291-293.
63. SPRENT, J.I. 1980. Root nodule anatomy, type of export product and evolutionary origin in some Leguminosae. Plant Cell Environ. 3: 35-43.
64. PATE, J.S., C.A. ATKINS, S.T. WHITE, R.M. RAINBIRD, K.C. WOO. 1980. Nitrogen nutrition and xylem transport of nitrogen in ureide-producing grain legumes. Plant Physiol. 65: 961-965.
65. McCLURE, P.R., D.W. ISRAEL. 1979. Transport of nitrogen in the xylem of soybean plants. Plant Physiol. 64: 411-416.
66. GOTO, S., S. INANGA, K. KUMAZAWA. 1987. Xylem sap composition of nodulated and non-nodulated groundnut plants. Soil Sic. Plant Nutr. 4: 619-627.
67. PEOPLES, M.B., J.S. PATE, C.A. ATKINS, F.J. BERGERSEN. 1986. Nitrogen nutrition and xylem sap composition of peanut (Arachis hypogaea L. cv Virginia Bunch). Plant Physiol. 82: 946-951.
68. SUTHERLAND, J.M., M. ANDREWS, S. McINROY, J.I. SPRENT. 1985. The distribution of nitrate assimilation between root and shoot in Vicia faba L. Ann. Bot. 56: 259-263.

69. WALLACE, W. 1986. Distribution of nitrate assimilation between the root and shoot of legumes and a comparison with wheat. Physiol. Plant. 66: 653-663.
70. WEBB, M.A., E.H. NEWCOMB. 1987. Cellular compartmentation of ureide biogenesis in root nodules of cowpea (Vigna unguiculata (L.) Walp.). Planta 172: 162-175.
71. PATE, J.S., C.A. ATKINS, K. HAMEL, D.L. McNEIL, D.B. LAYŻELL. 1979. Transport of organic solutes in phloem and xylem of a nodulated legume. Plant Physiol. 63: 1082-1088.
72. WEE, K.L., S. WANG. 1987. Effect of post-harvest treatment on the degradation of mimosine in Leucaena leucocephala leaves. J. Sci. Food Agric. 39: 195-201.
73. WINK, M., L. WITTE. 1984. Turnover and transport of quinolizidine alkaloids. Diurnal fluctuations of lupanine in the phloem sap, leaves and fruits of Lupinus albus L. Planta 161: 519-524.
74. GROBBELAAR, N. 1969. The isolation of amino acids from Pisum sativum. Identification of L(-)-Homoserine and L(+)-O-Acetylhomoserine and certain effects of environment upon their formation. Phytochemistry 8: 553-559.
75. JOY, K.W., C. PRABHA. 1986. The role of transamination in the synthesis of homoserine in peas. Plant Physiol. 82: 99-102.
76. HANSEN, A.P., J.S. PATE. 1987. Comparative growth and symbiotic performance of seedlings of Acacia spp. in defined pot culture or as natural understorey components of a eucalypt forest ecosystem in S.W. Australia. J. Exp. Bot. 38: 13-26.
77. DOWNUM, K.R., G.A. ROSENTHAL, W.S. COHEM. 1983. L-Arginine and L-canavanine metabolism in jack bean, Canavalia ensiformis (L) DC. and soybean, Glycine max (L.) Merr. Plant Physiol. 73: 965-968.
78. FUJIHARA, L.S., T. NAKASHIMA, Y. KUROGOCHI, M. YAMAGUCHI. 1986. Distribution and metabolism of sym-homospermidine and canavaline in the sword bean Canavalia gladiata cv Shironata. Plant Physiol. 82: 795-800.
79. ROSENTHAL, G.A., D. RHODES. 1984. L-Canavanine transport and utilization in developing jack

bean, Canavalia ensiformis (L.) DC. Leguminosae. Plant Physiol. 76: 541-544.
80. FOWDEN, L. 1954. The nitrogen metabolism of groundnut plants: the role of γ-methyleneglutamine and γ-methyleneglutamic acid. Ann. Bot. N.S. 72: 417-440.
81. THOMAS, R.J., J.I. SPRENT. 1984. The effects of temperature on vegetative and early reproductive growth of a cold-tolerant and a cold-sensitive line of Phaseolus vulgaris L. Ann. Bot. 53: 579-588.
82. GUNNING, B.E.S., J.S. PATE, F.R. MINCHIN, I. MARKS. 1974. Quantitative aspects of transfer cell structure in relation to vein loading in leaves and solute transport in legume nodules. Symp. Soc. Exp. Biol. 28: 87-125.
83. IRELAND, R.J., K.W. JOY. 1981. Two routes for asparagine metabolism in Pisum sativum L. Planta 151: 289-292.
84. TA, T.C., K.W. JOY, R.J. IRELAND. 1984. Utilization of the amide groups of asparagine and 2-hydroxysuccinamic acid by young pea leaves. Plant Physiol. 75: 527-530.
85. PATE, J.S. 1986. Xylem-to-phloem transfer - a vital component of the nitrogen-partitioning system of a nodulated legume. In Phloem Transport. (J. Cronshaw, W.J. Lucas, R.T. Giaquinta, eds.), A.R. Liss, Inc., New York, pp. 445-462.
86. LEA, P.J., J.S. HUGHES, B.J. MIFLIN. 1979. Glutamine- and asparagine-dependent protein synthesis in maturing legume cotyledons cultured in vitro. J. Exp. Bot. 30: 529-537.
87. GOMES, M.A.F., L. SODEK. 1984. Allantoinase and asparaginase activities in maturing fruits of nodulated and non-nodulated soybeans. Physiol. Plant. 62: 105-109.
88. PATE, J.S. 1984. The carbon and nitrogen nutrition of fruit and seed - case studies of selected grain legumes. In Seed Physiology - Development. (D.R. Murray, ed.), Academic Press, Australia, Vol. 1, pp. 41-81.
89. RATAJCZAK, W. 1986. Asparagine metabolism in developing seeds of Lupinus luteus L. Biochem. Physiol. Pflanz. 181: 17-22.
90. TA, T.C., K.W. JOY. 1986. Metabolism of some amino acids in relation to the photorespiratory

nitrogen cycle of pea leaves. Planta 169: 117-122.

91. SCHAEFER, J., T.A. SKOKUT, E.O. STEJSKAL, R.A. McKAY, J.E. VARNER. 1981. Asparagine amide metabolism in developing cotyledons of soybean. Proc. Nat. Acad. Sci. USA 78: 5978-5982.

92. ATKINS, C.A., J.S. PATE, P.J. SHARKEY. 1975. Asparagine metabolism - key to the nitrogen nutrition of developing legume seeds. Plant Physiol. 56: 807-812.

93. SODEK, L., P.J. LEA, B.J. MIFLIN. 1980. Distribution and properties of a potassium-dependent asparaginase isolated from developing seeds of Pisum sativum and other plants. Plant Physiol. 65: 22-26.

94. COKER, G.T., J. SCHAEFER. 1985. ^{15}N and ^{13}C NMR determination of allantoin metabolism in developing soybean cotyledons. Plant Physiol. 77: 129-135.

95. IMSANDE, J. 1986. Degradation and utilization of exogenous allantoin by intact soybean root. Physiol. Plant. 68: 685-688.

96. ATKINS, C.A., J.S. PATE, A. RITCHIE, M.B. PEOPLES. 1982. Metabolism and translocation of allantoin in ureide-producing grain legumes. Plant Physiol. 70: 476-482.

97. MATSUMOTO, T., M. YATAZAWA, Y. YAMAMOTO. 1978. Allantoin metabolism in soybean plants as influenced by grafts, a delayed inoculation with Rhizobium, and a late supply of nitrogen-compounds. Plant Cell Physiol. 19: 1161-1168.

98. COSTIGAN, S.A., V.R. FRANCESCHI, M.S.B. KU. 1987. Allantoinase activity and ureide content of mesophyll and paraveinal mesophyll of soybean leaves. Plant Sci. (Shannon) 50: 179-187.

99. YONEYAMA, T. 1984. Partitioning and metabolism of nitrate, asparagine, and allantoin in the soybean shoots at the grain-filling stage. Soil Sci. Plant Nutr. 30: 583-587.

100. SHELP, B.J., R.J. IRELAND. 1985. Ureide metabolism in leaves of nitrogen-fixing soybean plants. Plant Physiol. 77: 779-783.

101. WINKLER, R.G., J.C. POLACCO, D.G. BLEVINS, D.D. RANDALL. 1985. Enzymic degradation of allantoate in developing soybeans. Plant Physiol. 79: 787-793.

102. KERR, P.S., D.G. BLEVINS, B. RAPP, D.D. RANDALL. 1983. Soybean leaf urease: comparison with seed urease. Physiol. Plant. 57: 339-345.
103. POLACCO, J.C., R.G. WINKLER. 1984. Soybean leaf urease: a seed enzyme? Plant Physiol. 74: 800-803.
104. WINKLER, R.G., D.G. BLEVINS, J.C. POLACCO, D.G. RANDALL. 1987. Ureide catabolism of soybeans. Plant Physiol. 83: 585-591.
105. PATE, J.S., C.A. ATKINS, D.F. HERRIDGE, D.B. LAYZELL. 1981. Synthesis, storage, and utilization of amino compounds in white lupin (Lupinus albus L.). Plant Physiol. 67: 37-42.
106. PATE, J.S., P.J. SHARKEY, C.A. ATKINS. 1977. Nutrition of a developing legume fruit. Plant Physiol. 59: 506-510.
107. PATE, J.S. 1975. Exchange of solutes between phloem and xylem and circulation in the whole plant. In Encyclopedia of Plant Physiology New Series. (M.H. Zimmermann, J.A. Milburn, eds.), Springer-Verlag, Berlin, Vol. 1, pp. 451-473.
108. McNEIL, D.L., C.A. ATKINS, J.S. PATE. 1979. Uptake and utilization of xylem-borne amino compounds by shoot organs of a legume. Plant Physiol. 63: 1076-1081.
109. PATE, J.S., M.B. PEOPLES, C.A. ATKINS. 1984. Spontaneous phloem bleeding from cryopunctured fruits of a ureide-producing legume. Plant Physiol. 74: 499-505.
110. PEOPLES, M.B., C.A. ATKINS, J.S. PATE, D.R. MURRAY. 1985. Nitrogen nutrition and metabolic interconversions of nitrogenous solutes in developing cowpea fruits. Plant Physiol. 77: 382-388.
111. PEOPLES, M.B., J.S. PATE, C.A. ATKINS. 1985. The effect of nitrogen source on transport and metabolism of nitrogen in fruiting plants of cowpea (Vigna unguiculata (L.) Walp.). J. Exp. Bot. 36: 567-582.
112. HERRIDGE, D.F., C.A. ATKINS, J.S. PATE, R.M. RAINBIRD. 1978. Allantoin and allantoic acid in the nitrogen economy of the cowpea (Vigna unguiculata (L.) Walp.). Plant Physiol. 62: 495-498.

113. PATE, J.S., M.B. PEOPLES, A.J.E. VAN BEL, J. KUO, C.A. ATKINS. 1985. Diurnal water balance of the cowpea fruit. Plant Physiol. 77: 148-156.

114. LINDOO, S.J., L.D. NOODEN. 1976. The interrelation of fruit development and leaf senescence in 'anoka' soybeans. Bot. Gaz. 137: 218-223.

115. BENNETT, A.B., R.G. SPANSWICK. 1983. Derepression of amino acid-H^+ cotransport in developing soybean embryos. Plant Physiol. 72: 781-786.

116. FELLOWS, R.J., D.B. EGLI, J.E. LEGGETT. 1978. A pod leakage technique for phloem translocation studies in soybean (Glycine max Merr.). Plant Physiol. 62: 812-814.

117. HSU, F.C., A.B. BENNETT, R.M. SPANSWICK. 1984. Concentrations of sucrose and nitrogenous compounds in the apoplast of developing soybean seed coats and embryos. Plant Physiol. 75: 181-186.

118. VAN BEL, A.J.E., J.W. PATRICK. 1985. Proton extrusion in seed coats of Phaseolus vulgaris L. Plant Cell Environ. 8: 1-6.

119. WOLSWINKEL, P., A. AMMERLAAN. 1985. Characteristics of sugar, amino acid and phosphate release from the seed coat of developing seeds of Vicia faba and Pisum sativum. J. Exp. Bot. 36: 359-368.

120. RAINBIRD, R.M., J.H. THORNE, R.W.F. HARDY. 1984. Role of amides, amino acids, and ureides in the nutrition of developing soybean seeds. Plant Physiol. 74: 329-334.

121. THORNE, J.H., R.M. RAINBIRD. 1983. An in vivo technique for the study of phloem unloading in seed coats of developing soybean seeds. Plant Physiol. 172: 0268-0271.

122. PATE, J.S. 1983. Patterns of nitrogen metabolism in higher plants and their ecological significance. Op. cit. Reference 8, pp. 225-255.

123. PATE, J.S., D.B. LAYZELL. Energetics and biological costs of nitrogen assimilation. In The Biochemistry of Plants. A Comprehensive Treatise. (B.J. Miflin, ed.), Academic Press, Florida, USA, Vol. 16 (in press).

124. MAHON, J.D. 1977. Respiration and the energy requirement for nitrogen fixation in nodulated pea roots. Plant Physiol. 60: 817-821.

125. PATE, J.S., C.A. ATKINS, R.M. RAINBIRD. 1981. Theoretical and experimental costing of nitrogen fixation and related processes in nodules of legumes. In Current Perspectives in Nitrogen Fixation. (E.H. Gibson, W.E. Newton, eds.), Aust. Acad. Sci. Canberra, pp. 105-116.
126. RYLE, G.J.A., C.E. POWELL, A.J. GORDON. 1979. The respiratory costs of nitrogen fixation in soybean, cowpea and white clover. II. Comparisons of the costs of nitrogen fixation and the utilization of combined nitrogen. J. Exp. Bot. 30: 145-153.
127. SCHUBERT, K.R., G.J.A. ROYLE. 1980. The energy requirements for nitrogen fixation in nodulated legumes. In Advances in Legume Science. (R.J. Summerfield, A.H. Bunting, eds.), Royal Botanic Gardens, Kew, England, pp. 85-96.
128. DIXON, R.O.D., C.T. WHEELER. 1983. Biochemical, physiological and environmental aspects of symbiotic nitrogen fixation. In Biological nitrogen fixation in forest ecosystem. (J.C. Gordon, C.T. Wheeler, eds.), Martinus Nijhoff/ W. Junk Publishers, The Hague, pp. 107-171.
129. RAINBIRD, R.M., C.A. ATKINS, J.S. PATE, P. SANFORD. 1983. Significance of hydrogen evolution in the carbon and nitrogen economy of nodulated cowpea. Plant Physiol. 71: 122-127.
130. PHILLIPS, D.A. 1980. Efficiency of symbiotic nitrogen fixation in legumes. Annu. Rev. Plant Physiol. 31: 29-49.
131. BELL, E.A. 1980. The non-protein amino acids of higher plants. Endeavour 4: 102-107.
132. BERNAYS, E.A. 1983. Nitrogen in defense against insect. In Nitrogen as an Ecological Factor. (J.A. Lee, S. McNeill, I.H. Rorison, eds.), Blackwell Scientific Publication, Oxford, pp. 321-344.
133. ROSENTHAL, G.A., C.G. HUGHES, D.H. JANZEN. 1982. L-Canavanine, a dietary nitrogen source for the seed predator Caryedes brasiliensis (Bruchidae). Science 217: 353-355.
134. PATE, J.S. 1977. Nodulation and nitrogen metabolism. In Physiology of the Garden Pea. (J.F. Sutcliffe, J.S. Pate, eds.), Academic Press, London, pp. 349-383.

Chapter Four

GENETICS AND MOLECULAR BIOLOGY OF HIGHER PLANT NITRATE REDUCTASES

ANDRIS KLEINHOFS*, ROBERT L. WARNER*
AND JAMES M. MELZER+

*Department of Agronomy and Soils and
Program in Genetics and Cell Biology
Washington State University
Pullman, Washington 99164-6420

+Plant Genetic Engineering Laboratory
New Mexico State University
Las Cruces, New Mexico 88003

INTRODUCTION

Nitrate assimilation is the primary pathway through which most reduced nitrogen is accumulated in plants. The pathway consists of two well characterized enzymes, nitrate and nitrite reductases. Poorly characterized nitrate uptake and transport mechanisms are also involved. Nitrate reductase (NR) catalyzes the reduction of nitrate to nitrite which is then reduced to ammonium ion by nitrite reductase. Incorporation of ammonium ion into amino acids is carried out by the glutamine synthetase-glutamate synthase pathway. The first step in the nitrate assimilation pathway, nitrate reductase, is highly regulated and is believed to be rate limiting. There has been a great deal of research concerning the biochemistry, genetics and physiology of NR which has recently been reviewed.[1-3] Also a recent comprehensive conference was held on the topic of nitrate assimilation, and the proceedings will be published.[4] Readers are referred to these works and the references therein for a detailed review. Here we will concentrate on NR, especially the recent accomplishments.

Plant physiologists have been interested in NR for many years. Three major developments have led to the current interest in NR by a broad spectrum of biologists. These developments were: the isolation of NR-deficient mutants, the purification of the NR protein without major degradation, and the cloning of NR genes.

BIOCHEMISTRY OF NITRATE REDUCTASE

Existence of NR in higher plants has been known since Evans and Nason[5] demonstrated that extracts of soybean leaves contained an enzyme that could reduce nitrate using either NADPH or NADH as the electron donor. Thus, the first higher plant NR described was NAD(P)H bispecific. The most common form of higher plant NR, however, is specific for NADH as the electron donor.

Nitrate reductase was notoriously difficult to purify in an intact state until Solomonson[6] showed that the blue dye ligand Cibacron blue FG3-A provided a semi-specific affinity matrix for NR. This technique was rapidly applied to the purification of NR from a variety of higher plants.[7-13] These preparations provided valuable knowledge about higher plant NRs, but suffered from a nagging problem of inconsistent and heterogeneous subunit sizes. The presence of heterogeneous subunits in NR was contrary to the genetic evidence in fungi and higher plants indicating a single structural NR gene in haploid and true diploid organisms (discussed in more detail below). This problem was solved when Kuo and coworkers[11,12] demonstrated a single homogeneous subunit for the barley NR. The instability of NR was identified as primarily due to proteolytic degradation which could be managed by protease inhibitors,[14,15] buffer adjustments[16] and rapid purification. Other noteworthy developments were the use of zinc-chelate, 5'-AMP affinity and amphiphilic gel chromatography.[17-19] Even with these developments, NR has remained difficult to obtain in sufficient quantities for extensive biochemical analysis and has been highly purified from only two higher plant species, i.e., squash[15] and spinach.[18,19] The recent development of monoclonal NR antibody columns, permitting the purification of NR to apparent homogeneity in a single step, promises to overcome this last obstacle to the biochemical studies of NR from a broad range of species.[20-22]

Nitrate reductases of eukaryotic organisms catalyze the two electron reduction of nitrate to nitrite using reduced pyridine nucleotides. Three types of pyridine nucleotide-dependent NRs are recognized. Most higher plants contain a NADH-specific NR (EC 1.6.6.1) with a pH optimum around 7.5. A NAD(P)H-bispecific NR (EC 1.6.6.2) has also been identified in higher plants. This enzyme is known to occur in soybean,[23] *Erythrina senegalensis* D.C.,[24] and monocots such as barley,[25-27] maize[28] and rice.[8] It is not as extensively characterized as the NADH NR. The NADPH specific NR (EC 1.6.6.3), found in fungi, has not been reported from higher plants.

All eukaryotic NRs that have been investigated contain three prosthetic groups, namely FAD, cytochrome b_{557} and molybdenum. These probably occur in a ratio of one mole per subunit, although rigorous data have only been obtained with *Chlorella*,[29-32] *Neurospora crassa*,[33] and squash.[15] The NR-associated molybdenum is complexed within a pterin moiety[34] to form the molybdenum cofactor (MoCo). The NR MoCo is interchangeable with other known molybdo-enzymes except nitrogenase.[35] Enzymes that can reduce nitrate without the MoCo, or a different type of MoCo, have not been identified in higher plants. The b-type cytochrome component of NR, first identified by Garrett and Nason[36] in *N. crassa* NADPH NR, has also been found in higher plant NRs.[23,37-40]

Purified fungal NRs from *N. crassa*[33,41,42] and *Penicillium chrysogenum*[43,44] are termed "FAD-dependent" since they require the coenzyme in reaction mixtures for activity. In contrast, the FAD moiety in higher plant

NRs appears more tightly bound, and exogenous addition of FAD is not required for maximal activity. Recently, it has been demonstrated that partially purified preparations of both isozymes of barley NR can be inactivated by removal of FAD by ultrafiltration, and can subsequently be reactivated by addition of exogenous FAD (Kakefuda et al., personal communication). Similar to the fungal enzymes, FAD-free barley NR will bind tightly to a FAD-Sepharose column. There appears to be a great deal of variation in the degree to which FAD remains bound to NRs from various sources.

The NADH NRs from higher plants appear to be homodimers. The molecular weight of the NADH holoenzyme from barley cv Steptoe was estimated to be 220,000 to 230,000[11,12] and 202,000 for wheat.[45] Molecular weights for spinach NR of 200,000[46] and 270,000[19] have been reported, while that of the squash NR was estimated at 230,000.[15] Subunit molecular weights determined by SDS polyacrylamide gel electrophoresis range from 100,000 to 110,000 for barley,[11,12,14] 115,000 for squash[15] and corn,[47,48] 110,000 to 120,000 for spinach,[19,20] and 107,000 to 109,000 for the three soybean NRs.[23,49]

Although the NAD(P)H bispecific NR was the first one to be isolated from higher plants,[5] it is not as yet highly characterized. In soybean, the situation is complicated by the occurrence of three different types of NRs which have only recently been separated and partially purified from wild-type plants.[23] Earlier separation of the three types was achieved using a combination of induction conditions and the constitutive NR-deficient mutant nr1.[49] The three types include one inducible and two constitutive NRs. The inducible NADH-NR is similar to other higher plant NRs with a pH optimum of 7.5. Another NADH-NR, however, is constitutive and has an unusual pH optimum of 6.5. The NAD(P)H-NR in soybean is constitutive, has the unusual pH 6.5 optimum and appears to be different from the inducible NAD(P)H NR observed in barley (pH optimum 7.7)[26,27] and other monocots.[8,28] The NAD(P)H and NADH constitutive NRs from soybean have lower S values, measured in sucrose gradients, compared to the inducible NADH NR. The migration in native polyacrylamide gels is faster, suggesting a lower native molecular weight.[49] However, recent molecular weight measurements by SDS polyacrylamide gel electrophoresis indicate approximately the same subunit size (107-109,000) for all three NRs.[23] A NR isolated from norflurazan-treated, nitrate grown soybean cotyledons was reported to have a subunit molecular weight of 98,000.[50] The relationship of this NR to those described by Streit and coworkers[23,49] is not clear.

The tropical leguminous tree, E. senegalensis appears to be unique among higher plants in having only the NAD(P)H NR.[24] Thus E. senegalensis would be an ideal source of the NAD(P)H NR for further biochemical studies, particularly since high levels of NR activity were demonstrated. Unfortunately, we are not aware of any further studies of the NR from this or related species.

Mixtures of NADH and NAD(P)H NRs are found in several monocots including barley, maize and rice. In rice the two types of NR were found in the seedling[8] or shoots,[51] while in maize the NAD(P)H NR is found together with the NADH NR only in the roots and scutellum.[28-52] These enzymes remain to be characterized with respect to their molecular weights. In barley, the NAD(P)H NR is found in association with the NADH NR only in the roots while in the shoots only the NADH-specific NR can be demonstrated.[25-27,53] However, when the NADH NR activity is lost due to mutation, the NAD(P)H NR is induced (or derepressed) in the shoot. This was exploited to partially purify and characterize the NAD(P)H NR from barley and demonstrate a subunit molecular weight of 100,000, slightly smaller than the barley NADH NR.[27] Kinetic and physical properties of some higher plant NRs are summarized in Table 1.

The homodimer model of NR proposed by Pan and Nason[33] for the N. crassa NR has been adopted for the higher plant NR with only minor modifications.[12,15] Pan and Nason[33] proposed that the functional NR consisted of two identical subunits joined and held together by the MoCo. This model was based on the available biochemical and genetic evidence showing that NR subunits of the N. crassa mutant nit 1 could be dimerized and activated in vitro by an active MoCo from a variety of sources. A model reflecting our current understanding of the biochemistry and genetics of higher plant NR is presented in Figure 1. For the squash NR, Redinbaugh and

Table 1. Properties of Some Higher Plant Nitrate Reductases.

Source	MW (k daltons)		pH	K_m (μM)			References
	Native	Subunit	Optimum	NO_3^-	NADH	NADPH	
Barley	220	110	7.5	130	12		12,25,26
Spinach	270	114	7.5	50	5		3,19
Squash	230	115	7.5	40	5		3,15
Maize	160	115	7.5	70	--		28,48
Soybean	---	109	7.5	130	3		23,49
Soybean	---	107	6.5	190	3		49
Wheat	202	---	7.5	310	60		45
Barley	---	100	7.7	620		10	26,27
Soybean	---	107	6.5	4700		3	23,49
Maize	---	---	7.5	300		--	28
Erythrina	---	---	7.5	10000		14	24

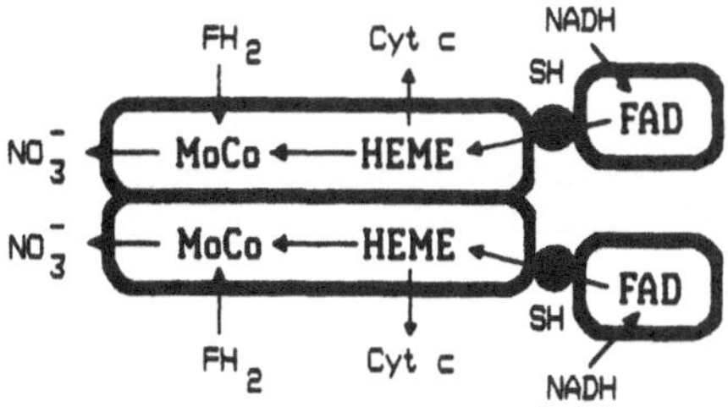

Fig. 1. Structure-function model of higher plant assimilatory nitrate reductase. The Chlorella nitrate reductase model, as proposed by Solomonson and coworkers,[56] was modified to reflect the two identical subunits and the elongated nature of the higher plant nitrate reductases. The model incorporates the information gained from sequencing of the tobacco[109] and Arabidopsis[110] nitrate reductases showing that the nitrate and NADH binding sites are located at the N- and C-terminal ends of the protein, respectively.

Campbell[15] obtained a value of 0.8 Mo per mole of subunit and concluded that each subunit contains one MoCo. This conclusion is in agreement with one MoCo per mole of subunit found in Chlorella vulgaris NR.[30,32,54-56] Additional evidence showed that each subunit behaved as an independent functional unit in radiation inactivation experiments.[31] Activity of the individual subunits in nitrate reduction, however, has not been demonstrated. Thus the MoCo role in the subunit dimerization or the dimerization role in catalytic activity remains to be resolved.

Information concerning the mechanism of electron transfer and the functional domains of NR has been obtained using inhibitors, mutants, and physical separation. Partial activities of NR are of two types, NADH-dehydrogenase such as cytochrome c or ferricyanide reductases and reduced flavins or viologens NR activity. NADH-dehydrogenase partial activities require a functional FAD but not MoCo and are inhibited by sulfhydryl reagents but not by cyanide. Nitrate reducing partial activities require a functional MoCo but not the FAD prosthetic group and are sensitive to cyanide and azide but not sulfhydryl reagents. Mutants that have lost the NADH, but retain the reduced flavin or viologen NR activities occur, but are relatively rare. Mutants with defective MoCo lack all ability to reduce nitrate but retain dehydrogenase activity. There is no evidence that these partial activities have physiological functions with the possible exception of iron-chelate reduction, which could play a part in iron assimilation.[57]

Treatment of the native *Chlorella* NR with a specific protease from corn or with *Staphylococcus aureus* V8 protease yielded a 30 kDa fragment containing FAD and a 69 kDa fragment containing heme and the MoCo.[56] The 30 kDa fragment retained the NADH-ferricyanide reductase activity of NR and the 69 kDa fragment retained the reduced flavin and viologen NR activities. These data are in good agreement with recent sequencing and homology analysis of the *Arabidopsis* and tobacco NRs (see Molecular Biology section below). Solomonson and coworkers[56,58] have determined that a single sulfhydryl group is involved in the binding of NADH necessary for NR activity. These data led to the proposal of a model for the *Chlorella* NR which contains two functional domains connected by a protease-sensitive hinge region. The FAD/NADH binding domain is exposed on the surface of the enzyme and has no apparent role in maintaining the quaternary structure, but is essential for NADH dehydrogenase activity. The quaternary structure of the enzyme is maintained through the heme/MoCo domain which is the site of nitrate reduction. The essential sulfhydryl group is closely associated with the NADH binding site, but is located on the heme/molybdenum domain.[58] While the details of this model remain to be confirmed for other NRs, the available evidence suggests that there will be great similarity among them.

GENETICS OF NITRATE REDUCTASE

Nitrate reductase-deficient mutants have been isolated from bacteria, fungi, algae, and higher plants. Here we will only deal with the higher plant mutants using examples from the other organisms only as needed for comparison. The first higher plant NR-deficient mutants were isolated in *Arabidopsis thaliana*,[59] followed by the identification of NR-deficient mutants in tobacco[60-62] and barley.[63,64] Currently, major efforts

Table 2. Summary of Nitrate Reductase Genes in Some Higher Plants.

Species	No. Loci	Function
Barley	1	NADH apoenzyme
	1	NAD(P)H apoenzyme
	5	MoCo
Arabidopsis	2	Apoenzyme
	1	Nitrate uptake
	3	MoCo
	3	Unidentified
Nicotiana	1	NADH apoenzyme/diploid
	6	MoCo
Soybean	1	Constitutive NADH NR (pH 6.5)
	1	Constitutive NADH & NAD(P)H NRs

to analyze the genetics of NR are under way in *Arabidopsis*, barley, soybean, and tobacco (Table 2) with more limited work in several other species. Cumulatively, this work has identified structural genes for the NADH-specific and the NAD(P)H-bispecific nitrate reductases and at least six different genes involved with the MoCo which is essential for the function of most, perhaps all, NRs. Specific NR regulatory genes have not been identified in any higher plant species, and nitrate uptake mutants have been reported only in *Arabidopsis*.

The two major methods for mutant selection are based on chlorate resistance[59] and rapid assay *in vivo* of individual seedlings[63] in segregating mutant (M2) populations. The chlorate selection has the advantage of speed and ability to process large populations, but it is not effective for selecting structural gene mutants in species that have more than one NR structural gene.

Structural Gene Mutants

In barley, twenty-two NR-deficient mutants have been found to be alleles of the narl locus. This locus has been rigorously identified as the NADH-NR structural gene based on: 1) variation in NR-associated activities of different narl mutants;[65,66] 2) the presence in some narl mutants of modified NR antigen;[67,68] 3) failure of functional MoCo to reconstitute NR activity in vitro from extracts of narl mutants;[69] 4) a single peptide difference in Cleveland maps of wild-type and mutant narld NR protein;[70] and 5) cosegregation of the narl locus with a locus giving restriction fragment length polymorphism when hybridized with the barley NADH-NR cDNA clone bNRp10 (Kleinhofs, unpublished).

The NR-deficient barley mutants that are alleles of the narl locus all have residual NR activity averaging about 5% of the wild type when assayed in young seedlings grown under growth chamber conditions. This activity is primarily due to a NAD(P)H bispecific NR.[25,26] In a few mutants, this activity appears to be a mixture of the NADH-NR produced by a leaky mutation and the NAD(P)H-NR. The NAD(P)H-NR is not found in the seedling leaves of wild-type barley plants, but is apparently turned on in leaves of narl mutants. The NAD(P)H-NR activity is present in the wild-type roots together with the NADH-NR. A single NAD(P)H-NR deficient mutant has been isolated and the gene and allele designated nar7w.[53] The narl and nar7 loci are not linked. The nar7w mutant is deficient in NAD(P)H-NR activity, but normal with respect to the NADH-NR activity, therefore it can only be detected by assaying the roots. The double mutant, defective at both narl and nar7 loci, has essentially zero NR activity and does not produce significant growth with nitrate as the sole nitrogen source.

Müller and coworkers[60-62,71,72] isolated and characterized numerous NR-deficient cell lines in Nicotiana tabacum allodihaploid cells. Thirty-six mutant lines were characterized as NR structural gene alleles and designated nia locus. Plants regenerated from the nia cell cultures and analyzed by classical genetic methods were shown to have mutations in each of the duplicate NR structural genes.[73] Numerous other NR-deficient cell lines have been isolated using chlorate selection on haploid protoplasts of Nicotiana plumbaginifolia.[74-79] These have been studied using somatic cell fusion or classical genetics of regenerated plants. All of the mutants that are not also deficient in xanthine dehydrogenase fall into a single complementation group designated nia. Several lines of evidence suggest that the nia locus encodes the NADH-NR structural gene:[79] 1) nia is the only complementation group with normal MoCo function; 2) immunologically altered NR antigen is present in some nia alleles; 3) nia alleles have different partial NR catalytic activities; and 4) a nia allele has been transformed to wild-type phenotype by a cloned NR gene (Caboche, personal communication).

Eleven NR-deficient mutants have been isolated in Arabidopsis.[59,80-82] These mutants fall into 7 complementation groups and all but three grow reasonably well on nitrate as the sole nitrogen source. The chl-2 and chl-3 loci, represented by mutants B2-1 and B29, have been proposed as encoding the structural genes. Due to limited biochemical characterization of these mutants, identification of the NR structural gene loci is tentative.

Soybean NRs are unique among higher plant NRs that have been studied. Two constitutive NRs and one inducible NR have been described.[49,83] Among the constitutive NRs, one is NADH specific while the other one is NAD(P)H bispecific. The inducible NR is NADH specific and similar to other higher plant NRs.[23,49,84] Seven mutants have been isolated that result in the loss of the constitutive NRs, but no mutants for the inducible NR have been found.[83,85-87] Three mutants selected by resistance to chlorate were deficient in both constitutive NR activities, allelic and controlled by a single recessive nuclear gene designated nr1.[49,84-86] Two other mutants isolated by this group using in vivo assays were deficient only in the NADH constitutive NR.[83] This suggests that nr1 probably is not the structural gene locus, but might represent a regulatory or a MoCo locus specific to the constitutive NRs. Two additional mutants isolated by Carroll and Gresshoff[87] have not been characterized at the biochemical level.

Molybdenum Cofactor Mutants

The NR-deficient mutants that show pleiotropic loss of xanthine dehydrogenase activity, also a molybdoenzyme, are assumed to be defective in the MoCo functions.[60,88] In *Nicotiana*, these mutants are designated *cnx*. Six *cnx* loci have been identified by complementation studies using somatic cell fusion or genetic crosses of regenerated plants. The NR activity of *cnxA* locus mutants can be partially restored by unphysiologically high levels of molybdenum in the growth medium.[71,89] Thus this locus is probably involved in molybdenum incorporation into the cofactor.[90] The *cnxB* and *cnxC* loci were identified in a series of experiments with mutants isolated in *N. tabacum* and *N. plumbaginifolia* and allelism testing by somatic cell fusion.[76-78,91] Mutants at these loci are unable to dimerize the NR subunits and therefore are probably deficient in the MoCo moiety itself.[90] More recently the *cnxD* locus was identified among a series of mutants isolated in *N. plumbaginifolia* some of which could be regenerated into plants.[75,78] Plant regeneration from *N. plumbaginifolia* protoplasts has been greatly improved[92] resulting in the regeneration and genetic analysis of numerous mutants and the identification of *cnxE* and *cnxF* loci.[93,94] It is not clear if all of the possible *cnx* loci have been identified in *Nicotiana*. In the extensively studied genetic organism *Aspergillus nidulans*, six molybdenum cofactor genes (*cnx* ABC,E,F,G,H,J) have been identified.[95,96] The *cnx* ABC locus is complex, perhaps encoding three functions. Thus, a total of eight functions would be required for the MoCo assembly and interaction with the apoprotein. In *Escherichia coli* five genes (*chl*, A,B,D,E,G) are known to be involved, but *chlA* and *chlE* are probably operons encoding three and two functions, respectively.[97] Thus, again a total of eight functions seem to be involved.

In barley five loci involved with the MoCo functions have been identified. These have been designated *nar2* through *nar6*.[66] Since two of these loci are represented by only a single mutant, it is likely that more MoCo loci will be uncovered.[66]

In *Arabidopsis* the mutants B25 (*rgn*) and B73 (*cnx*) exhibit low xanthine dehydrogenase activity and are presumed to be MoCo mutants.[80,98] A third MoCo locus was identified by Scholten and coworkers[99] among chlorate resistant cell lines. This mutant, designated G1, was shown to be deficient in xanthine dehydrogenase activity and not allelic to the mutants B25 and B73. Another mutant, designated G3, was suggested as a possible MoCo mutant, but the evidence presented was not convincing. The number of loci identified as having MoCo functions in *Arabidopsis* seems small considering the number known in *Nicotiana* and barley. This may be a reflection of the small number of mutants that have been analyzed, although, paradoxically, the largest number of loci involved with higher plant NR activity have been identified in this species (see Other Mutants).

Mutants deficient in NR activity due to MoCo defects have been identified in most species examined, except soybean. The lack of MoCo mutants in soybean may be due to the small number of mutants examined to date or to the possibility that soybean is a polyploid species.[100]

Other Mutants

The genetics of nitrate reductase in eukaryotes has been most extensively investigated in *A. nidulans*.[95] Three major classes of NR mutants are found: 1) structural gene (*niaD*); 2) MoCo genes (*cnx* ABC,E,F,G,H,J); and 3) regulatory genes (*nirA* and *areA*). Structural and MoCo gene mutants similar to those identified in *Aspergillus* have been found in higher plants, but clear-cut regulatory mutants are yet to be described. The sophistication of NR genetic analysis in higher plants is now high enough to expect that regulatory gene mutants would be identified if they exist. The most likely explanation is that these mutants, perhaps due to lethality, are not identified by current selection procedures. NR-deficient mutants in higher plants that do not fit into either the structural or MoCo gene categories are known. Some of these may be regulatory gene mutants although their characterization at this time is not complete enough for a reasonably sound assignment of function.

In *Arabidopsis* 11 mutants were assigned to seven complementation groups.[80] Of these, two were presumed to be structural gene loci and two others MoCo loci, leaving 3 complementation groups without a function assignment. One or more of these may be regulatory gene mutants;

unfortunately direct evidence for or against this hypothesis is not available.

The soybean *nr1* locus mutants (described under Structural Gene Mutants) are possible regulatory gene mutants, but restricted to the regulation of the constitutive NR expression. In this case a single recessive gene results in the loss of the NAD(P)H and NADH constitutive NR activities, but without effect on the inducible NADH NR. Other mutants have been identified that are constitutive NADH NR-deficient but constitutive NAD(P)H NR positive, indicating that the two NRs are controlled by separate loci. Thus the *nr1* locus must control some function which affects both constitutive NRs. The most probable suggestions are a regulatory role or a cofactor that is specific to the constitutive NRs.

A very commonly occurring mutation in *Arabidopsis* gives rise to mutants of the B1 type (*chl*-1) which are affected in nitrate uptake. These mutants are interesting for physiological studies of nitrate uptake. It is also interesting to note that similar uptake mutants have not been described in any of the other higher plant systems where NR mutants have been investigated.

Reverse mutants of the probable MoCo mutant B25 (*rgn*) have been isolated in *A. thaliana*.[82] All seven independently arisen revertants were found to be mutations in the same suppressor gene designated *su*, which is not linked to the originally mutated gene *rgn*. Other NR-deficient mutants tested, B1 (*chl*-1) and B2-1 (*chl*-2), were not suppressed by this locus. Unfortunately the other MoCo mutant B73 (*cnx*) was not tested.[82]

Genetic Mapping of Nitrate Reductase Genes

The NR genes in *Arabidopsis* have been mapped.[80] The genes *chl*-1, *chl*-3 and *rgn* all map to chromosome 1. The *rgn* locus is independent of *chl*-3. The *chl*-2 locus maps to chromosome 2 and is closely linked to mutant B40 (gene symbol not assigned) and weakly linked to mutants B31-1 and B36 (gene symbols not assigned). The *cnx* locus maps to chromosome 5 as does the suppressor *su*.

In barley the narl locus was mapped 44.2 cM distal to the ribosomal RNA (Rrnl) locus on the short arm of chromosome 6.[101,102] The nar2 locus was mapped 54.7 cM from the short rachilla (s) locus near the centromere of chromosome 7. The other nar loci have not yet been mapped.

MOLECULAR BIOLOGY OF NITRATE REDUCTASE

Gene Structure

The first eukaryotic NR to be cloned was from barley.[103] This work was made possible by the development of monospecific antibodies to the barley NR[68] and the expression vector lambda gtll.[104,105] Total poly(A)+RNA from nitrate-induced barley leaves was fractionated according to size on a native sucrose gradient. The fractions containing the NR mRNA were identified by translation *in vitro* and used to construct a cDNA library. Two positive clones were selected from among twenty-five thousand recombinants screened. The cDNA inserts in these clones represented overlapping clones from a single transcript and were used to screen a second cDNA library constructed in pUC12. Several clones were isolated and the longest among these containing a 1.1 kb insert was designated bNRp10. This clone was identified as NR specific by hybrid selection followed by translation *in vitro*.[103] The identity of this clone with the barley NADH NR was confirmed by demonstrating that a locus identified by restriction fragment length polymorphism with the bNRp10 probe cosegregates with the narla phenotype (Kleinhofs, unpublished). The bNRp10 cDNA insert has been used to isolate additional cDNA and genomic clones from barley which are being characterized.

A similar experimental approach was used to clone the squash NR where one positive clone (lambda Cmcl) was found among 250,000 recombinant phage carrying cDNA synthesized from unfractionated poly(A)+RNA.[106] This clone was shown to encode nitrate reductase by demonstrating that antibodies bound to lambda Cmcl lysate protein were inhibitory to squash NR activity and recognized NR on western blots.

Tobacco cDNA clones of NR were also obtained from a lambda gt11 expression library.[107] These clones ranged from 1.6 to 2.1 kb and shared sequence homologies. The identity of the clones was confirmed by demonstrating that a fusion protein from a recombinant phage was recognized by polyclonal and monoclonal NR antibodies. Antibodies bound by the fusion protein could also be eluted and shown to inhibit tobacco NR activity. A 1.6 kb insert was sequenced and shown to have strong homology at the amino acid level with proteins of the cytochrome b5 superfamily and with human erythrocyte cytochrome b5 reductase. Le and Lederer[108] previously reported that the heme-binding domain of the _N. crassa_ NR is a member of the cytochrome b5 superfamily. The results with the tobacco NR show that this protein is also a member of that family and has homology with the heme binding domain. Homology was also found with the human cytochrome b5 reductase, which is known to catalyze the reduction of cytochrome b5 using NADH as an electron donor and FAD as a redox intermediate. This is functionally comparable to the NR catalytic activity involving NADH, FAD and cytochrome b5 as cofactors. Therefore, this homology defines the NADH/FAD domain of the tobacco NR.[107]

The complete amino acid sequence translated from the nucleotide sequence of the _Nicotiana sylvestris_-type NR was recently published.[109] The protein was found to be 904 amino acids long with a calculated molecular weight of 102,000. In this coding sequence the cytochrome b5 domain is located between amino acids 526 and 615. The cytochrome b5 reductase domain, by homology with the human b5 reductase, is located in the C-terminal part of the protein between the amino acid residues 646 and 904. An unusual stretch of eight amino acids containing six aspartic and two glutamic acid residues is located at the N-terminal end of the protein between residues 58-65. The function of this region is not known, but a similar highly acidic region is also found in the _A. thaliana_ NR sequence.[110] This region, encompassing residues 56 to 60, contains four aspartic and one glutamic acid residues.

The _Arabidopsis_ NR was cloned by using the squash cDNA as a hybridization probe[111] and by using barley NR antibodies (Cheng, personal communication) to screen

Arabidopsis cDNA libraries. One recombinant phage (10Atc-22), carrying a 3.1 kb cDNA insert, was subcloned into a plasmid designated pNN351 and sequenced.[110] Homology of the amino acid sequence was compared to sequences of other proteins. Homology was found between residues 540-620 of the A. thaliana NR and proteins of the cytochrome b5 superfamily. These proteins bind heme and include microsomal, mitochondrial and erythrocyte cytochrome b5, yeast flavocytochrome b2, and sulfite oxidase. A second group of proteins had homology to residues 640-917 of A. thaliana NR. These proteins include the microsomal and erythrocyte NADH cytochrome b5 reductases which bind FAD.[110] Two small oligopeptide sequences of the sulfite oxidase molybdenum domain shared sequence homology with the A. thaliana NR from residues 110-125 and 270-300.[110] Thus the three redox domains of the plant NRs are assigned as follows: molybdenum binding domain to the N-terminal half of the protein, the heme binding domain to the central region, and the FAD binding domain to the C-terminal portion of the protein. The functional domains and sequence homologies are summarized in Figure 2 for Arabidopsis, tobacco and barley NR sequences.

A single cysteine residue reacting with N-ethyl maleimide was identified by Barber and Solomonson[58] in Chlorella NR and by Hackett and coworkers[112] in the microsomal NADH cytochrome b5 reductase. The cysteine was localized to residue 283 of the microsomal NADH cytochrome b5 reductase. Crawford and coworkers[110] compared the Arabidopsis NR sequence to the cytochrome b5 reductase sequence and determined that this cysteine was not conserved between the two proteins. A nearby cysteine at residue 273 in the cytochrome b5 reductase and cysteine residue 889 of the Arabidopsis NR are, however, conserved. Crawford and coworkers[110] suggested that cys-889 in the Arabidopsis NR may be the critical residue. A cysteine residue is also found at this position in the N. sylvestris NR,[109] Hordeum vulgare NR (Dewdney, personal communication), and A. nidulans NR (Johnstone and Kinghorn, personal communication). In fact, the septapeptide CGPPPMI is conserved in all of the plant NR sequences available, and a similar sequence, CGPEAME, is found in A. nidulans NR. Thus, this cysteine residue may be the one that has been proposed to be involved in NADH binding.[58,112,113]

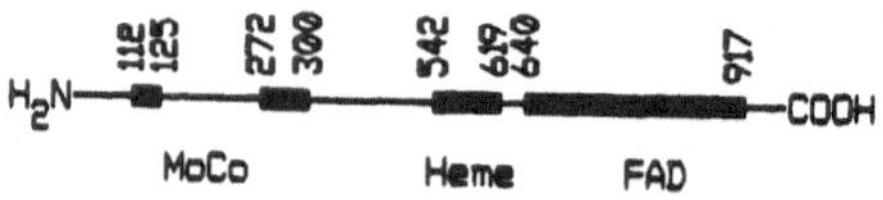

	% Homology			
Nitrate reductase				
N. sylvestris	93	83	86	73
H. vulgare	-	-	-	62
Sulfite oxidase				
Rat	35	55	-	-
Cytochrome b5				
Chicken	-	-	49	-
Cytochrome b5 reductase				
Bovine	-	-	-	47
Human	-	-	-	44

Fig. 2. Amino acid sequence homology of regions of nitrate reductase and other protein domains to *A. thaliana* nitrate reductase.

Barber and Solomonson[58] determined that the 4-maleimido-2,2,6,6-tetramethylpiperidiooxyl (spin-labeled analog of N-ethylmaleimide) labeled sulfhydryl group was located within the heme/molybdenum-containing fragment and not in the FAD-containing fragment which carries the NADH activity. Several other cysteine residues in the FAD-binding domain are conserved in the *Arabidopsis* and the tobacco NRs. These residues, however, were not present in the cytochrome b5 reductase.[110] The only cysteine residue conserved between the *Arabidopsis* and tobacco NRs in the heme domain is at cys-592 (*Arabidopsis* NR) and cys-581 (tobacco NR). This cysteine residue is not conserved in the chicken microsomal cytochrome b5.[110] Thus it seems probable that the cys-889 residue in the *Arabidopsis* NR and the homologous cysteine residues in other NRs are the critical residues involved in NADH-binding. Since the cys-889 residue is located very close to the C-terminal end of the NR protein, the NR structure model may have to be modified to show the C-terminal end folding back to associate with the heme/MoCo domain. Such a model would require at least two protease sensitive sites to accommodate the limited proteolysis data of Solomonson and coworkers.[56]

Additional cloning of NR genes has been carried out in rice[51,114] and maize.[115] Two different genomic clones were recovered from rice. These were subcloned into pUC8 and designated pHBH1 and pHBH2. Analysis of mRNA induction kinetics suggests that the pHBH1 clone codes for a NADH NR while the pHBH2 clone codes for a NAD(P)H NR.[51] Further analysis, however, is required to fully identify the pHBH2 clone as the NAD(P)H NR, and to characterize and compare it with the NADH NRs.

Regulation

The regulation of higher plant NR has received extensive attention and, although much has been learned, we are far from understanding how it works. The pre-molecular work has been extensively reviewed.[2,3,116-126] Therefore, we will limit this discussion to a brief summary of the current molecular work, which is just getting started.

Higher plant NR activity increases in response to nitrate. This was first demonstrated by Hageman and Flesher[116] and seems to be the closest approximation to a universal truth about NR regulation. However, a few higher plant NRs are known to be constitutive. The response of NR activity to nitrate induction can be broken down into three phases: 1) lag phase, lasting up to a few hours; 2) log phase, lasting 8 to 40 hours; and 3) steady state phase.

In barley,[67,127] maize,[128] spinach cells,[129] tobacco XD culture cells[130] and wheat,[131] increases in NR activity after induction by nitrate result from *de novo* synthesis of NR protein. This was first demonstrated by Zielke and Filner[130] in tobacco XD cells using an elegant triple-labeling technique. Somers and coworkers[127] used immunological techniques to demonstrate that in barley NR activity and NR protein increased from time zero and in parallel upon nitrate induction. When nitrate was removed, NR activity and protein declined, with activity declining slightly faster than NR protein. Similar experiments have been recently reported with NR induction in the leaves of maize,[128] except that a more sensitive enzyme-linked immunosorbent assay (ELISA) was used.[47] They found that etiolated and nitrate-starved maize seedlings, when fed nitrate and exposed to light,

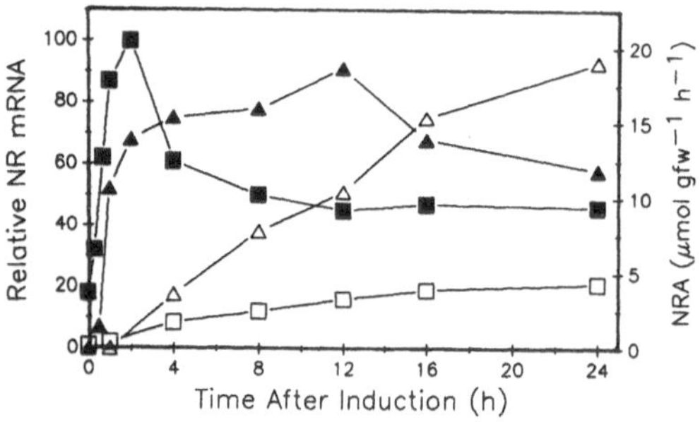

Fig. 3. Accumulation of nitrate reductase activity and mRNA in barley seedling roots and shoots during induction by nitrate. NR mRNA in roots (■) and shoots (▲); NADH NR activity in roots (□) and shoots (△).

responded by synthesizing NR protein and acquiring NR activity. The appearance of the NR protein slightly preceded the appearance of NR activity. These results are in accordance with observations of NR induction in N. crassa,[132] but activation of inactive NR protein is well documented in Chlorella.[133,138] Inactive NR protein appears to be induced in maize by low concentrations of nitrate.[139] Most previous NR induction experiments were conducted at relatively high levels of nitrate.

Induction of NR by nitrate has been demonstrated at the mRNA level in barley,[103,140] squash,[106] rice,[51] Arabidopsis,[111] and tobacco.[109] Detailed time course induction studies have been carried out in barley and rice.[51,140] Barley seedlings grown in the absence of nitrate lacked detectable NR activity, antigen, and NR mRNA.[127,140] Approximately 40 min after addition of nitrate to the seedling roots, NR mRNA was detected in the total RNA isolated from the shoots. A rapid accumulation of NR mRNA resulted in near maximum levels in 2 h, although a slower increase continued for up to about 12 h (Fig. 3). This peak was followed by a steady decline of the NR mRNA levels to about 60% of peak levels by 24 h. NR mRNA increased slightly to steady state levels on continuous nitrate (Fig. 4). Low levels of NR mRNA appeared to be present in uninduced barley roots. Induction by nitrate resulted in a rapid increase in NR mRNA

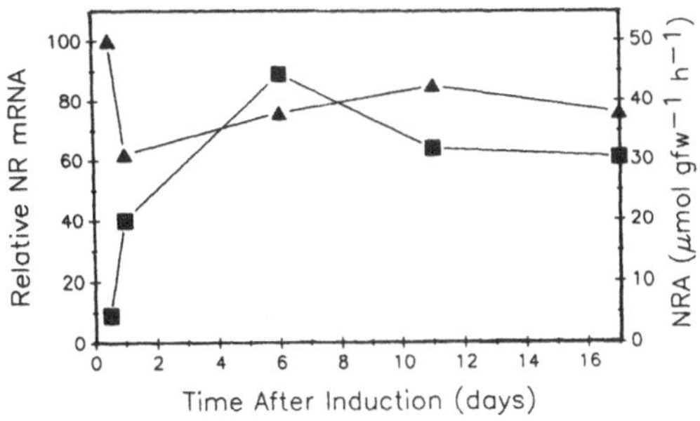

Fig. 4. Steady state nitrate reductase activity and mRNA in barley seedling leaves. NADH NR activity (■) and NR mRNA (▲).

levels with the peak occurring approximately 2 h after induction (Fig. 3). After the peak there was a rapid decline in the NR mRNA levels to steady state levels of about 45% of peak levels. Accumulation of root NR mRNA peaked and declined much more rapidly than shoot NR mRNA.[140]

The nitrate induction of rice NR mRNA was very similar to the barley NR when the presumed NADH NR specific clone (pHBH1) was used as the hybridization probe.[51] The NR mRNA levels increased very rapidly in leaves, reaching peak levels about 5 h after addition of nitrate to the roots and decreased to low levels by 24 h (Fig. 5). The NADH NR activity also increased rapidly, especially in the first 5 h after nitrate addition. The NADH NR activity continued to increase throughout the 24 h experiment, but at slower rate. The NAD(P)H NR activity induction kinetics, however, were different. The NAD(P)H NR activity increased very rapidly, reaching a peak in leaves 3 h after addition of nitrate to the roots (Fig. 5). Thereafter, NR activity declined very rapidly reaching very low levels 5 h after nitrate addition and continuing to decline slowly to the end of the experiment at 12 h. The NR mRNA levels assayed with the presumed NAD(P)H NR clone (pHBH2) showed a pattern similar to the NAD(P)H NR activity induction curve. The NAD(P)H NR mRNA increased very rapidly reaching a peak

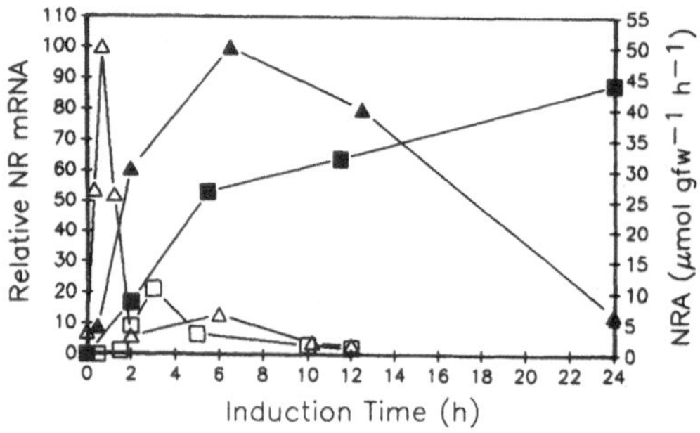

Fig. 5. Induction of NADH and NAD(P)H nitrate reductases and NR mRNA by nitrate in *Oryza sativa* L., sub-species *japonica* cv. M201. Nitrate reductase mRNA was measured by hybridization with fragment BglII/EcoRI from clone pHBH1 (▲) or with fragment ApaI/XhoI from clone pHBH2 (△). Nitrate reductase activity was determined with NADH (■) or with NADPH (□) as the electron donor.

in leaves in less than 1 h after addition of nitrate to the roots and then declined equally rapidly reaching very low levels by 2 h (Fig. 5).

The regulation of NR by light has been extensively investigated (see reviews).[124,141] However, only recently has it become possible to attack this problem at the molecular level. In young barley seedlings grown in the light, Somers and coworkers[127] showed that induction of NR by nitrate in the dark resulted in the accumulation of low levels of NR activity and protein. The accumulation of NR activity and protein was, however, strongly stimulated by subsequent exposure of the seedlings to white light. The authors concluded that nitrate was the primary inducer and that light provided a secondary stimulating effect, perhaps an increased energy status. Similar results were obtained with the induction of NR mRNA.[140] Barley seedlings grown in continuous light without nitrate and induced with nitrate in the dark for 12 h accumulated detectable levels of NR mRNA and NR activity. Exposure of the seedlings to white light at this point resulted in an accelerated

Table 3. Nitrate Reductase Activity and mRNA in Etiolated or Light-Grown Seedling Leaves of Steptoe Barley Subjected to Various Light Treatments During Induction by Nitrate.

Light Treatment**	Light-Grown Seedlings		Etiolated Seedlings	
	NR Activity*	Relative NR mRNA	NR Activity	Relative NR mRNA
Dark	0.0	100	0.0	100
White	0.3	1274	0.2	250
Red	0.0	121	0.4	2011
Red/Far Red	0.0	168	0.1	111
Blue	0.1	163	0.3	2061
Continuous Light	3.2	5263	---	---

*µmol NO_2^-/gfw/h.

**Light-grown seedlings were grown in continuous light (300 µE m^{-2} s^{-1}) for five days, given an 8 min far-red light treatment and moved to the dark for 15 h. The seedlings were then watered with a nutrient solution containing 15 mM nitrate, held in the dark for an additional 2 h and then subjected to the following light treatments: 2 h dark; 2 h white light (300 µE m^{-2} s^{-1}); 5 min red light, 115 min dark; 5 min red light, 5 min far-red light, 110 min dark; 1 h blue light, 1 h dark; 4 h continuous white light. Etiolated seedlings were grown in the dark for 8 days on a nutrient solution containing 15 mM nitrate. Light treatments were as described above.

accumulation of NR mRNA and NR activity without a detectable lag. This response was also detected in light-grown seedlings induced with nitrate for two hours in the dark and for two hours in white light (Table 3). Stimulation of NR mRNA accumulation and NR activity in light-grown seedlings by white light appears to be fluence dependent.

Red, far red and blue light had little effect on NR mRNA or NR activity. These results suggested that the primary effect of light on the accumulation of NR mRNA and NR activity was an increased energy status of the seedlings.

The response of etiolated barley seedlings to light was different from seedlings grown in light. Both red and blue light significantly enhanced NR mRNA accumulation and NR activity in etiolated seedlings during induction by nitrate but had little or no effect in light-grown seedlings (Table 3). The response to red light was reversed by far-red light. The low irradiance phytochrome response has been shown to enhance NR activity in several species, especially in etiolated tissues.[141] Accumulation of NR mRNA in etiolated seedlings during induction by nitrate appears to be modulated by phytochrome, perhaps at the transcriptional level. The reason for the lack of a phytochrome response in light-grown barley seedlings even after a 15 h dark period is not clear.

In tobacco and tomato, remarkably different results were obtained. When plants were grown on a 16 h light/ 8 h dark regime, the NR mRNA levels decreased progressively during the day to virtually undetectable levels and increased during the dark period.[109,142] The levels of NR activity and protein were only decreased by about a factor of two during the light period. A similar change in the NR mRNA levels was also observed in the roots. Tobacco and tomato plants starved for nitrogen for one week expressed approximately the same level of NR mRNA as plants grown continuously on nitrate, while the NR activity and protein were about ten-fold less. Nitrogen-starved plants, when exposed to nitrate, responded by rapidly accumulating NR mRNA.[109,142] These data indicate that NR activity in tobacco and tomato is regulated at the translation as well as the transcription level.

The role of light in the regulation of nitrate assimilation and NR activity is complex. In most cases, nitrate appears to be required for the induction of NR and NR mRNA. The degree of induction can be modulated by light, perhaps by enhancing the energy status of the plant and/or the participation of a photoreceptor such as phytochrome. Additional studies of NR gene structure and transcription will be required to identify the role(s) of light on regulation of NR.

CONCLUSIONS

Interest in NR has rapidly accelerated during the past few years. This has led to a remarkable new understanding of the genetics, biochemistry and molecular biology of NR. Much, however, remains to be learned. The stated aim of much of the research on NR is the improvement in the efficiency of nitrogen use. While it is self-evident that such improvement is not possible without a thorough knowledge of NR and its regulation at all levels, it is not equally obvious how the improved efficiency will be achieved. Approaches worth investigating include engineering of the NR protein and manipulating NR regulation. Engineering of NR will be difficult due to its very large size and our limited understanding of the three-dimensional structure, kinetics of substrate binding, electron flow and product dissociation. In other words, we still do not know what limits the efficiency of NR at the biochemical level. Manipulating NR regulation provides another possibility for achieving improved N-use efficiency. The complexities involved may be as great as with protein engineering. The major problem is that we do not have an assay for efficiency at the enzyme or plant level. The Kms for nitrate and the physiological electron donor could provide useful parameters. The values currently available in the literature, however, do not appear to be very reliable due to the different levels of purification and states of inactivation of the various preparations used. Rates of induction and turnover *in vivo* have not been determined and will not be easy to estimate. The ability to manipulate and use NR genes to produce transgenic plants should, however, provide a means for testing their effects on N-use efficiency *in vivo*. In the meantime, we can expect that the pace of NR research will continue to increase. Consequently, a clear understanding of the gene structure, function, and regulation at the molecular level should emerge. Comparable understanding at the plant level will take longer to achieve.

ACKNOWLEDGMENTS

The authors' research is supported in part by National Science Foundation Grant DMB 85-05095, U.S. Department of Agriculture Competitive Research Grant 86-CRCR-1-2004, and the Washington Technology Center.

This review was completed while AK was a visiting scientist at the Institute of Plant Science Research, Cambridge Laboratory, United Kingdom. Gratitude is expressed to the Institute for providing space, facilities and support for this work.

REFERENCES

1. KLEINHOFS, A., R.L. WARNER, K.R. NARAYANAN. 1985. Current progress towards an understanding of the genetics and molecular biology of nitrate reductase in higher plants. In Oxford Surveys of Plant Molecular and Cell Biology. (B.J. Miflin, ed.), Oxford University Press, Oxford, Vol. 2, pp. 91-121.
2. WRAY, J.L. 1986. The molecular genetics of higher plant nitrate assimilation. In A Genetic Approach to Plant Biochemistry. (A. Blomstein, P. King, eds.), Springer-Verlag, Vienna, pp. 101-157.
3. CAMPBELL, W.H., J. SMARRELLI. 1985. Nitrate reductase: Biochemistry and regulation. In Biochemical Basis of Plant Breeding. (C. Neyra, ed.), CRC Press, Boca Raton, Vol. II, pp. 1-39.
4. WRAY, J.L., J.R. KINGHORN, eds. 1988. Molecular and genetic aspects of nitrate assimilation. Oxford University Press, Oxford (in press).
5. EVANS, H.J., A. NASON. 1953. Pyridine nucleotide-nitrate reductase from extracts of higher plants. Plant Physiol. 28: 233-254.
6. SOLOMONSON, L.P. 1975. Purification of NADH-nitrate reductase by affinity chromatography. Plant Physiol. 56: 853-855.
7. CAMPBELL, W.H. 1976. Separation of soybean leaf nitrate reductases by affinity chromatography. Plant Sci. Lett. 7: 239-247.
8. SHEN, T.-C., E.A. FUNKHOUSER, M.G. GUERRERO. 1976. NADH- and NAD(P)H-nitrate reductases in rice seedlings. Plant Physiol. 58: 292-294.
9. CAMPBELL, W.H., J. SMARRELLI, JR. 1978. Purification and kinetics of higher plant NADH:nitrate reductase. Plant Physiol. 61: 611-616.
10. SHERRARD, J.H., M.J. DALLING. 1979. In vitro stability of nitrate reductase from wheat leaves. I. Stability of highly purified enzyme and its component activities. Plant Physiol. 63: 346-353.

11. KUO, T.M., A. KLEINHOFS, R.L. WARNER. 1980. Purification and partial characterization of nitrate reductase from barley leaves. Plant Sci. Lett. 17: 371-381.
12. KUO, T.M., D.A. SOMERS, A. KLEINHOFS, R.L. WARNER. 1982. NADH-nitrate reductase in barley leaves. Identification and amino acid composition of subunit protein. Biochim. Biophys. Acta 708: 75-81.
13. SMALL, I.S., J.L. WRAY. 1980. NADH nitrate reductase and related NADH cytochrome c reductase species in barley. Phytochemistry 19: 387-394.
14. CAMPBELL, J.M., J.L. WRAY. 1983. Purification of barley nitrate reductase and demonstration of nicked subunits. Phytochemistry 22: 2375-2382.
15. REDINBAUGH, M.G., W.H. CAMPBELL. 1985. Quaternary structure and composition of squash NADH-nitrate reductase. J. Biol. Chem. 260: 3380-3385.
16. KUO, T.M., R.L. WARNER, A. KLEINHOFS. 1982. In vitro stability of nitrate reductase from barley leaves. Phytochemistry 21: 531-533.
17. REDINBAUGH, M.G., W.H. CAMPBELL. 1983. Purification of squash NADH:nitrate reductase by zinc chelate affinity chromatography. Plant Physiol. 71: 205-207.
18. FIDO, R.J., B.A. NOTTON. 1984. Spinach nitrate reductase: further purification and removal of nicked sub-units by affinity chromatography. Plant Sci. Lett. 37: 87-91.
19. NAKAGAWA, H., Y. YONEMURA, H. YAMAMOTO, T. SATO, N. OGURA, R. SATO. 1985. Spinach nitrate reductase. Purification, molecular weight and subunit composition. Plant Physiol. 77: 124-128.
20. FIDO, R.J. 1987. Purification of nitrate reductase from spinach (Spinacea oleracea L.) by immunoaffinity chromatography using a monoclonal antibody. Plant Sci. (Shannon) 50: 111-115.
21. CAMPBELL, W.H. 1988. Structure and regulation of nitrate reductase in higher plants. Op. cit. Reference 4,
22. NOTTON, B.A. 1988. Immunology of nitrate reductase with special reference to higher plants. Op. cit. Reference 4,
23. STREIT, L., B.A. MARTIN, J.E. HARPER. 1987. A method for the separation and partial purification of the three forms of nitrate reductase present in

wild-type soybean leaves. Plant Physiol. 84: 654-657.

24. STEWART, G.R., J.O. OREBAMJO. 1979. Some unusual characteristics of nitrate reduction in *Erythrina senegalensis* DC. New Phytol. 83: 311-319.

25. DAILEY, F.A., T.M. KUO, R.L. WARNER. 1982. Pyridine nucleotide specificity of barley nitrate reductase. Plant Physiol. 69: 1196-1199.

26. DAILEY, F.A., R.L. WARNER, D.A. SOMERS, A. KLEINHOFS. 1982. Characteristics of a nitrate reductase in a barley mutant deficient in NADH nitrate reductase. Plant Physiol. 69: 1200-1204.

27. HARKER, A.R., K.R. NARAYANAN, R.L. WARNER, A. KLEINHOFS. 1986. NAD(P)H bispecific nitrate reductase in barley leaves: partial purification and characterization. Phytochemistry 25: 1275-1279.

28. REDINBAUGH, M.G., W.H. CAMPBELL. 1981. Purification and characterization of NAD(P)H:nitrate reductase and NADH:nitrate reductase from corn roots. Plant Physiol. 68: 115-120.

29. GEWITZ, H.-S., J. PIEFKE, B. VENNESLAND. 1981. Purification and characterization of demolybdo nitrate reductase (NADH-cytochrome c oxidoreductase) of *Chlorella vulgaris*. J. Biol. Chem. 256: 11527-11531.

30. GIRI, L., C.S. RAMADOSS. 1979. Physical studies on assimilatory nitrate reductase from *Chlorella vulgaris*. J. Biol. Chem. 254: 11703-11712.

31. SOLOMONSON, L.P., M.J. McCREERY. 1986. Radiation inactivation of assimilatory NADH:nitrate reductase from *Chlorella*. Catalytic and physical sizes of functional units. J. Biol. Chem. 261: 806-810.

32. SOLOMONSON, L.P., G.H. LORIMER, R.L. HALL, R. BORCHERS, J.L. BAILEY. 1975. Reduced nicotinamide adenine dinucleotide-nitrate reductase of *Chlorella vulgaris*. J. Biol. Chem. 250: 4120-4127.

33. PAN, S.-S., A. NASON. 1978. Purification and characterization of homogeneous assimilatory reduced nicotinamide adenine dinucleotide phosphate-nitrate reductases from *Neurospora crassa*. Biochim. Biophys. Acta 523: 297-313.

34. RAJAGOPALAN, K.V. 1988. Chemistry and biology of the molybdenum cofactor. *Op. cit*. Reference 4,

35. PIENKOS, T., V.K. SHAH, W.J. BRILL. 1977. Molybdenum cofactors from molybdoenzymes and *in vitro* reconstitution of nitrogenase and nitrate reductase. Proc. Natl. Acad. Sci. USA 74: 5468-5471.
36. GARRETT, R.H., A. NASON. 1967. Involvement of a b-type cytochrome in the assimilatory nitrate reductase of *Neurospora crassa*. Proc. Natl. Acad. Sci. USA 58: 1603-1610.
37. NOTTON, B.A., R.J. FIDO, E.J. HEWITT. 1977. The presence of functional haem in a higher plant nitrate reductase. Plant Sci. Lett. 8: 165-170.
38. MENDEL, R.R., A.J. MÜLLER. 1980. Comparative characterization of nitrate reductase from wild type and molybdenum cofactor-defective cell cultures of *Nicotiana tabacum*. Plant Sci. Lett. 18: 277-288.
39. SMARRELLI, J., JR., W.H. CAMPBELL. 1983. Heavy metal inactivation and chelator stimulation of higher plant nitrate reductase. Biochim. Biophys. Acta 742: 435-445.
40. SOMERS, D.A., T. KUO, A. KLEINHOFS, R.L. WARNER. 1982. Barley nitrate reductase contains a functional cytochrome b_{557}. Plant Sci. Lett. 24: 261-265.
41. GARRETT, R.H., A. NASON. 1969. Further purification and properties of *Neurospora* nitrate reductase. J. Biol. Chem. 244: 2870-2882.
42. HORNER, R.D. 1983. Purification and comparison of *nit-1* and wild-type NADPH:nitrate reductases of *Neurospora crassa*. Biochim. Biophys. Acta 744: 7-15.
43. RENOSTO, F., N.D. SCHMIDT, I.H. SEGEL. 1982. Nitrate reductase from *Penicillium chrysogenum*: the reduced flavin-adenine dinucleotide-dependent reaction. Arch. Biochem. Biophys. 219: 12-20.
44. RENOSTO, F., D.M. ORNITZ, D. PETERSON, I.H. SEGEL. 1981. Nitrate reductase from *Penicillium chrysogenum*. Purification and kinetic mechanism. J. Biol. Chem. 256: 8616-8625.
45. JONES, P.W., M.N. MHUIMHUEACHAIN. 1985. The activity and stability of wheat nitrate reductase *in vitro*. Phytochemistry 24: 385-392.
46. NOTTON, B.A., E.J. HEWITT. 1979. Structure and properties of higher plant nitrate reductase, especially *Spinacea oleracea*. *In* Nitrogen Assimilation in Plants. (E.J. Hewitt, C.V.

Cutting, eds.), Academic Press, New York, pp. 227-244.

47. CAMPBELL, W.H., J.L. REMMLER. 1986. Regulation of corn leaf nitrate reductase. I. Immunochemical methods for analysis of the enzyme's protein component. Plant Physiol. 80: 435-441.

48. NAKAGAWA, H., M. POULLE, A. OAKS. 1984. Characterization of nitrate reductase from corn leaves (Zea mays cv W64A x W182E). Two molecular forms of the enzyme. Plant Physiol. 75: 285-289.

49. STREIT, L., R.S. NELSON, J.E. HARPER. 1985. Nitrate reductases from wild-type and nr_1-mutant soybean (Glycine max [L.] Merr.) leaves. Plant Physiol. 78: 80-84.

50. VAUGHN, K.C., S.O. DUKE, E.A. FUNKHOUSER. 1984. Immunochemical characterization and localization of nitrate reductase in non-flurazon-treated soybean cotyledons. Physiol. Plant 62: 481-484.

51. HAMAT, H.B. 1987. Molecular characterization of nitrate reductase genes in rice (Oryza sativa L.). Dissertation, Washington State University, Pullman, 98 pp.

52. CAMPBELL, W.H. 1978. Isolation of NAD(P)H:nitrate reductase from the scutellum of maize. Z. Pflanzenphysiol. 88: 357-361.

53. WARNER, R.L., K.R. NARAYANAN, A. KLEINHOFS. 1987. Inheritance and expression of NAD(P)H nitrate reductase in barley. Theor. Appl. Genet. 74: 714-717.

54. HOWARD, W.D., L.P. SOLOMONSON. 1981. Kinetic mechanism of assimilatory NADH:nitrate reductase from Chlorella. J. Biol. Chem. 256: 12725-12730.

55. HOWARD, W.D., L.P. SOLOMONSON. 1982. Quaternary structure of assimilatory NADH:nitrate reductase from Chlorella. J. Biol. Chem. 257: 10243-10250.

56. SOLOMONSON, L.P., M.J. BARBER, A.P. ROBBINS, A. OAKS. 1986. Functional domains of assimilatory NADH: nitrate reductase from Chlorella. J. Biol. Chem. 261: 11290-11294.

57. CAMPBELL, W.H., M.G. REDINBAUGH. 1984. Ferric-citrate reductase activity of nitrate reductase and its role in iron assimilation by plants. J. Plant Nutr. 7: 799-806.

58. BARBER, M.J., L.P. SOLOMONSON. 1986. The role of the essential sulfhydryl group in assimilatory NADH: nitrate reductase of Chlorella. J. Biol. Chem. 261: 4562-4567.

59. OOSTINDIER-BRAAKSMA, F.J., W.J. FEENSTRA. 1973. Isolation and characterization of chlorate-resistant mutants of Arabidopsis thaliana. Mutat. Res. 19: 175-185.
60. MENDEL, R.R., A.J. MÜLLER. 1976. A common genetic determinant of xanthine dehydrogenase and nitrate reductase in Nicotiana tabacum. Biochem. Physiol. Pflanz. 170: 538-541.
61. MENDEL, R.R., A.J. MÜLLER. 1979. Nitrate reductase deficient mutant cell lines of Nicotiana tabacum. Mol. Gen. Genet. 177: 145-153.
62. MÜLLER, A.J., R. GRAFE. 1978. Isolation and characterization of cell lines of Nicotiana tabacum lacking nitrate reductase. Mol. Gen. Genet. 161: 67-76.
63. WARNER, R.L., C.J. LIN, A. KLEINHOFS. 1977. Nitrate reductase-deficient mutants in barley. Nature 269: 406-407.
64. TOKAREV, B.I., V.K. SHUMNY. 1977. Detection of barley mutants with low level of nitrate reductase activity after the seed treatment with ethylmethanesulphonate. Genetika 13: 2097-2103.
65. KLEINHOFS, A., T.M. KUO, R.L. WARNER. 1980. Characterization of nitrate reductase-deficient barley mutants. Mol. Gen. Genet. 177: 421-425.
66. KLEINHOFS, A., R.L. WARNER, J.M. LAWRENCE, J.M. MELZER, J.M. JETER, D.A. KUDRNA. 1988. Molecular genetics of nitrate reductase in barley. Op. cit. Reference 4,
67. KUO, T.M., A. KLEINHOFS, D. SOMERS, R.L. WARNER. 1981. Antigenicity of nitrate reductase-deficient mutants in Hordeum vulgare L. Mol. Gen. Genet. 181: 20-23.
68. SOMERS, D.A., T.M. KUO, A. KLEINHOFS, R.L. WARNER. 1983. Nitrate reductase-deficient mutants in barley: immunoelectrophoretic characterization. Plant Physiol. 71: 145-149.
69. NARAYANAN, K.R., A.J. MÜLLER, A. KLEINHOFS, R.L. WARNER. 1984. In vitro reconstitution of NADH: nitrate reductase in nitrate reductase-deficient mutants of barley. Mol. Gen. Genet. 197: 358-362.
70. KUO, T.M., A. KLEINHOFS, D.A. SOMERS, R.L. WARNER. 1984. Nitrate reductase-deficient mutants in barley: enzyme stability and peptide mapping. Phytochemistry 23: 229-232.

71. MENDEL, R.R., Z.A. ALIKULOV, N.P. LVOV, A.J. MÜLLER. 1981. Presence of the molybdenum-cofactor in nitrate reductase-deficient mutant cell lines of Nicotiana tabacum. Mol. Gen. Genet. 181: 395-399.
72. MÜLLER, A.J., R.R. MENDEL. 1982. Nitrate reductase-deficient tobacco mutants and the regulation of nitrate assimilation. In Plant Tissue Culture 1982. (A. Fujiwara, ed.), Abe Photo Printing Co., Ltd., Tokyo, pp. 233-234.
73. MÜLLER, A.J. 1983. Genetic analysis of nitrate reductase-deficient tobacco plants regenerated from mutant cells. Evidence for duplicate structural genes. Mol. Gen. Genet. 192: 275-281.
74. DE VRIES, S.E., R. DIRKS, R.R. MENDEL, J.G. SCHAART, W.J. FEENSTRA. 1986. Biochemical characterization of some nitrate reductase deficient mutants of Nicotiana plumbaginifolia. Plant Sci. (Shannon) 44: 105-110.
75. DIRKS, R., I. NEGRUTIU, V. SIDOROV, M. JACOBS. 1985. Complementational analysis by somatic hybridization and genetic crosses of nitrate reductase-deficient mutants of Nicotiana plumbaginifolia. Evidence for a new category of cnx mutants. Mol. Gen. Genet. 201: 339-343.
76. MARTON, L., T.M. DUNG, R.R. MENDEL, P. MALIGA. 1982. Nitrate reductase deficient cell lines from haploid protoplast cultures of Nicotiana plumbaginifolia. Mol. Gen. Genet. 186: 301-304.
77. MARTON, L., V. SIDOROV, G. BAISINI, P. MALIGA. 1982. Complementation in somatic hybrids indicates four types of nitrate reductase deficient lines in Nicotiana plumbaginifolia. Mol. Gen. Genet. 187: 1-3.
78. NEGRUTIU, I., R. DIRKS, M. JACOBS. 1983. Regeneration of fully nitrate reductase-deficient mutants from protoplast culture of Nicotiana plumbaginifolia (Viviani). Theor. Appl. Genet. 66: 341-347.
79. GABARD, J., A. MARION-POLL, I. CHEREL, C. MEYER, A. MÜLLER, M. CABOCHE. 1987. Isolation and characterization of Nicotiana plumbaginifolia nitrate reductase-deficient mutants: genetic and biochemical analysis of the NIA complementation group. Mol. Gen. Genet. 209: 596-606.
80. BRAAKSMA, F.J., W.J. FEENSTRA. 1982. Isolation and characterization of nitrate reductase-deficient

mutants of *Arabidopsis thaliana*. Theor. Appl. Genet. 64: 83-90.

81. BRAAKSMA, F.J., W.J. FEENSTRA. 1982. Nitrate reduction in the wild type and a nitrate reductase deficient mutant of *Arabidopsis thaliana*. Physiol. Plant. 54: 351-360.

82. BRAAKSMA, F.J., W.J. FEENSTRA. 1982. Reverse mutants of the nitrate reductase-deficient mutant B25 of *Arabidopsis thaliana*. Theor. Appl. Genet. 61: 263-271.

83. STREIT, L., J.E. HARPER. 1986. Biochemical characterization of soybean mutants lacking constitutive NADH:nitrate reductase. Plant Physiol. 81: 593-596.

84. ROBIN, P., L. STREIT, W.H. CAMPBELL, J.E. HARPER. 1985. Immunochemical characterization of nitrate reductase forms from wild-type (cv Williams) and nr_1 mutant soybean. Plant Physiol. 77: 232-236.

85. RYAN, S.A., R.S. NELSON, J.E. HARPER. 1983. Soybean mutants lacking constitutive nitrate reductase activity. II. Nitrogen assimilation, chlorate resistance and inheritance. Plant Physiol. 72: 510-514.

86. NELSON, R.S., S.A. RYAN, J.E. HARPER. 1983. Soybean mutants lacking constitutive nitrate reductase activity. I. Selection and initial plant characterization. Plant Physiol. 72: 503-509.

87. CARROLL, B.J., P.M. GRESSHOFF. 1986. Isolation and initial characterization of constitutive nitrate reductase-deficient mutants NR328 and NR345 of soybean (*Glycine max*). Plant Physiol. 81: 572-576.

88. PATEMAN, J.A., D.J. COVE, B.M. REVER, D.B. ROBERTS. 1964. A common co-factor for nitrate reductase and xanthine dehydrogenase which also regulates the synthesis of nitrate reductase. Nature 201: 58-60.

89. MENDEL, R.R., A.J. MÜLLER. 1985. Repair *in vitro* of nitrate reductase-deficient tobacco mutants (*cnxA*) by molybdate and by molybdenum cofactor. Planta 163: 370-375.

90. MENDEL, R.R., L. MARTON, A.J. MÜLLER. 1986. Comparative biochemical characterization of mutants at the nitrate reductase/molybdenum

cofactor loci cnxA, cnxB and cnxC of Nicotiana plumbaginifolia. Plant Sci. (Shannon) 43: 125-129.
91. XUAN, L.T., R. GRAFE, A.J. MÜLLER. 1983. Complementation of nitrate reductase-deficient mutants in somatic hybrids between Nicotiana species. In Protoplasts 1983. (I. Potrykus, C.T. Harms, A. Hinnen, P.S. King, R.D. Shillito, eds.), Berkhauser Verlag, pp. 75-76.
92. CABOCHE, M. 1987. Nitrogen, carbohydrate and zinc requirements for the efficient induction of shoot morphogenesis from protoplast-derived colonies of Nicotiana plumbaginifolia. Plant Cell Tissue Organ Cult. 8: 197-206.
93. CABOCHE, M., I. CHEREL, F. GALANGAU, M.A. GRANDBASTIEN, C. MEYER, T. MOUREAUX, F. PELSY, P. ROUZE, H. VAUCHERET, F. VEDELE, M. VINCENTZ. 1988. Molecular genetics of nitrate reduction in Nicotiana. Op. cit. Reference 4
94. GABARD, J., F. PELSY, A. MARION-POLL, M. CABOCHE, I. SAALBACH, R. GRAFE, A.J. MÜLLER. 1988. Genetic analysis of nitrate reductase deficient mutants of Nicotiana plumbaginifolia: evidence for six complementation groups among 70 classified molybdenum cofactor deficient mutants. Mol. Gen. Genet. 213: 206-213.
95. COVE, D.J. 1979. Genetic studies of nitrate assimilation in Aspergillus nidulans. Biol. Rev. 54: 291-327.
96. ARST, H.N., JR., D.W. TOLLERVEY, H.M. SEALY-LEWIS. 1982. A possible regulatory gene for the molybdenum-containing cofactor in Aspergillus nidulans. J. Gen. Microbiol. 128: 1083-1093.
97. REISS, J., A. KLEINHOFS, W. KLINGMÜLLER. 1987. Cloning of seven differently complementing DNA fragments with chl functions from Escherichia coli K12. Mol. Gen. Genet. 206: 352-355.
98. JACOBSEN, E., F.J. BRAAKSMA, W.J. FEENSTRA. 1984. Determination of xanthine dehydrogenase activity in nitrate reductase deficient mutants of Pisum sativum and Arabidopsis thaliana. Z. Pflanzenphysiol. 113: 183-188.
99. SCHOLTEN, J.H., S.E. DE VRIES, H. NIJDAM, W.J. FEENSTRA. 1985. Nitrate reductase deficient cell lines from diploid cell cultures and lethal

mutant M_2 plants of Arabidopsis thaliana. Theor. Appl. Genet. 71: 263-271.

100. HADLEY, H.H., T. HYMOWITZ. 1973. Speciation and cytogenetics. In Soybeans: Improvement, Production and Uses. (B.E. Caldwell, ed.), American Society of Agronomy, Inc., Madison, pp. 97-116.

101. MELZER, J.M., A. KLEINHOFS, D.A. KUDRNA, R.L. WARNER, T.K. BLAKE. 1988. Genetic mapping of the barley nitrate reductase-deficient nar1 and nar2 loci. Theor. Appl. Genet. 75: 767-771.

102. KLEINHOFS, A., S. CHAO, P.J. SHARP. 1988. Mapping of nitrate reductase genes in barley and wheat. In Proceedings of the Seventh International Wheat Genetics Symposium, Cambridge (in press).

103. CHENG, C.L., J. DEWDNEY, A. KLEINHOFS, H.M. GOODMAN. 1986. Cloning and nitrate induction of nitrate reductase mRNA. Proc. Natl. Acad. Sci. USA 83: 6825-6828.

104. YOUNG, R.A., R.W. DAVIS. 1983. Efficient isolation of genes by using antibody probes. Proc. Natl. Acad. Sci. USA 80: 1194-1198.

105. HUYNH, T.V., R.A. YOUNG, R.H. DAVIS. 1985. Construction and screening cDNA libraries in lambda gt 10 and lambda gt 11. In DNA Cloning: A Practical Approach. (D.M. Glover, ed.), IRL Press, Washington, D.C., Vol. 1, pp. 49-78.

106. CRAWFORD, N.M., W.H. CAMPBELL, R.W. DAVIS. 1986. Nitrate reductase from squash: cDNA cloning and nitrate regulation. Proc. Natl. Acad. Sci. USA 83: 8073-8076.

107. CALZA, R., E. HUTTNER, M. VINCENTZ, P. ROUZE, F. GALANGAU, H. VAUCHERET, I. CHEREL, C. MEYER, J. KRONENBERGER, M. CABOCHE. 1987. Cloning of DNA fragments complementary to tobacco nitrate reductase mRNA and encoding epitopes common to the nitrate reductases from higher plants. Mol. Gen. Genet. 209: 552-562.

108. LE, K.H.D., F. LEDERER. 1983. On the presence of a heme-binding domain homologous to cytochrome b_5 in Neurospora crassa assimilatory nitrate reductase. EMBO J. 2: 1909-1914.

109. GALANGAU, F., I. CHEREL, M. DENG, C. MEYER, T. MAUREAUX, P. ROUZE, H. VAUCHERET, F. VEDELE, M. VINCENTZ, M. CABOCHE. 1988. Nitrate reductase expression in tobacco and tomato. Curr. Top. Plant Biochem. Physiol. 7: 26-34.

110. CRAWFORD, N.M., M. SMITH, D. BELLISSIMO, R.W. DAVIS. 1988. Sequence and nitrate regulation of the Arabidopsis thaliana mRNA encoding nitrate reductase: a metalloflavoprotein with three functional domains. Proc. Natl. Acad. Sci. USA 85: 5006-5010.

111. CRAWFORD, N.M., R.W. DAVIS. 1988. Molecular analysis of nitrate regulation of nitrate reductase in squash and Arabidopsis. Op. cit. Reference 4

112. HACKETT, C.S., W.B. NOVAO, J. OZOLS, P. STRITTMATTER. 1986. Identification of the essential cysteine residue of NADH cytochrome b_5 reductase. J. Biol. Chem. 261: 9854-9857.

113. GUIARD, B., F. LEDERER. 1979. Amino acid sequence of the 'b_5-like' heme-binding domain from chicken sulfite oxidase. Eur. J. Biochem. 100: 441-453.

114. KLEINHOFS, A., R.L. WARNER, H.B. HAMAT, M. JURICEK, C. HUANG, K. SCHNORR. 1988. Molecular genetics of barley and rice nitrate reductases. Curr. Top. Plant Biochem. Physiol. 7: 35-42.

115. GOWRI, G., V.K. RAJASEKHAR, W.H. CAMPBELL. 1987. Isolation of cDNA clones for corn leaf NADH nitrate reductase. Plant Physiol. 83: s17.

116. HAGEMAN, R.H., D. FLESHER. 1960. Nitrate reductase activity in corn seedlings as affected by light and nitrate content of nutrient media. Plant Physiol. 35: 700-708.

117. BEEVERS, L., R.H. HAGEMAN. 1969. Nitrate reduction in higher plants. Annu. Rev. Plant Physiol. 20: 495-522.

118. BEEVERS, L., R.H. HAGEMAN. 1980. Nitrate and nitrite reduction. In The Biochemistry of Plants. (P.K. Stumpf, E.E. Conn, eds.), Amino Acids and Derivatives, B.J. Miflin, ed., Academic Press, New York, Vol. 5, pp. 115-168.

119. BEEVERS, L., R.H. HAGEMAN. 1983. Uptake and reduction of nitrate: bacteria and higher

plants. In Inorganic Plant Nutrition. (A. Lauchli, R.L. Bielesk, eds.), Springer-Verlag, Vienna, pp. 351-375.

120. HEWITT, E.J. 1975. Assimilatory nitrate-nitrite reduction. Annu. Rev. Plant Physiol. 26: 73-100.

121. HEWITT, E.J., D.P. HUCKLESBY, B.A. NOTTON. 1976. Nitrate assimilation. In Plant Biochemistry. (J. Bonner, J.E. Varner, eds.), Academic Press, New York, pp. 633-681.

122. GUERRERO, M.G., J.M. VEGA, M. LOSADA. 1981. The assimilatory nitrate-reducing system and its regulation. Annu. Rev. Plant Physiol. 32: 169-204.

123. LOSADA, M., M.G. GUERRERO. 1979. The photosynthetic reduction of nitrate and its regulation. In Photosynthesis in Relation to Model Systems. (J. Barber, ed.), Elsevier, Amsterdam, pp. 365-408.

124. SRIVASTAVA, H.S. 1980. Regulation of nitrate reductase activity in higher plants. Phytochemistry 19: 725-733.

125. ULLRICH, W.R. 1983. Uptake and reduction of nitrate: algae and fungi. Op. cit. Reference 119, pp. 376-397.

126. WALLACE, W., A. OAKS. 1986. Role of proteinases in the regulation of nitrate reductase. In Plant Proteolytic Enzymes. (M. Dalling, ed.), CRC Press, Boca Raton, Vol. II, pp. 81-89.

127. SOMERS, D.A., T.-M. KUO, A. KLEINHOFS, R.L. WARNER, A. OAKS. 1983. Synthesis and degradation of barley nitrate reductase. Plant Physiol. 72: 949-952.

128. REMMLER, J.L., W.H. CAMPBELL. 1986. Regulation of corn leaf nitrate reductase. II. Synthesis and turnover of the enzyme's activity and protein. Plant Physiol. 80: 442-447.

129. MAKI, H., K. YAMAGISHI, T. SATO, N. OGURA, H. NAKAGAWA. 1986. Regulation of nitrate reductase activity in cultured spinach cells as studied by an enzyme-linked immunosorbent assay. Plant Physiol. 82: 739-741.

130. ZIELKE, H.R., P. FILNER. 1971. Synthesis and turnover of nitrate reductase induced

by nitrate in cultured tobacco cells. J. Biol. Chem. 246: 1772-1779.

131. SOUALMI-BONJEMAA, K., A. MOYSE, M.-L. CHAMPIGNY. 1985. Modulation of nitrate reductase in wheat shoot and root by nitrate. Physiol. Veg. 23: 869-875.

132. AMY, N.K., R.H. GARRETT. 1979. Immunoelectrophoretic determination of nitrate reductase in *Neurospora crassa*. Anal. Biochem. 95: 97-107.

133. LORIMER, G.H., H.S. GEWITZ, W. VOLKER, L.P. SOLOMONSON, B. VENNESLAND. 1974. The presence of bound cyanide in the naturally inactivated form of nitrate reductase of *Chlorella vulgaris*. J. Biol. Chem. 249: 6074-6079.

134. VENNESLAND, B., C. JETSCHMANN. 1971. Nitrate reductase of *Chlorella pyrenoidosa*. Biochim. Biophys. Acta 227: 554-564.

135. FUNKHOUSER, E.A., T.-C. SHEN, R. ACKERMANN. 1980. Synthesis of nitrate reductase in *Chlorella*. I. Evidence for an inactive protein precursor. Plant Physiol. 65: 939-943.

136. FUNKHOUSER, E.A., C.S. RAMADOSS. 1980. Synthesis of nitrate reductase in *Chlorella*. II. Evidence for synthesis in ammonia-grown cells. Plant Physiol. 65: 944-948.

137. SOLOMONSON, L.P., K. JETSCHMANN, B. VENNESLAND. 1973. Reversible inactivation of the nitrate reductase of *Chlorella vulgaris* Beijerinck. Biochim. Biophys. Acta 309: 32-43.

138. PISTORIUS, E.K., H.-S. GEWITZ, H. VOSS, B. VENNESLAND. 1976. Reversible inactivation of nitrate reductase in *Chlorella vulgaris in vivo*. Planta 128: 73-80.

139. OAKS, A., M. POULLE, J.V.J. GOODFELLOW, L.A. CASS, H. DEISING. 1988. The role of nitrate and ammonium ions and light on the induction of nitrate reductase in maize leaves. Plant Physiol. 88: 1067-1072.

140. MELZER, J.M. 1987. Effect of nitrate, light and mutation on nitrate reductase messenger RNA in barley. Dissertation, Washington State University, Pullman, 118 pp.

141. DUKE, S.H., S.O. DUKE. 1984. Light control of extractable nitrate reductase activity in higher plants. Physiol. Plant. 62: 485-493.
142. GALANGAU, F., F. DANIEL-VEDELE, T. MOUREAUX, M.-F. DORBE, M.-T. LEYDECKER, M. CABOCHE. 1988. Expression of the nitrate reductase genes from tobato and tobacco in relation to light-dark regimes and the nitrate supply. Plant Physiol. 88: 383-388.

Chapter Five

THE USE OF MUTANTS LACKING GLUTAMINE SYNTHETASE AND GLUTAMATE SYNTHASE TO STUDY THEIR ROLE IN PLANT NITROGEN METABOLISM

PETER J. LEA, RAY D. BLACKWELL,
ALAN J.S. MURRAY* AND KENNETH W. JOY†

Division of Biological Sciences
University of Lancaster
Lancaster, LA1 4YQ
United Kingdom

*William Grants and Sons Ltd.
The Girvan Distillery
Girvan, Ayrshire
United Kingdom

†Department of Biology
Carleton University
Ottawa, Canada K1S 5B6

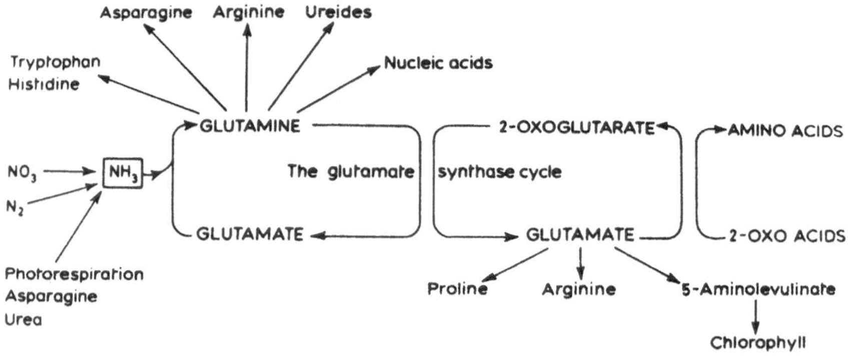

Fig. 1. The glutamate synthase cycle.

INTRODUCTION

It is now generally accepted that at least 90% of the ammonia in higher plants is assimilated via the glutamate synthase cycle as outlined in Figure 1.[1,2] Following the incorporation of ammonia into glutamine by the enzyme glutamine synthetase (GS), the nitrogen can be transferred to a wide range of different compounds. Nitrate or ammonia taken up through the roots is eventually assimilated through the glutamate synthase pathway. In C_3 plants during photorespiration, ammonia released in the mitochondrion from the conversion of glycine to serine is recycled into the chloroplast for reassimilation (Fig. 2).[3] Estimates of the flux through the photorespiratory pathway are varied and values ranging from 15 to 75% of the rate of carbon assimilation have been obtained.[4] More recent evidence suggests that the photorespiration rate is closer to 40% of the rate of net CO_2 fixation[5,6] (see later section). On the other hand, the rate of primary nitrogen assimilation in barley can be as low as 3.0 mol gFW^{-1} h^{-1} or approximately 1% of the net rate of CO_2 fixation.[7] The amount of NH_3 derived from photorespiration within C_3 plant leaves is, therefore, quantitatively the major flux of nitrogen within plant metabolism.

The glutamate synthase cycle was first presented in 1974,[8] and since that time there have been a number of authors who have proposed a role in ammonia assimilation for the ubiquitous enzyme glutamate dehydrogenase.[9,11]

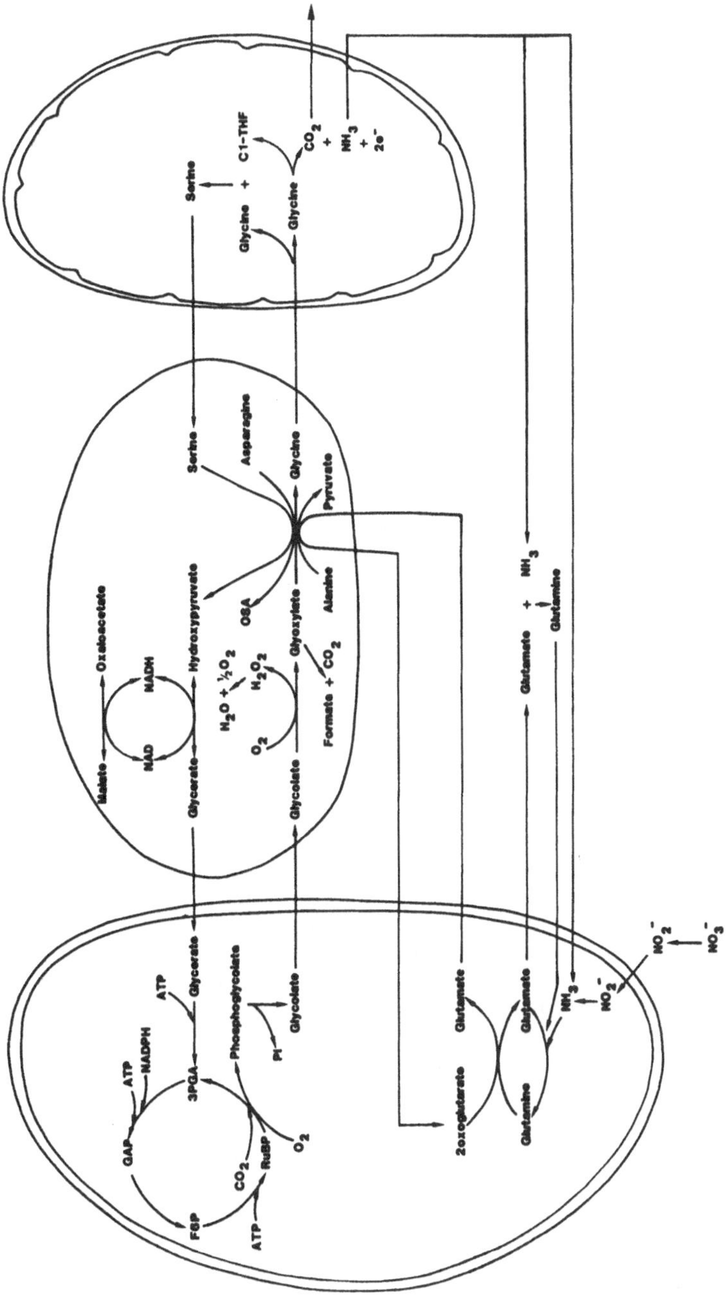

Fig. 2. The photorespiratory carbon and nitrogen cycle.

In the following chapter Rhodes *et al.*[12] present a detailed analysis of the flux of nitrogen through the two possible pathways. A major part of the evidence used to favor the establishment of the glutamate cycle, has been acquired with inhibitors such as methionine sulphoximine and azaserine.[13-15] The use of such compounds has been criticized on the grounds that no inhibitors are entirely specific and they may be having secondary, hitherto unknown, effects on metabolism. A detailed discussion of the mode of action of glutamine synthetase inhibitors on plant metabolism has recently been published.[16]

A method of circumventing the problem of non-specific inhibitors is to isolate conditional lethal mutants of higher plants lacking the enzymes under study. This principle has been particularly useful in studies on nitrate reductase, as a range of plants lacking the active enzyme and only able to grow on a reduced nitrogen source have been obtained.[17,18] Amino acid requiring mutants of tissue culture have been known for some time,[19] but more recently mutant plants of *Arabidopsis thaliana* that require tryptophan for growth have been shown to be lacking the enzyme anthranilate phosphoribosyl transferase.[20] In this article we will establish the importance of mutants of barley lacking key nitrogen metabolizing enzymes in improving our understanding of the process of photorespiration.

ISOLATION OF MUTANTS OF THE PHOTORESPIRATORY PATHWAY

To date, all the plants lacking either glutamine synthetase or glutamate synthase have been recovered using a screen specifically designed to reveal plants deficient in photorespiratory metabolism (Fig. 2). The screen was devised by Somerville and Ogren[21-23] and exploits the competition between CO_2 and O_2 for the bifunctional enzyme ribulose-1,5-bisphosphate carboxylase-oxygenase (rubisco).[24] At high levels (>0.2%) of CO_2, the flow of carbon is exclusively through the Calvin cycle. As the CO_2 concentration is lowered a greater proportion of the carbon is diverted to phosphoglycolate, the initial substrate of the photorespiratory cycle, under the influence of the oxygenase activity of rubisco. At ambient levels of CO_2 and O_2, plants with defects in photorespiration develop characteristic lesions and in most cases senesce prematurely. Lea and Ridley[16] have discussed the possible explanations for the appearance of these visible symptoms in detail.

The photorespiratory pathway contains two interdependent component cycles. One channels carbon from phosphoglycolate, the initial product of rubisco, back into the Calvin cycle, while the other, the nitrogen cycle, regenerates the necessary aminodonors. Mutant lines deficient in a number of the enzymes of this cycle have now been identified (see Blackwell *et al.*[25] for review). The initial mutants recovered using this screen included individual plants lacking phosphoglycolate phosphatase, catalase, glycine decarboxylase, serine transhydroxymethylase and serine:glyoxylate aminotransferase.[26] These plants were then used to clarify confusing aspects of the photorespiratory cycle. For example, the isolation of a plant lacking phosphoglycolate phosphatase which accumulated high levels of the toxic organic acid phosphoglycolate in air, indicated, beyond all doubt, that the initial precursor of the photorespiratory cycle was phosphoglycolate.[27] Similarly, until 1981, two possible sources of CO_2 existed in the cycle even though the mitochondrial conversion of glycine to serine seemed more likely than the decarboxylation of glyoxylate. Plants lacking either glycine decarboxylase,[28] serine transhydroxymethylase[29] or catalase[30] were used to show that the decarboxylation of glycine is the major source of photorespiratory-derived CO_2 in plants. The original experiments used the small, but easily studied, crucifer *A. thaliana*, while more recently a number of plants which lack these enzymes have been isolated from barley. As yet the only mutation available in *A. thaliana*, but not in barley, is that leading to a loss of mitochondrial serine transhydroxymethylase.[29] In addition mutants lacking ferredoxin-dependent glutamate synthase have now been isolated from pea.[25] Barley mutants lacking either NADH-dependent hydroxypyruvate reductase or glutamate:glyoxylate aminotransferase[31] have also been isolated. Mutant lines deficient in glutamine synthetase are frequently recovered and have been characterized in barley.[32,33] By crossing individuals lacking either glutamine synthetase or glutamate synthase, plants deficient in both enzymes of ammonia assimilation have now been produced.[34] The ways in which examination of these three types of mutant plants has clarified a number of problems associated with the operation of the photorespiratory nitrogen cycle and the relationship between nitrogen and carbon metabolism in the plant are detailed in the following sections.

ROLE OF GLUTAMATE SYNTHASE AND GLUTAMINE SYNTHETASE IN PHOTORESPIRATION

The isolation of air-sensitive mutants of *A. thaliana*, barley and pea lacking glutamate synthase provided unequivocal evidence that the major role for this enzyme *in vivo* was the reincorporation of ammonia released in the photorespiratory nitrogen cycle, as originally suggested by Keys *et al.*[3] The subsequent isolation of barley plants containing only low levels of glutamine synthetase[32,33] has in our opinion confirmed that any role for glutamate dehydrogenase in the reincorporation of ammonia must be minimal. The plants deficient in glutamine synthetase contained wild-type levels of glutamate dehydrogenase yet accumulated large amounts of ammonia when placed in air.

Ferredoxin-dependent glutamate synthase is located in the chloroplast in leaves[35] and in the plastid in the root and other non-chlorophylous tissues.[36] In the mutant lines of barley isolated at both Rothamsted and Lancaster, the ferredoxin-dependent enzyme is virtually absent in both the leaves and roots.[33,37,38] However an NADH-dependent enzyme (>5% of the leaf ferredoxin-dependent activity)[39] is present in the all mutant lines so far tested at normal wild-type levels. It must be assumed that there is sufficient activity of this enzyme to carry out the normal process of nitrate assimilation under non-photorespiratory conditions. In contrast to the work to be described later on glutamine synthetase, little is known about the genes regulating glutamate synthase synthesis. However, work with the mutant lines has clearly shown that the ferredoxin-dependent and NADH-dependent enzymes are under separate gene control and that the chloroplast ferredoxin-dependent enzyme is coded for by a nuclear gene.

The 2-oxoglutarate required for the glutamate synthase reaction must be transported across the chloroplast envelope into the stroma. Mutants of both

Arabidopsis[40] and barley[41] have been isolated that are deficient in this uptake process. In both species these mutants behave in a manner similar to the mutants deficient in glutamate synthase. The uptake protein, which has been identified on the chloroplast envelope,[42] is regulated by a single nuclear gene. The presence of two dicarboxylate carriers with different but overlapping specificities has been proposed for chloroplasts.[43] The existence of these mutants suggests that one of the carriers is involved primarily in the recycling of the 2-oxoglutarate required for the synthesis of glutamate in the photorespiratory nitrogen cycle.

The first mutant of *A. thaliana* deficient in glutamate synthase was isolated in 1980,[44] but no corresponding mutants lacking glutamine synthetase have been isolated from this plant. Two laboratories at Rothamsted Experimental Station and the University of Lancaster have now isolated mutant lines of barley lacking the chloroplast form of glutamine synthetase.[32,33] In leaves of many species, two isoenzymes of glutamine synthetase are present.[45,46] In barley one form is localized in the cytoplasm and is termed GS_1 and one in the chloroplast termed GS_2. The two forms can be readily separated by ion exchange chromatography (see Fig. 3) and GS_1 usually accounts for only 10% of the total glutamine synthetase activity in the leaf. At least three different protein subunits of cytosolic glutamine synthetase have been identified along with one major subunit located in the chloroplast.[47-49] The separate genes for these subunits have been cloned.[50,51] The chloroplast GS_2 gene has been shown to synthesize a precursor protein with an attached transit peptide which allows the protein to enter the chloroplast.[50,51]

Photorespiratory ammonia is released from glycine in the mitochondria, an organelle which contains no glutamine synthetase activity.[52] Ammonia must, therefore, leave the mitochondria to be reassimilated. Logically, the cytoplasmic form of GS_1 might be expected to be responsible for the assimilation, given the possible toxic effects of an accumulation of ammonia. This was, in fact, proposed in the original description of the photorespiratory nitrogen cycle.[3] However, every line of air sensitive plant deficient in glutamine synthetase that has been isolated so far has lacked only the chloroplastic

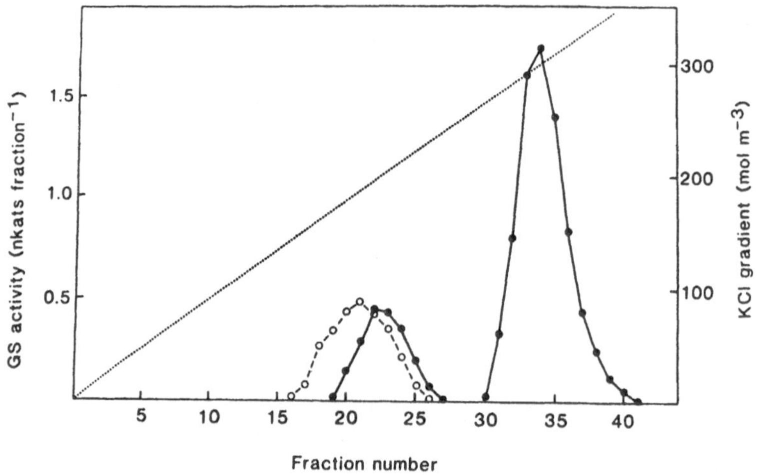

Fig. 3. Separation of isoenzymes of glutamine synthetase from leaves of wild-type barley (●) and LaPr GM31 (o), a mutant deficient in both glutamine synthetase and glutamate synthase activity by FPLC using a Pharmacia mono Q column.

form of the enzyme, retaining wild-type levels of the minor cytoplasmic form (Fig. 3).[33] This demonstrates unequivocally that it is GS_2, located in the chloroplast, along with ferredoxin-dependent glutamate synthase that is responsible for the assimilation of photorespiratory ammonia. As mentioned earlier, lines deficient in glutamine synthetase are only available in barley and not in _A. thaliana_. Unlike barley, _A. thaliana_ contains only the single chloroplast glutamine synthetase isoform[45] and it must be assumed that the loss of the enzyme would produce a glutamine auxotroph, inviable even in an atmosphere of high CO_2.

Plants which are deficient in GS_2 accumulate high levels of ammonia in their leaves. These plants have been used to demonstrate directly the high flux of nitrogen through the photorespiratory cycle. As can be seen in Table 1, only under conditions which promote photorespiration does ammonia accumulate in the mutant plants.

Table 1. Effect of ambient O_2 concentration on ammonia accumulation by leaves of barley deficient in glutamine synthetase.

O_2 Concentration	Wild Type	Deficient Plants
2	1.04	3.78
21	2.31	18.09
50	2.18	45.71

All values are quoted as μmol min^{-1} g^{-1}FW and were made at a PFR of 1300 μmol m^{-2} s^{-1} at a gas flow rate of 0.4 l min^{-1} using the equipment of Blackwell *et al.*[34]

During the study of glutamine synthetase activity in a population of barley derived from mutagenized seed, Blackwell *et al.*[33] related the total activity to the ability of each plant in air to accumulate ammonia. Only when the plants contained less than 40% of the wild type level of glutamine synthetase activity did ammonia accumulate. This demonstrates that barley contains at least a 2-fold excess of glutamine synthetase activity *in vivo*. Given this large excess, accumulation of ammonia in plants infected with biotrophic pathogens such as powdery mildew is unexpected.[7,53] Presumably, the accumulation reflects an imbalance in the availability of substrates for glutamine synthetase, probably ATP.

A replot of the data (Fig. 4) relating ammonia accumulation by mutant barley plants to glutamine synthetase activity on a semi-log scale gives a straight line. Extrapolating to zero glutamine synthetase activity reveals a maximum rate of ammonia production of 120 μmol h^{-1} g^{-1}FW. This value should theoretically be equivalent to the rate of photorespiration and was 40% of the net rate of CO_2 fixation in air. This is in good agreement with current estimates of the photorespiratory rate in air obtained by other methods[5,6] and reinforces the earlier statement (see Introduction) that quantitatively

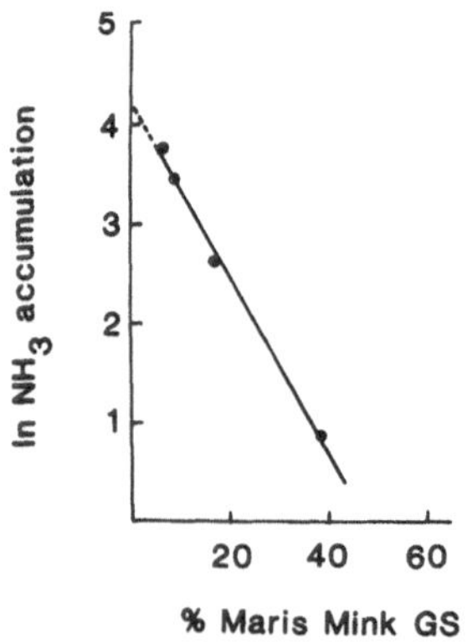

Fig. 4. A semi-log plot of ammonia accumulation by leaves of mutants of barley containing various levels of glutamine synthetase activity as compared to the wild-type variety cv. Maris mink.

the nitrogen flux through photorespiration is the most important flux of ammonia in plants.

Classical genetic studies on the inheritance of the deficiency by all the mutants have demonstrated that the lack of GS_2 activity behaves as a single nuclear recessive gene,[32,33] in a way similar to the glutamate synthase mutants. All of the lines tested so far have been shown to be allelic,[32] reconfirming that only one gene is responsible for the chloroplastic isoenzyme.[50,51] Furthermore, of the plants studied so far, all except one have been shown to lack a protein component present in the wild type capable of reacting with antibody to the enzyme[25,32] (R.D. Blackwell, unpublished results). The mutants contain wild-type levels of glutamine synthetase activity in the roots, further emphasizing the current view that this enzyme is also coded for by different gene(s) from the one coding for GS_2 in leaves.[49-51] Barley, unlike pea and alfalfa,[54] contains only one form of glutamine synthetase in the roots.

Following the isolation of mutant lines of barley deficient in glutamine synthetase and glutamate synthase, individuals lacking the two different enzymes were crossed. The progeny, which were not sensitive to air, were allowed to grow through one complete generation and the F_2 seedlings were screened for the two enzyme activities. Approximately 6% of the plants were found to lack both enzymes of ammonia assimilation. Such plants were extremely sensitive to air and were very difficult to grow even at high levels of CO_2.[34]

INTERACTIONS OF NITROGEN METABOLISM WITH CO_2 FIXATION

Ammonia Accumulation and the Rate of CO_2 Fixation

Plants lacking glutamine synthetase and/or glutamate synthase demonstrate the very close interdependence of CO_2 fixation with nitrogen metabolism and have been used to examine the biochemical basis of this relationship. In common with all photorespiratory mutants isolated so far, except catalase deficiency, plants lacking glutamine synthetase and/or glutamate synthase are unable to maintain wild-type rates of CO_2 uptake in air. Although the CO_2 fixation rates of all three mutants were indistinguishable from that of the wild type under non-photorespiratory conditions, a dramatic decrease in the rate could be seen when the mutant plants were transferred to air (Fig. 5). A similar situation was found in plants treated with the glutamine synthetase inhibitor, methionine sulphoximine (MSO), where there is a fall in the rate of photosynthetic CO_2 assimilation and an accumulation of ammonia.[55,56] Similarly, plants lacking either glutamine synthetase or glutamate synthase accumulate ammonia after exposure to air (Fig. 5). Since plants deficient in glutamate synthase are still capable of synthesizing glutamine using the small amounts of glutamate available, these plants accumulate less ammonia than plants lacking either glutamine synthetase alone or both enzymes. In fact, the levels of ammonia that have been detected in these plants in air are much higher than those known to uncouple photophosphorylation in isolated chloroplast preparations.[57,58] However, over the last few years a body of evidence has been collected, firstly from studies with inhibitors such as MSO[55,56] and more recently using mutant plants, which shows that high concentrations of ammonia do not directly inhibit photosynthetic CO_2 fixation. Close inspection of the change in the rate of CO_2 fixation

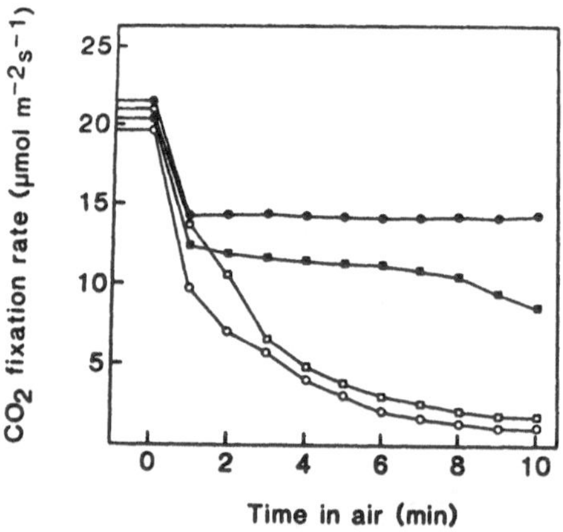

Fig. 5. Rate of CO_2 fixation by detached leaves of wild-type barley (●), LaPr 85/80 (■) (glutamine synthetase deficient mutant), LaPr 85/73 (□) (glutamate synthase deficient mutant) and LaPr GM31 (○) (mutant deficient in both enzyme activities), after transfer to air from 1% O_2. At the end of the experiment the ammonia concentrations in the leaves were 1.5, 20, 6 and 19 μmol g^{-1} fresh weight, respectively.

with time exhibited by the mutants on transfer to air (Fig. 5) revealed that the rate detected in plants lacking glutamine synthetase fell more slowly than in plants lacking glutamate synthase. Simultaneous measurements of the ammonia content indicated that plants deficient in glutamine synthetase accumulated much more of this potentially toxic compound than similarly treated plants lacking glutamate synthase. Further investigation indicated that in air the fixation rate was progressively inhibited by increasing the light intensity (PFR) above 400 μmol m^{-2} s^{-1}, while the content of ammonia remained virtually unchanged. We have concluded that at least initially the ammonia may accumulate in a compartment, perhaps the vacuole, away from the chloroplast and has no direct effect on the photosynthetic machinery.

Recently, Sivak _et al._[59] have thoroughly investigated the photosynthetic characteristics of the photorespiratory mutants using the powerful technique of monitoring chlorophyll fluorescence and have found no evidence supporting uncoupling of electron transport in the mutants by high levels of ammonia. Further analysis of the dark decay of fluorescence quenching in mutants deficient in glutamine synthetase showed a large decrease in F_V indicating that photoinhibition probably accompanied the sharp decline in photosynthesis.

Importance of Photorespiratory Aminodonors to CO_2 Fixation

For the continued operation of the photorespiratory pathway, the plant requires a supply of glutamate and serine to transaminate glyoxylate in the peroxisome. However, the production of these aminodonors is also dependent on the operation of the cycle. Serine is produced in the mitochondria directly from 2 molecules of glycine. During the conversion, one molecule of glycine is decarboxylated and deaminated. The ammonia released is then used to produce glutamate. Photorespiratory mutants are analogous to plants fed the glutamine synthetase inhibitor MSO. As with the mutants, the presence of this inhibitor caused the fixation rate to decline quickly, even though MSO has no direct effect on photosynthesis by isolated chloroplast.[60] Walker et al.[56] proposed that a lack of aminodonors was responsible for the decline in the rate of CO_2 fixation in plant leaves fed MSO by preventing the return of carbon to the Calvin cycle, thus suggesting a very close relationship between CO_2 fixation and nitrogen metabolism. The availability of mutants lacking glutamine synthetase and glutamate synthase activities has allowed this relationship to be studied further, without any doubt concerning the specificities of the inhibitors. The lack of aminodonors was demonstrated in an extensive investigation of the amino acid content of the mutants following exposure to air (Fig. 6).[34] When the plants were grown under non-photorespiratory conditions, the content of amino acids was essentially similar to the pool of amino acids present in the wild type, although the plant deficient in glutamine synthetase obviously contained very low levels of glutamine. On transfer to air, the wild-type plants accumulated glycine and serine reflecting the onset of photorespiration, but this was not seen in the leaves of the mutant plants. In fact, with the exception of the glutamate synthase deficient plant, very low levels of amino acids were detected. The plant lacking glutamate synthase accumulated large quantities of

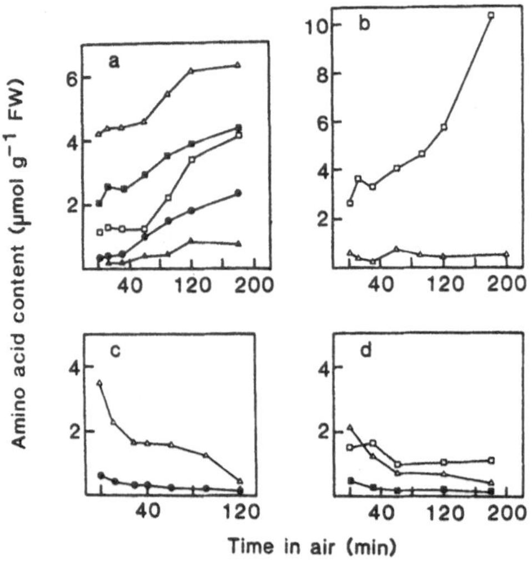

Fig. 6. Change of leaf amino acid content with time following transfer of the wild-type (a), LaPr 85/73 (b), LaPr 85/80 (c) and LaPr GM31 (d) from 0.7% CO_2 to air. Levels of glutamic acid (△), glutamine (□), alanine (■), glycine (▲) and serine (●). Amino acids not shown in the mutants were below the levels of detection.

nitrogen in glutamine which could not be metabolized further. The plants lacking both enzyme activities were in a particularly difficult position. Unable to synthesize glutamine, these plants could not accumulate the high levels of glutamine detected in the glutamate synthase deficient plant. Similarly, their inability to synthesize glutamate at the rate of the wild type was reflected in a reduced content of glutamate on transfer to air.[34]

The direct involvement of aminodonors in preventing the fall in the CO_2 fixation rate of plants deficient in glutamine synthetase has also been shown. Detached leaves of such plants fed glutamine under non-photorespiratory conditions maintained the wild-type rate of CO_2 fixation on transfer to air for much longer than the mutants receiving distilled water at the same pH. As yet, no unequivocal demonstration that a supply of

Table 2. Glutamate, alanine and ammonia content of detached leaves of LaPr 85/73 fed distilled water, pH 6 or 40 mM glutamate or 40 mM alanine in the xylem stream.

Solution Fed	Glutamate	Alanine	Ammonia
	nmol 30 min^{-1} gFW^{-1}		
Distilled water, pH 6.5	1157	135	5300
40 mM Glutamate, pH 6.5	6511	319	8800
40 mM Alanine, pH 6.5	3724	10858	8900

Amino acids were fed for 1 hour in non-photorespiratory conditions (1% O_2, 0.034% CO_2) before the O_2 content of the gas phase flowing over the leaf was increased to 21% to start photorespiration. For comparison leaves of wild-type plant fed distilled water would contain approximately 3652 nmol glutamate, 1774 nmol alanine and 1800 nmol ammonia.

glutamate can prevent the fall in the fixation rate has been achieved in glutamate synthase deficient plants. Kendall et al.[37] demonstrated that 10 mM glutamate fed to the leaves could prevent, to a slight extent, the inhibition of photosynthesis, but the rate soon fell to that of the control. A plant lacking both glutamine synthetase and glutamate synthase is analogous to one lacking only glutamate synthase for this purpose since glutamine synthetase is dependent on glutamate for a supply of glutamate. When glutamate (40 mM) was fed to such a plant it was slightly more effective at preventing the fall in the fixation rate, but the drop was still considerable.[34] The lack of success is difficult to comprehend. One possible explanation is that the rate of uptake could not match the large demand for nitrogen by the aminotransferase enzymes. Analysis of the amino acid content of detached glutamate synthase deficient leaves fed glutamate during the measurement of CO_2 fixation in air indicated a large accumulation of glutamate (Table 2). The limitation on CO_2 fixation in this mutant, therefore, does not involve the actual uptake of glutamate by the

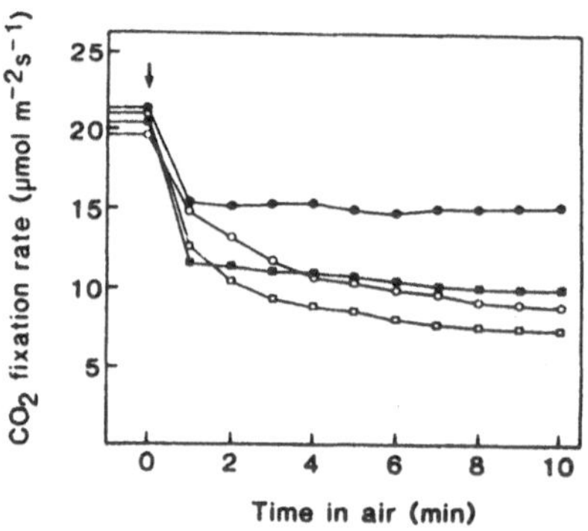

Fig. 7. Rate of CO_2 fixation by wild-type (●), LaPr 85/80 (■), LaPr 85/73 (□) and LaPr GM31 (○) fed 40 mol m^{-3} alanine for 40 min, whilst in 1% O_2 prior to the transfer to air. This figure should be compared to Figure 5 in the absence of alanine.

leaves. A further more probable limitation is the uptake of glutamate by the peroxisome in which the aminotransferase enzymes are located. After careful isolation of intact peroxisomes using Percoll density gradients, these organelles were found to be much less permeable to the ingress of glutamate compared with other amino acids such as serine.[61,62]

After feeding $^{14}CO_2$ to plants deficient in glutamate synthase, a higher proportion of ^{14}C was recovered in the mutant in 6-phosphogluconate, a known effector of rubisco. Subsequently, Hall et al.[63] demonstrated a considerable accumulation of 6-phosphogluconate in leaves of barley plants lacking glutamate synthase after being exposed to air for only 10 minutes. They suggested that the rapid decline in CO_2 fixation exhibited by this mutant on transfer to air may be mediated through accumulation of such an inhibitor rather than lack of aminodonors. However, it is possible to reduce considerably the extent of the fall in the fixation rate on transfer to air by supplying 40 mM alanine to detached leaves of a glutamate synthase deficient plants (Fig. 7).[34] Although providing alanine did not circumvent the lesion completely, the fixation rate was maintained at 60% of the wild-type rate for at least one hour. It appears, therefore, that the fall in the fixation rate may have more than one component.

Chastain and Ogren[64] examined the effect on rubisco activity of placing photorespiratory mutants of *A. thaliana* in air. They concluded that the block in the photorespiratory cycle caused a decrease in the rubisco activation level, possibly due to the action of glyoxylate. An effect on the rubisco activity would be reflected in the content of Calvin cycle intermediates such as ribulose-1,5-bisphosphate (RuBP). Leegood *et al.*[65] have measured the levels of a number of intermediates in photorespiratory mutants of barley. When these mutants were transferred to photorespiratory conditions, a remarkable stability of RuBP content was detected. In fact, both the glutamine synthetase and glutamine synthetase/glutamate synthase deficient mutants accumulated RuBP over the initial 15 min in air, suggesting that in these plants, rubisco activity was being regulated. Only the glutamate synthase deficient plant contained less than 50% of the wild-type level of RuBP after being exposed to air for 15 minutes.

EFFECT OF SUPPRESSING PHOTORESPIRATORY NITROGEN CYCLING ON SUCROSE SYNTHESIS AND THE PHOTOSYNTHETIC MECHANISMS

Mutant plants containing lesions in the nitrogen cycle of photorespiration have also been used to study the mechanisms of photosynthesis and aspects of the synthesis of sucrose and starch. When the plants lacking either glutamine synthetase, glutamate synthase or both enzymes were fed $^{14}CO_2$, the mutants contained a lower proportion of ^{14}C in sucrose indicating the reduced ability of the mutants to export carbon from the chloroplast. This was also reflected in the lower fresh weight of roots recovered from plants lacking glutamine synthetase after growth for three weeks in air (Table 3). Similarly, the plants were unable to produce wild-type levels of sucrose or starch when transferred to high light in air. Barley stores only 30% of its soluble carbohydrate as high molecular weight storage compounds, but the presence of the mutations did not affect this balance suggesting that the lack of sucrose and starch accumulation reflects only the lack of carbon supply.[34]

Table 3. Total root and shoot fresh weights (mg) of wild-type plants and plants deficient in glutamine synthetase grown hydroponically for three weeks in air.

	Wild Type	Deficient Plants
Shoot	446 ± 33	372 ± 51
Root	319 ± 24	227 ± 20
Total Plant	765 ± 58	599 ± 69
S:R Ratio	1.39	1.64

Seeds were germinated for seven days in high CO_2 before transfer to air. At the time of transfer, wild-type plants and plants deficient in glutamine synthetase were indistinguishable in size. Values are means of four replicates.

When plants start to fix CO_2 after a period of darkness, they exhibit a distinct induction phase during which the Calvin cycle builds up adequate levels of intermediates before photosynthate becomes available for export to the cytoplasm. When $^{14}CO_2$ was supplied during induction of CO_2 fixation in air to plants lacking glutamine synthetase and/or glutamate synthase, the proportion of ^{14}C recovered in the mutant as sucrose was higher than in the wild type indicating a possible effect of the mutation on some aspect of the induction process. Furthermore, the proportion of ^{14}C detected as 3-phosphoglyceric acid in the mutant was lower during induction than the corresponding fraction extracted from the wild type. Since the production of triose phosphate for sucrose synthesis from 3-phosphoglycerate is dependent on the energy state of the chloroplast, these results may indicate an altered energy status in the mutants compared to the wild type during the induction period. Recent measurements of the Calvin cycle intermediates have also suggested an increased energization in the mutants.[65]

ALTERNATIVE PHOTORESPIRATORY AMINODONORS

The photorespiratory nitrogen cycle includes two aminotransferase steps, both transferring amino groups in the peroxisome to glyoxylate derived from the initial carbon component of the cycle, namely phosphoglycolate. The nitrogen cycle balances the production of aminodonors exactly with the flow of carbon through the photorespiratory C cycle.[3] The cycle would, therefore, appear to have no need for a further input of nitrogen. However, recent evidence suggests that the nitrogen component cycle is not closed, but can involve an input of nitrogen from amino acids other than glutamate and serine. Ireland and Joy[66] showed that asparagine could donate amino groups to glyoxylate in place of serine. This step was, however, shown to be catalyzed by the same peroxisomal enzyme responsible for transferring amino groups from serine to glyoxylate. Work with a photorespiratory mutant lacking serine:glyoxylate aminotransferase activity (see later section) has added further evidence to this by demonstrating that the lack of both enzyme activities are inherited as a single recessive mutation.[67] Furthermore, supplying asparagine to the cut end of a leaf taken from a mutant which lacks glutamine synthetase can almost totally prevent the fall in the CO_2 fixation rate normally exhibited by this mutant on transfer to air (Fig. 8). Under normal circumstances, it appears that asparagine only supplies approximately 7% of the total nitrogen donated to photorespiration,[68] but our knowledge of asparagine as a photorespiratory aminodonor is still far from complete. The fate of the hydroxysuccinamic acid produced as a result of the deamination of asparagine is particularly fragmentary (see reference 69 for a review) and may well be more easily studied in a system such as that offered by glutamine synthetase mutant containing very low amounts of all the amino acids.

Evidence, based on analysis of the transfer of ^{15}N, that alanine can supply a large proportion of nitrogen to the photorespiratory cycle has been collected over the last few years.[68-70] The peroxisomal enzyme alanine: glyoxylate aminotransferase is currently thought to be an alternative activity carried on the glutamate:glyoxylate aminotransferase enzyme.[71,72] In mutants lacking glutamine synthetase, glutamate synthase or both enzymes, alanine was able to prevent the drop in the fixation

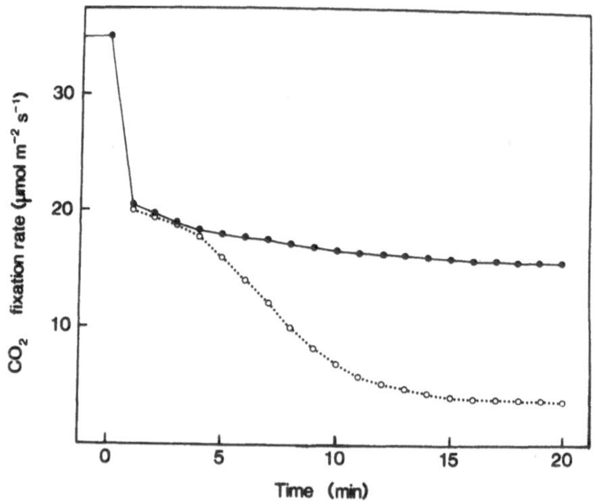

Fig. 8. The rate of CO_2 fixation by leaves of a mutant of barley deficient in glutamine synthetase (LaPr 85/80) in the presence of 40 mM asparagine (●) and distilled water (o), following the transfer from 1% O_2 to air.

rate in air, as shown in Fig. 7.[34] These experiments not only strengthened the evidence that alanine can enter the photorespiratory cycle, but also showed directly that lack of aminodonors was at least partially responsible for the reduction of the rate of CO_2 fixation in a plant lacking glutamate synthase activity. However, it is difficult to envisage a situation where the photorespiration cycle does not produce sufficient aminodonors to fulfill its operation. It, therefore, remains uncertain to what extent the nitrogen cycle is open under normal conditions. It seems highly likely that in a healthy plant the proportions derived from each amino acid will depend on the concentrations of each at the active site of the aminotransferase enzymes and their relative Michaelis constants. A scheme of the role of asparagine and alanine in photorespiration is shown in Figure 9.

The use of alternative amino donors may, however, be more important in a stressed plant. For example the obligate biotroph Erysiphe graminis (mildew) draws a supply of nitrogen from the amino acid pools of its

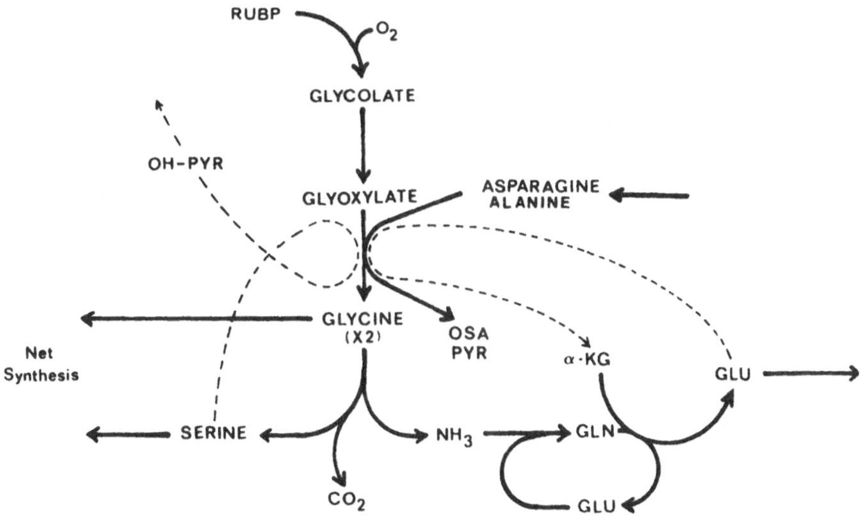

Fig. 9. Possible role of asparagine and alanine in photorespiratory nitrogen metabolism. OSA, 2-oxosuccinamic acid; PYR, pyruvic acid.

host although the actual form withdrawn is not known.[7,73] If it were a photorespiratory amino acid, then the balance of photorespiration would be disturbed and alternative aminodonors would be required. The effect of the mutations on the CO_2 fixation rate is only detected above a PFR of 400 μmol m^{-2} s^{-1}. Photorespiratory mutants can therefore be grown at a low PFR in air for at least 3 months. Preliminary experiments have demonstrated that under such conditions it is impossible to infect plants lacking glutamine synthetase, glutamate synthase or serine:glyoxylate aminotransferase with mildew. On the other hand, extensive coverings of sporulating mildew colonies develop on the leaves of wild-type plants under the same conditions. At high levels of CO_2, however, the mutants are readily infected with mildew. Interestingly, another type of photorespiratory mutant, lacking phosphoglycolate phosphatase is capable of supporting sporulating colonies of mildew in air. It is clear from Table 4, that the inability of the mildew to infect is not linked to the ammonia accumula-

Table 4. Ammonia content of leaves of photorespiratory mutants of barley infected with *Erysiphe graminis* (powdery mildew) grown at a low PFR (100 μmol m^{-2} s^{-1}) for six weeks.

Mutation	Ammonia Content
	nmol g^{-1}FW
Uninfected wild type	80
Infected wild type	200
Glutamine synthetase	7900
Glutamate synthase	2100
Serine:glyoxylate aminotransferase	100

tion since the serine:glyoxylate aminotransferase deficient mutant does not show the characteristic lesions.

THE USE OF MUTANTS LACKING OTHER ENZYMES OF NITROGEN METABOLISM

In the previous sections we have concentrated on mutant plants lacking the two key enzymes of ammonia assimilation. However, information on ammonia assimilation can also be derived from two other mutant lines.

Serine:Glyoxylate Aminotransferase

Mutants of *A. thaliana* lacking this peroxisomal enzyme were first isolated by Somerville and Ogren[74] and more recently in barley by Murray *et al.*[67] but as mentioned previously the activity of asparagine:glyoxylate aminotransferase is also absent in the mutant barley plant. The barley mutants accumulate large quantities of serine upon transfer to air (Fig. 10). The second enzyme glutamate:glyoxylate aminotransferase is, however, able to evolve ammonia at half the rate of the wild-type

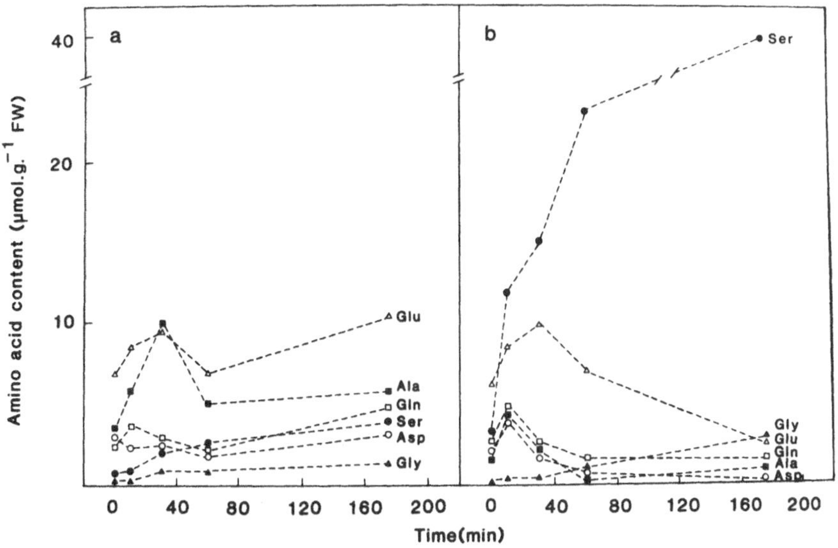

Fig. 10. Amino acid content of leaves from wild-type barley (a) and LaPr 85/84 deficient in serine:glyoxylate aminotransferase (b) following transfer to air.

plant (Fig. 11). The surplus glyoxylate was released as CO_2, by a mechanism that had been previously proposed to be involved in the normal process of photorespiration (see Singh et al.[27] for a review).

Mitochondrial Glycine-Serine Conversion

The ammonia released during photorespiration is formed by the conversion of two molecules of glycine to one molecule of serine in a two-step process in the mitochondria. The CO_2 released in the same reaction is that evolved by the C_3 plants. The first reaction, the glycine decarboxylase (cleavage) system, involves four proteins and catalyses the oxidation of glycine yielding CO_2, NH_3 and 5,10-methylene tetrahydrofolate. The mitochondrial isoenzyme of serine transhydroxymethylase then transfers the C_1 group to a second molecule of glycine to produce serine.[75,76] The complex that carries out both reactions has now been isolated in an active form from spinach mitochondria.[77] Mutant lines

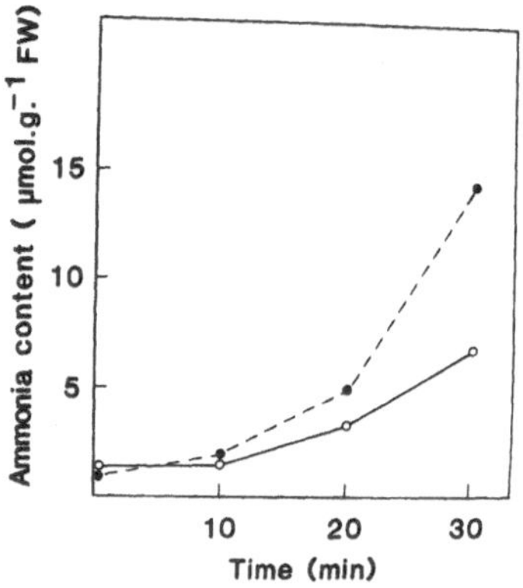

Fig. 11. Accumulation of ammonia by wild-type barley leaves (●----●) and LaPr 85/84 (o——o) in air in the presence of methionine sulphoximine.

of *A. thaliana* have been isolated that lack the two enzyme activities and have been used to demonstrate that the major source of photorespiratory CO_2 loss does in fact come from the conversion of glycine to serine.

Two mutant lines of barley have been isolated that accumulate glycine when exposed to air (Fig. 12).[78] These plants have normal serine transhydroxymethylase activity but are deficient in glycine decarboxylase activity to different extents. If MSO is fed to wild-type leaves ammonia is rapidly evolved as has frequently been demonstrated previously. The two mutants deficient in glycine decarboxylase only produced very small amounts of ammonia, as shown in Figure 13. A combination of Figure 11 and Figure 13 clearly indicates that the major source of ammonia evolution in C_3 plants is derived from the metabolism of glycine during the operation of the photorespiratory nitrogen cycle.

CONCLUSIONS

The availability of mutants lacking key enzymes of nitrogen metabolism have clearly indicated the dominant role played by photorespiration in C_3 plants. The

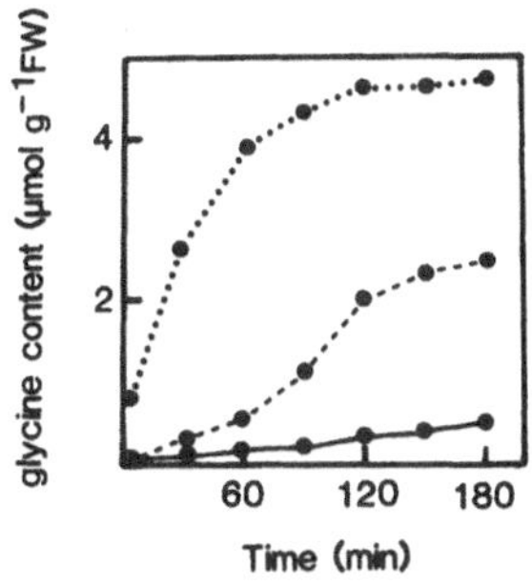

Fig. 12. Accumulation of glycine in leaves of wild-type barley (●——●) and two mutants lacking glycine decarboxylase activity LaPr 85/85 (●----●), LaPr 87/30 (●······●) on transfer to air.

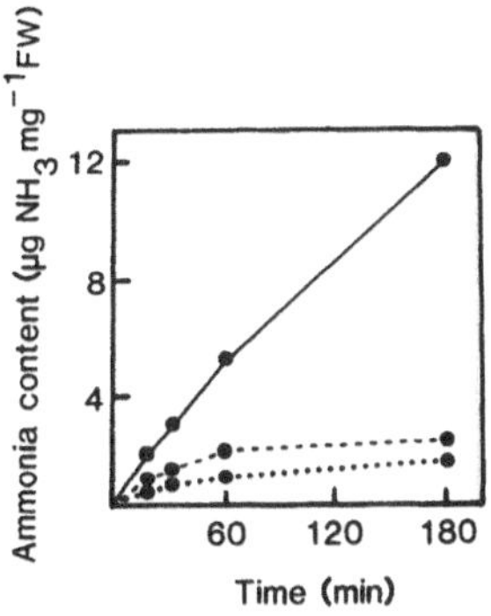

Fig. 13. Accumulation of ammonia by wild-type barley leaves (●——●), LaPr 85/55 (●----●) and LaPr 87/30 (●······●) in the presence of methionine sulphoximine in air.

mutants also allow detailed studies of metabolism to be carried out without the side effects of non-specific inhibitors. The characterization of the mutants has allowed us to be certain that the pathway outlined in Figure 2 is now correct. The recent identification of a rather sickly mutant plant deficient in glutamate: glyoxylate aminotransferase insures that all possible mutants of nitrogen metabolism are now available. The question that we now ask is, given our thorough knowledge of the photorespiration process, is it now possible to circumvent it and prevent the wasteful loss of CO_2?

REFERENCES

1. MIFLIN, B.J., P.J. LEA. 1976. The pathway of nitrogen assimilation in plants. Phytochemistry 15: 873-885.
2. MIFLIN, B.J., P.J. LEA. 1980. Ammonia assimilation. In The Biochemistry of Plants: A Comprehensive Treatise. (B.J. Miflin, ed.), Academic Press, New York, Vol. 5, pp. 169-202.
3. KEYS, A.J., I.F. BIRD, M.J. CORNELIUS, P.J. LEA, R.M. WALLSGROVE, B.J. MIFLIN. 1978. The photorespiratory nitrogen cycle. Nature 275: 741-743.
4. ZELITCH, I. 1979. Photorespiration studies with whole tissue. In Encyclopaedia of Plant Physiology. (M. Gibbs, E. Latzko, eds.), Springer Verlag, Berlin, Vol. 6, pp. 353-367.
5. SOMERVILLE, S.C., C.R. SOMERVILLE. 1983. Effect of oxygen and carbon dioxide on photorespiratory flux determined from glycine accumulation in a mutant of *Arabidopsis*. J. Exp. Bot. 34: 415-421.
6. GERBAUD, A., M. ANDRÉ. 1987. An evaluation of recycling in measurements of photorespiration. Plant Physiol. 83: 933-937.
7. MURRAY, A.J.S., P.G. AYRES. 1986. Studies on nitrate reductase activity and *in vitro* nitrate reduction in barley leaves infected by *Erysiphe graminis*. New Phytol. 104: 367-372.
8. LEA, P.J., B.J. MIFLIN. 1974. An alternative route for nitrogen assimilation in higher plants. Nature 251: 614-616.
9. LOYOLA-VARGAS, V.M., E.S. JIMENEZ. 1984. Differential role of glutamate dehydrogenase in nitrogen metabolism of maize tissues. Plant Physiol. 76, 536-540.
10. SRISTAVA, H.S., R.P. SINGH. 1987. Role and regulation of L-glutamate dehydrogenase activity in higher plants. Phytochemistry 26: 597-610.
11. YAMAYA, T., A. OAKS. 1987. Synthesis of glutamate by mitochondria - an anaplerotic function for

glutamate dehydrogenase. Physiol. Plant. 70: 749-756.

12. RHODES, D., D.G. BRUNK. J.R. MAGALHAES. 1989. In Plant Nitrogen Metabolism. (E.E. Conn, ed.), Plenum Press, New York, pp. 191-226.

13. STEWART, G.R., D. RHODES. 1976. Evidence for the assimilation of ammonia via the glutamine pathway in nitrate-grown *Lemna minor* L. FEBS Lett. 64: 296-299.

14. FENTEM, P.A., P.J. LEA, G.R. STEWART. 1983. Action of inhibitors of ammonia assimilation on amino acid metabolism in *Hordeum vulgare* L. (Golden Promise). Plant Physiol. 71: 502-506.

15. RHODES, D., L. DEAL, P. HAWORTH, G.C. JAMIESON, G.C. REUTER, M.C. ERICSON. 1986. Amino acid metabolism of *Lemna minor* L. I. Responses to methionine sulphoximine. Plant Physiol. 82: 1057-1062.

16. LEA, P.J., S.M. RIDLEY. 1989. Glutamine synthetase and its inhibition. In Herbicides and Plant Metabolism. (A.D. Dodge, ed.), Cambridge University Press, Cambridge, pp. 137-167.

17. WRAY, J.L. 1986. The molecular genetics of higher plant nitrate assimilation. In A Genetic Approach to Plant Biochemistry. (A.D. Blonstein, P.J. King, eds.), Springer Verlag, New York, pp. 101-157.

18. KLEINHOFFS, A. 1989. Genetics and molecular biology of higher plant nitrate reductases. In Plant Nitrogen Metabolism. (E.E. Conn, ed.), Plenum Press, New York, pp. 117-155.

19. McCOURT, P., C.R. SOMERVILLE. 1987. The use of mutants for the study of plant metabolism. In The Biochemistry of Plants. (D.D. Davies, ed.), Academic Press, New York, Vol. 13, pp. 33-64.

20. LAST, R.L. G.R. FINK. 1988. Tryptophan-requiring mutants of the plant *Arabidopsis thaliana*. Science 240: 305-310.

21. SOMERVILLE, C.R., W.L. OGREN. 1973. A phosphoglycolate phosphatase-deficient mutant of *Arabidopsis*. Nature 280: 833-836.

22. SOMERVILLE, C.R., W.L. OGREN. 1982. Genetic modification of photorespiration. Trends Biochem. Sci. 7: 171-174.

23. SOMERVILLE, C.R. 1986. Analysis of photosynthesis and photorespiration using mutants of higher

plants and algae. Annu. Rev. Plant Physiol. 37: 467-507.

24. KEYS, A.J. 1986. Rubisco, its role in photorespiration. Philos. Trans. R. Soc. Lond. B. Biol. 313: 325-336.

25. BLACKWELL, R.D., A.J.S. MURRAY, P.J. LEA, A.C. KENDALL, N.P. HALL, J.C. TURNER, R.M. WALLSGROVE. 1988. The value of mutants unable to carry out photorespiration. Photosyn. Res. 16: 155-176.

26. SOMERVILLE, C.R. 1984. The analysis of photosynthetic carbon dioxide fixation and photorespiration by mutant selection. In Oxford Surveys of Plant Molecular and Cell Biology. (B.J. Miflin, ed.), Oxford University Press, Oxford, Vol. 1, pp. 103-131.

27. SINGH, P., P.A. KUMAR, Y.P. ABROL, M.S. NAIK. 1985. Photorespiratory nitrogen cycle -- a critical evaluation. Physiol. Plant. 66: 169-176.

28. SOMERVILLE, C.R., W.L. OGREN. 1981. Mutants of the cruciferous plant Arabidopsis thaliana lacking glycine decarboxylase. Biochem. J. 202: 373-380.

29. SOMERVILLE, C.R., W.L. OGREN. 1981. Photorespiration-deficient mutants of Arabidopsis thaliana lacking mitochondrial serine transhydroxymethylase activity. Plant Physiol. 67: 666-671.

30. KENDALL, A.C., A.J. KEYS, J.C. TURNER, P.J. LEA, B.J. MIFLIN. 1983. The isolation and characterization of a catalase-deficient mutant of barley. Planta 159: 505-511.

31. LEA, P.J., R.D. BLACKWELL, A.J.S. MURRAY. 1988. The isolation of mutants of barley lacking hydroxypyruvate reductase and glutamate:glyoxylate aminotransferase activity. AFRC Meeting on Photosynthesis, AFRC, London, p. 84.

32. WALLSGROVE, R.M., J.C. TURNER, N.P. HALL, A.C. KENDALL, S.W.J. BRIGHT. 1987. Barley mutants lacking chloroplast glutamine synthetase - biochemical and genetical analysis. Plant Physiol. 83: 155-158.

33. BLACKWELL, R.D., A.J.S. MURRAY, P.J. LEA. 1987. Inhibition of photosynthesis in barley with decreased levels of glutamine synthetase activity. J. Exp. Bot. 38: 1799-1809.

34. BLACKWELL, R.D., A.J.S. MURRAY, P.J. LEA, K.W. JOY. 1988. Sucrose synthesis and N metabolism in a photorespiratory mutant of barley deficient in both chloroplastic glutamine synthetase and ferredoxin-dependent glutamate synthase. J. Exp. Bot. 39: 845-858.
35. WALLSGROVE, R.M., P.J. LEA, B.J. MIFLIN. 1979. Distribution of the enzymes of nitrogen assimilation within the pea leaf. Plant Physiol. 63: 232-236.
36. OAKS, A., B. HIREL. 1985. Nitrogen metabolism in roots. Annu. Rev. Plant Physiol. 36: 345-365.
37. KENDALL, A.C., R.M. WALLSGROVE, N.P. HALL, J.C. TURNER, P.J. LEA. 1986. Carbon and nitrogen metabolism in barley (Hordeum vulgare) mutants lacking ferredoxin-dependent glutamate synthase. Planta 168: 316-323.
38. BLACKWELL, R.D., A.J.S. MURRAY, P.J. LEA. 1987. The isolation and characterization of photorespiratory mutants of barley and pea. In Progress in Photosynthesis Research. (J. Biggins, ed.), Martinus Nijhoff, Drodrecht, Vol. 3, pp. 625-628.
39. WALLSGROVE, R.M., P.J. LEA, B.J. MIFLIN. 1982. The development of NAD(P)H-dependent and ferredoxin-dependent glutamate synthase in greening pea and barley leaves. Planta 154: 473-476.
40. SOMERVILLE, S.C., W.L. OGREN. 1983. An Arabidopsis thaliana mutant defective in chloroplast dicarboxylate transport. Proc. Nat. Acad. Sci. USA 80: 1290-1294.
41. WALLSGROVE, R.M., A.C. KENDALL, N.P. HALL, J.C. TURNER, P.J. LEA. 1986. Carbon and nitrogen metabolism in a barley (Hordeum vulgare) mutant with impaired chloroplast dicarboxylate transport. Planta 168: 324-329.
42. SOMERVILLE, S.C., SOMERVILLE, C.R. 1985. A mutant of Arabidopsis thaliana deficient in chloroplast dicarboxylate transport is missing an envelope protein. Plant Sci. Lett. 37: 317-320.
43. WOO, K.C., U.I. FLUGGE, H.W. HELDT. 1987. A two-translocator model for the transport of 2-oxoglutarate and glutamate in chloroplasts during ammonia assimilation in the light. Plant Physiol. 84: 624-632.

44. SOMERVILLE, C.R., W.L. OGREN. 1980. The inhibition of photosynthesis in Arabidopsis mutants lacking in glutamate synthase activity. Nature 286: 257-259.

45. McNALLY, S.F., B. HIREL, P. GADAL, A.F. MANN, G.R. STEWART. 1983. Glutamine synthetases of higher plants. Evidence for specific isoform content related to their possible physiological role and their compartmentation within the plant. Plant Physiol. 72: 22-25.

46. McNALLY, S.F., B. HIREL. 1983. Glutamine synthetase isoforms in higher plants. Physiol. Veg. 21: 761-774.

47. LARA, M., H. PORTA, J. PADILLA, J. FOLCH, F. SANCHEZ. 1984. Heterogeneity of glutamine synthetase polypeptides in Phaseolus vulgaris L. Plant Physiol. 76: 1019-1023.

48. CULLIMORE, J.V., C. GEBHARDT, R. SAARELAINEN, B.J. MIFLIN, K.B. IDLER, R.F. BARKER. 1984. Glutamine synthetase of Phaseolus vulgaris L.: Organ-specific expression of a multigene family. J. Mol. Appl. Genet. 2: 589-599.

49. GEBHARDT, C., J.E. OLIVER, B.G. FORDE, R. SAARELAINEN, B.J. MIFLIN. 1986. Primary structure and differential expression of glutamine synthetase genes in nodules, roots and leaves of Phaseolus vulgaris. EMBO J. 5: 1429-1435.

50. TINGEY, S.V., G.M. CORUZZI. 1987. Glutamine synthetase of Nicotiana plumbaginifolia: cloning and in vivo expression. Plant Physiol. 84: 366-373.

51. BENNETT, M.J., D.A. LIGHTFOOT, J.V. CULLIMORE. 1988. Glutamine synthetase cDNA sequences and subunit identification of the Phaseolus vulgaris L. multigene family. Mol. Gen. Genet. (in press).

52. WALLSGROVE, R.M. A.J. KEYS, I.F. BIRD, M.J. CORNELIUS, P.J. LEA, B.J. MIFLIN. 1980. The location of glutamine synthetase in leaf cells and its role in the reassimilation of ammonia released in photorespiration. J. Exp. Bot. 31: 1005-1017.

53. WALTERS, D.R., P.G. AYRES. 1983. Changes in nitrogen utilization and enzyme activities associated with CO_2 exchanges in healthy leaves

of powdery mildew infected barley. Plant Path. 23: 447-459.

54. VEZINA, L., H.J. HOPE, K.W. JOY. 1987. Isoenzymes of glutamine synthetase in roots of pea (Pisum sativum L. cv. Little Marvel) and alfalfa (Medicago media Pers cv. Saranac). Plant Physiol. 83: 58-62.

55. IKEDA, M., OGREN. W.L. HAGEMAN, R.H. 1984. Effect of methionine sulphoximine on photosynthetic carbon metabolism in wheat (Triticum aestivum) cv. Poland) leaves. Plant Cell Physiol. 25: 447-452.

56. WALKER, K.A., A.J. KEYS, C.V. GIVAN. 1984. Effect of L-methionine sulphoximine on the products of photosynthesis in wheat (Triticum aestivum) leaves. J. Exp. Bot. 35, 1800-1810.

57. KROGMAN, D.W., A.T. JAGENDORF, M. AVRON. 1959. Uncouplers of spinach chloroplast photosynthetic phosphorylation. Plant Physiol. 34: 272-277.

58. SLOVACEK, R.E., G. HIND. 1981. Correlation between photosynthesis and the transthylakoid proton gradient. Biochim. Biophys. Acta 635: 393-404.

59. SIVAK, M.N., P.J. LEA, R.D. BLACKWELL, A.J.S. MURRAY, N.P. HALL, A.C. KENDALL, J.C. TURNER, R.M. WALLSGROVE. 1988. Some effects of oxygen on photosynthesis by photorespiratory mutants of barley (Hordeum vulgare L.). J. Exp. Bot. 39: 655-666.

60. MUHITCH, M., J.S. FLETCHER. 1983. Influence of methionine sulphoximine on photosynthesis in isolated chloroplasts. Photosynth. Res. 4: 241-244.

61. LIANG, Z., A.H.C. HUANG. 1983. Metabolism of glycolate and glyoxylate in intact spinach leaf peroxisomes. Plant Physiol. 73: 147-152.

62. ANDERSON, I.W., V.S. BUTT. 1986. Permeability of lettuce leaf peroxisomes to photorespiratory metabolites. Biochem. Soc. Trans. 14: 106-107.

63. HALL, N.P., A.C. KENDALL, P.J. LEA, J.C. TURNER, R.M. WALLSGROVE. 1987. Characteristics of a photorespiratory mutant of barley (Hordeum vulgare) deficient in phosphoglycolate phosphatase. Photosynth. Res. 11: 89-96.

64. CHASTAIN, C.J., W.L. OGREN. 1985. Photorespiration-induced reduction of ribulose bisphosphate

carboxylase activation level. Plant Physiol. 77: 851-856.
65. LEEGOOD, R.C., M.J. ADCOCK, C.A. LABATE, A.J.S. MURRAY, R.D. BLACKWELL, P.J. LEA. 1987. Regulation of photosynthetic carbon metabolism following the transition from 2% to 20% oxygen in photorespiratory mutants of barley. AFRC Meeting on Photosynthesis, AFRC, London, p. 82.
66. IRELAND, R.J., K.W. JOY. 1983. Purification and properties of an asparagine aminotransferase from Pisum sativum leaves. Arch. Biochem. Biophys. 223: 291-296.
67. MURRAY, A.J.S., R.D. BLACKWELL, P.J. LEA. 1987. Photorespiratory donors aminotransferase specificity and photosynthesis in a mutant of barley deficient in serine:glyoxylate aminotransferase activity. Planta 172: 106-113.
68. TA, T.C., K.W. JOY. 1986. Metabolism of some amino acids in relation to the photorespiration cycle of pea leaves. Planta 169: 117-122.
69. SIECIECHOWICZ, K.A., K.W. JOY, R.J. IRELAND. 1988. The metabolism of asparagine in plants. Phytochemistry 27: 663-671.
70. BETSCHE, T. 1983. Aminotransfer from alanine and glutamate to glycine and serine during photorespiration in oat leaves. Plant Physiol. 71: 961-965.
71. TOLBERT, N.E. 1980. Microbodies - peroxisomes and glyoxysomes. In The Biochemistry of Plants: A comprehensive treatise. (N.E. Tolbert, ed.), Academic Press, New York, Vol. 1, pp. 359-388.
72. REHFIELD, D.W., N.E. TOLBERT. 1970. Aminotransferases in peroxisomes from spinach leaves. J. Biol. Chem. 247: 4803-4811.
73. MURRAY, A.J.S., P.G. AYRES. 1986. Uptake and translocation of nitrogen by mildewed barley seedlings. New Phytol. 104: 355-365.
74. SOMERVILLE, C.R., W.L. OGREN. 1980. Photorespiration mutants of Arabidopsis thaliana deficient in serine-glyoxylate aminotransferase activity. Proc. Nat. Acad. Sci. USA 77: 2684-2687.
75. WALKER, J.L., D.J. OLIVER. 1986. Glycine decarboxylase multienzyme complex: Purification and partial characterization from pea leaf mitochondria. J. Biol. Chem. 261: 2214-2221.

76. NEUBURGER, M., J. BOURGUIGNON, R. DOUCE. 1986. Isolation of a large complex from the matrix of pea leaf mitochondria involved in rapid transformation of glycine into serine. FEBS Lett. 207: 18-22.
77. BOURGUIGNON, J., M. NEUBURGER, R. DOUCE. 1988. Resolution and characterization of the glycine cleavage reaction in pea leaf mitochondria. Biochem. J. 255: 169-178.
78. BLACKWELL, R.D., A.J.S. MURRAY, P.J. LEA. 1988. Mutations of photorespiratory CO_2 evolution in the mitochondrial conversion of glycine to serine. AFRC Meeting on Photosynthesis, AFRC, London, p. 78.

Chapter Six

ASSIMILATION OF AMMONIA BY GLUTAMATE DEHYDROGENASE?

DAVID RHODES,* DENNIS G. BRUNK*
AND JOSÉ R. MAGALHÃES+

*Department of Horticulture
Purdue University
West Lafayette, Indiana 47907

+EMBRAPA
Centro Nacional de Pesquisa
de Hortaliças - C.N.P.H.
Brasilia - DF - 70,000
Brazil

INTRODUCTION

Prior to 1974, glutamate dehydrogenase (GDH; EC 1.4.1.2) was considered to be the major route of ammonia assimilation in higher plants,[1-3] as in yeast, where clear kinetic evidence (specifically, the direct incorporation of $^{15}NH_4^+$ into glutamic acid) exists for the operation of a GDH pathway of glutamate biosynthesis.[4,5] However, since the discovery of glutamate synthase in higher plant tissues[1-3] and the realization that the glutamine synthetase/glutamate synthase (GS/GOGAT) cycle may represent a major alternative pathway of ammonia assimilation into glutamic acid in plants,[1-3,6,7] controversy has raged over the precise role of GDH in ammonia metabolism. As discussed by Lea[8] elsewheres in this volume, considerable evidence, both biochemical and genetic, has accrued in the last decade to indicate that the GS/GOGAT cycle is the principal, if not the sole, pathway of ammonia assimilation in higher plants. This leaves the NAD-specific GDH as an enigma, since it is present often at extremely high levels in higher plant tissues, particularly in roots,[9] but with no obvious major biosynthetic role (see reviews [6,7,10]). The current concensus appears to be that GDH occupies either a small, minor or negligible role in plant ammonia assimilation, relative to the flux via the GS/GOGAT cycle.[6,7,10] Although several relatively recent articles[11-14] continue to note that in certain plant tissues a contribution of GDH to net ammonia assimilation cannot be ruled out, there appears to be little doubt that the GS/GOGAT cycle may carry a large percentage of the ammonia assimilation of the higher plant cell. Joy[15] has recently reviewed aspects of ammonia assimilation and related amino acid metabolism in plants, and concludes that there is little evidence for the participation of GDH in these processes.

Several recent studies have demonstrated ^{15}N-glutamate synthesis in isolated mitochondria supplied with ^{15}N-glycine,[14,31,38] indicating that mitochondrial GDH may reassimilate a fraction of the ammonia released by mitochondria during photorespiration. Isolated pea mitochondria typically synthesize 5 nmol of ^{15}N-glutamate/h/mg of mitochondrial protein from 2 mM ^{15}N-glycine, presumably via mitochondrial GDH.[31] However, this rate represents only 0.2% of the rate of ammonia production in the photorespiratory pathway.[31] Hartman and Ehmke[38] have reported rates of ammonia assimilation into glutamate of between 1.6 and 14.2% of the photorespiratory flux in isolated pea mitochondria. With this low and variable apparent flux via GDH *in vitro*, it clearly becomes technically challenging to attempt to conclusively demonstrate the operation of this pathway *in vivo*.

This chapter will attempt to highlight some of the technical difficulties associated with assigning upper and lower bounds to the flux via GDH *in vivo*, when superimposed upon an admittedly much larger flux via the GS/GOGAT pathway. Much of the work to be described herein focuses on the use of ^{15}N tracer technology alone or in combination with inhibitors of amino acid metabolism to address this question in order to define limitations of the techniques and needs for alternative experimental approaches. Preliminary $^{15}NH_4^+$ tracer and inhibitor studies with a mitochondrial GDH-deficient mutant of *Zea mays* are reported. These results help focus attention on the validity of the underlying assumption that GS inhibitors such as methionine sulfoximine (MSX) are without effects on the carbon economy and the supply of 2-oxoglutarate to mitochondrial GDH *in vivo*. We describe possible alternative experimental approaches to distinguishing between the operation of the GS/GOGAT and GDH pathways which do not employ inhibitors of the GS/GOGAT cycle, and which rely upon quantitative kinetic analysis of $^{15}NH_4^+$ assimilation.

PRELIMINARY EVIDENCE FOR $^{15}NH_4^+$ ASSIMILATION VIA GDH IN TOMATO ROOTS

Traditionally, various inhibitors, usually of GS and GOGAT, have been employed to attempt to assess the relative contributions of the GS/GOGAT and GDH pathways.[7,16,22,24-27,35] Thus, methionine sulfoximine (MSX), an inhibitor of GS, has been widely used to demonstrate that the GS/GOGAT cycle is the major route of ammonia assimilation in plants.[16,18,22,24-27] This inhibitor appears to block almost all $^{15}NH_4^+$ assimilation in most plant tissues that have been examined.[7,22,26] The absence of detectable ^{15}N-labeling of glutamate (or any other amino acid) following inhibition of GS by MSX *in vivo* is frequently cited as evidence against a significant role of GDH in ammonia assimilation,[7,22,26,35] despite the high activity of GDH in certain tissues.[26,35]

One exception to these generalizations appears to be tomato (*Lycopersicon esculentum* cv. Campbell 1327). The labeling kinetics of amino acids in roots of tomato plants supplied with 1 mM MSX for 2 h and then with 5 mM 99% $(^{15}NH_4)_2SO_4$ for the subsequent 24 h (pH of medium adjusted to pH 5.7 with $CaCO_3$), clearly show that continued assimilation of $^{15}NH_4^+$ can occur when the GS/GOGAT cycle is inhibited (Fig. 1) (Magalhães and Rhodes, unpublished). In the presence of MSX, three

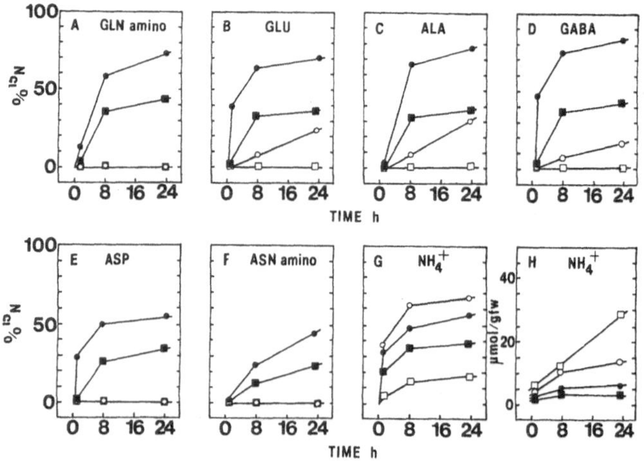

Fig. 1. Labeling behavior of amino acid and free ammonia pools of root and shoot tissue of 2-week old tomato seedlings grown in hydroponic culture and supplied with 5 mM 99% $(^{15}NH_4)_2SO_4$ via the aerated root medium (pH 5.7) in the presence and absence of a 2 h preincubation with 1 mM MSX (preincubation with or without MSX in minus N medium) (Magalhães and Rhodes, unpublished). Plants were pregrown on 5 mM NH_4NO_3 for one week in vermiculite and one week in modified Hoagland's nutrient solution containing 5 mM NH_4NO_3 prior to the labeling experiment. Circles refer to root metabolites, squares refer to shoot metabolites. Closed symbols refer to control (- MSX) treatments; open symbols refer to 1 mM MSX treatments. Panels A - G show changes in ^{15}N abundance of glutamine-amino N (A), glutamate (B), alanine (C), γ-aminobutyrate (GABA) (D), aspartate (E), asparagine-amino N (F) and free NH_4^+ (G). Panel H shows the changes in free NH_4^+ pool size (µmol/g f.w.).

amino acids [glutamate, alanine and γ-aminobutyrate (GABA)] of the root tissue continue to become labeled with ^{15}N under conditions where labeling of the amino-N moiety of glutamine is completely inhibited (Fig. 1A-D). The most logical explanation for these observations is that GDH catalyzes some synthesis of glutamate in tomato roots and that this glutamate can be used both for transamination with pyruvate to yield alanine, and decarboxylation to yield GABA. It is notable that in roots in the absence of MSX, glutamate and GABA are much more heavily labeled than glutamine-amino N after 1 h of exposure to $^{15}NH_4^+$ (Fig. 1A,B,D); this would not be inconsistent with some primary ammonia assimilation via GDH. An alternative explanation for these results is that, in the presence of MSX, direct assimilation of ammonia into alanine occurs via an alanine dehydrogenase in tomato roots, with alanine being transaminated to glutamate and/or GABA. However, in the absence of MSX, glutamate and GABA are much more heavily labeled than alanine at early time points (Fig. 1B,C,D), suggesting that it is unlikely that alanine is a precursor of glutamate and GABA.

It is evident that there is an MSX-insensitive pathway of ammonia assimilation in operation in tomato roots, with GDH remaining a leading candidate for this pathway. In tomato roots we estimate this flux to be approximately 30 to 50 nmol/h/g f.w.; this is perhaps no more than 1% of total NH_4^+ assimilation in the absence of MSX. However, as discussed below, this rate could be underestimated as a result of protein turnover and the associated isotopic dilution of the free glutamate and alanine pools by ^{14}N-amino acids released from protein. It may be further underestimated if MSX has secondary effects on photosynthesis and carbon economy which in turn affect the supply of 2-oxoglutarate to GDH *in vivo*.

Free ammonia accumulates rapidly in both shoots and roots of tomato in response to MSX (Fig. 1H). In shoots, this free ammonia is poorly labeled with ^{15}N (Fig. 1G) and is probably derived from photorespiratory catabolism of unlabeled storage pools of amino acids as well as catabolism of amino acids derived from protein turnover.[27,30] The low ^{15}N abundance of the ammonia accumulated in shoots (<15% ^{15}N) (Fig. 1G) means that continued glutamate synthesis from NH_4^+ in the shoot tissue of MSX-treated plants would be difficult to detect because of isotope dilution from $^{14}NH_4^+$. In roots, the ammonia accumulated in the presence of MSX is fairly heavily labeled (65 - 70% ^{15}N) (Fig. 1G), and must be largely derived from the medium. The pools of glutamine, glutamate, alanine, aspartate and asparagine after 24 h exposure to $^{15}NH_4^+$ were, on average, 5- to 10-fold lower in the MSX-treated plants than in the controls (- MSX) (results not shown).

In contrast, the pools of valine, leucine, isoleucine, proline, threonine, phenylalanine, lysine and tyrosine increased 5- to 10-fold above control values in the shoots of MSX-treated tomato plants, and 2- to 4-fold above control values in the roots of MSX-treated plants after 24 h (results not shown). The latter amino acids all exhibited low isotope abundance (<3% ^{15}N in roots; <1% ^{15}N in shoots after 24 h) (results not shown), and are presumably derived from protein turnover (cf. *Lemna minor*[27,30]). These amino acids could possibly be sequestered in the vacuoles and therefore rendered inaccessible to catabolism.[27] Alternatively, reduced availability of 2-oxoglutarate, caused by MSX, may restrict the transamination of amino acids such as valine, leucine and isoleucine derived from protein hydrolysis with glutamate. The equilibria of these transamination reactions would be shifted in favor of glutamate production by the rapid decline in glutamate level in response to MSX *only* if 2-oxoglutarate was freely available:

amino acid + 2-oxoglutarate ------> glutamate + 2-oxo acid

If 2-oxoglutarate becomes limiting for these transamination reactions, then could 2-oxoglutarate also become limiting for GDH activity in the presence of MSX? It is often assumed that specific inhibitors of GS (such as MSX) are without effects on 2-oxoglutarate availability, but data to support this assumption are limited.

Lewis *et al.*[26] have suggested that 5 mM MSX does not interfere to any perceptible extent with the metabolism of glutamine-amido ^{15}N in barley roots, implying that 2-oxoglutarate is not limiting for metabolism of glutamine via GOGAT in MSX-treated roots. Fentem *et al.*[36] have further argued that since MSX does not inhibit glutamate synthesis in fungi where GDH operates (see e.g. the lichen *Peltigera aphthosa*[37]), it is unlikely that MSX has secondary inhibitory action on either the production of NADH or 2-oxoglutarate. However, it is debatable as to whether or not this latter evidence from fungal systems can be readily extrapolated to higher plant photosynthetic systems. MSX has been reported to rapidly inhibit photosynthesis,[17,28,29] and it therefore remains possible that MSX may influence the transport of photosynthetically derived carbon from the shoot to the root; such transport processes are likely required to sustain root ammonia assimilation by providing carbon and energy for nitrogen metabolism.[11] Walker *et al.*[18] have shown that MSX increases the incorporation of U-^{14}C-glutamate into malate and succinate, but decreases incorporation into 2-oxoglutarate in illuminated wheat leaves.

Given the possibility that MSX may have secondary effects on carbon metabolism affecting the supply of 2-oxoglutarate to GDH and leading to underestimation of flux via the GDH pathway in roots of MSX-treated plants, it would be useful to develop alternative experimental approaches to distinguishing between the GDH and GS/GOGAT pathways. One possible approach would be to compare and contrast the kinetics of $^{15}NH_4^+$ assimilation of higher plant mutants lacking GDH, with the corresponding isogenic GDH-positive wildtypes. If GDH catalyzes a significant ammonia flux into glutamate, then one would predict the following: (i) mutants lacking GDH should exhibit a significantly lower rate of $^{15}NH_4^+$ assimilation into total N; (ii) mutants lacking GDH should exhibit correspondingly reduced rates of ^{15}N labeling of glutamate (and products derived therefrom) in comparison to the wild-type. Moreover, if MSX has no secondary effects on carbon metabolism which interfere with GDH activity *in vivo*, one might predict (iii) that the wild-type should exhibit continued ^{15}N-glutamate synthesis in the presence of MSX under conditions where ^{15}N-glutamate synthesis is completely inhibited in the GDH mutant. Although no such mutants have been identified for tomato, a GDH-deficient mutant of maize has been isolated and is available for such investigations (see below).

STUDIES WITH A MITOCHONDRIAL GDH-DEFICIENT MUTANT OF MAIZE

A mutant of *Z. mays* L. deficient in mitochondrial GDH (genotype 82-137S), and the GDH-positive wild-type from which this mutant was derived (a derivative of N6) (genotypes provided by Dr. Anthony J. Pryor, CSIRO, Canberra, Australia), present intriguing opportunities for investigations of the relative contributions of the GS/GOGAT and GDH pathways to $^{15}NH_4^+$ assimilation. If mitochondrial GDH contributes to net ammonia assimilation in maize, this might be revealed by differences in

Table 1. Growth and GDH Activities of a GDH Mutant and Wild-type of *Zea mays* as Influenced by Nitrogen Source.

		Nitrogen Source[a]			
		NH_4^+	NH_4^+ + $CaCO_3$	NH_4NO_3	NO_3^-
Mean shoot weight (g f.w.)[c]	W	5.60	5.55	7.75	6.72
	M	5.30	5.40	5.42	5.40
Mean root weight (g f.w.)[c]	W	3.10	2.90	3.40	3.02
	M	3.55	3.45	3.20	3.05
Shoot/root ratio	W	1.81	1.91	2.21	2.23
	M	1.49	1.57	1.69	1.77
GDH activity of root[b]	W	384	384	361	298
	M	20	18	13	14

[a] Nitrogen sources were supplied at 10 mM (total N) for 10 days after planting in sand culture. After 10 days, N supply was reduced to 5 mM for a further 7 days, and then increased to 10 mM for a further 10 days; plants were harvested 27 days after planting.

[b] GDH activities expressed as nmol NADH consumed/min/g f.w. of root extract with NADH, NH_4^+ and 2-oxoglutarate as substrates using the assay conditions given in Reference 39 (means of 2 independent extracts).

[c] Means of 4 plants. W = GDH wildtype; M = GDH-deficient mutant. Unpublished results of Magalhães and Rhodes.

rates of assimilation of total $^{15}NH_4^+$, and differences in the labeling kinetics of glutamate and its products in the wild-type and mutant [predictions (i) to (iii) as outlined above].

Three to four weeks after planting, root GDH activities per g f.w. differ by a factor of up to 20-fold between these genotypes, whether plants are grown on NO_3^-, NH_4^+ or NH_4NO_3 as nitrogen sources (Table 1) (Magalhães and Rhodes, unpublished). An interesting feature of the GDH mutant is that it exhibits a 20 to 30% lower shoot/root ratio than the wild-type, regardless of N source at the 3- to 4-week growth stage in sand culture (Table 1). It is not yet known whether this difference in shoot/root ratio is due to the GDH mutation _per se_, but it may be pertinent to the interpretation of the ^{15}N tracer studies which follow, where comparisons were made between the kinetics of $^{15}NH_4^+$ assimilation of wild-type and mutant plants grown in hydroponic culture for 2 weeks initially on 5 mM NH_4NO_3 as nitrogen source. In these specific cultures the wild-type exhibited an average shoot weight of 6.44 g (f.w.), an average root weight of 1.896 g (f.w.) and a shoot/root ratio of 3.396. The GDH mutant plants exhibited an average shoot weight of 5.40 g (f.w.), an average root weight of 2.335 g (f.w.), and a shoot/root ratio of 2.313. Shoot/root ratios were significantly different at the $P = 0.05$ level.

Plants were transferred to a minus N medium for 2 h in the presence or absence of 1 mM MSX, and then were supplied with 99% 5 mM $(^{15}NH_4)_2SO_4$ (plus $CaCO_3$ to prevent medium acidification due to ammonia assimilation) via the aerated root medium. Glutamine synthetase assays showed that within 2 h, >99% of the GS activity had been eliminated from both the roots and shoots of both genotypes in response to MSX, but that GDH activities were unaffected (results not shown).

Measurements of total ^{15}N assimilation (subtracting ^{15}N recovered as free NH_4^+) revealed that in the absence of MSX the wild-type had a substantially higher $^{15}NH_4^+$ assimilation rate than the GDH mutant (Fig. 2) in accordance with prediction (i), above (Magalhães and Rhodes, unpublished). Total NH_4^+ assimilation rates correspond to 10.75 µmol/h plant for the GDH mutant (Fig. 2C). Assuming that the root carries all of this ammonia assimilation flux, and that all N assimilated in the shoot is derived from root metabolism, then these values translate to rates of root ammonia assimilation of 5.7 µmol/h/g f.w. for the wild-type and 3.122 µmol/h/g

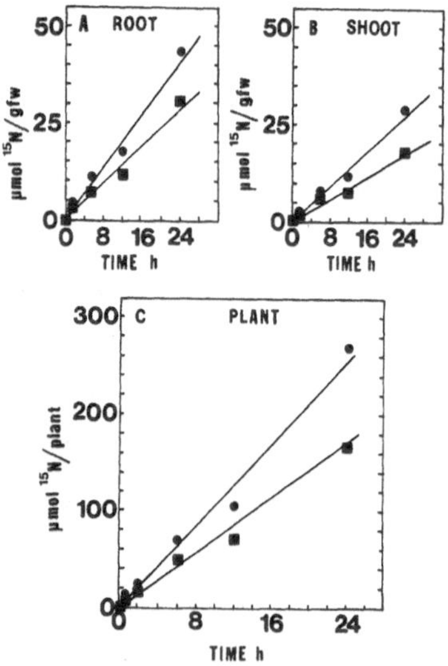

Fig. 2. $^{15}NH_4^+$ assimilation (total reduced ^{15}N minus ^{15}N present as free $^{15}NH_4^+$) by root and shoot tissue of 2-week old plants of the GDH wild-type and mutant of *Zea mays* in the presence of 5 mM 99% $(^{15}NH_4)_2SO_4$ supplied to the aerated root medium (pH 5.7) at t = 0 h in the absence of inhibitors (Magalhães and Rhodes, unpublished). Seeds were germinated in silica sand, and after 2 days were transplanted to aerated modified Hoagland's solution containing 5 mM NH_4NO_3. Plants were transferred to medium containing no nitrogen for 2 h prior to supplying $^{15}NH_4^+$. Circles, wild-type; squares, GDH-deficient mutant. Panel A, root assimilated ^{15}N; panel B, shoot assimilated ^{15}N; panel C, total plant assimilated ^{15}N.

f.w. for the GDH mutant (mean root weights per plant = 1.896 g (f.w.) and 2.335 g (f.w.), respectively). Figures 3 and 4 show kinetics of ^{15}N incorporation into several amino acids in root and shoot tissue of the wild-type and GDH mutant during this experiment, both in the presence and absence of MSX. It is notable that in the absence of MSX, virtually all amino acids become more heavily labeled in the wild-type than in the mutant, in accordance with prediction (ii) above. The labeling behavior of total N (Fig. 2), and amino acids (Figs. 3

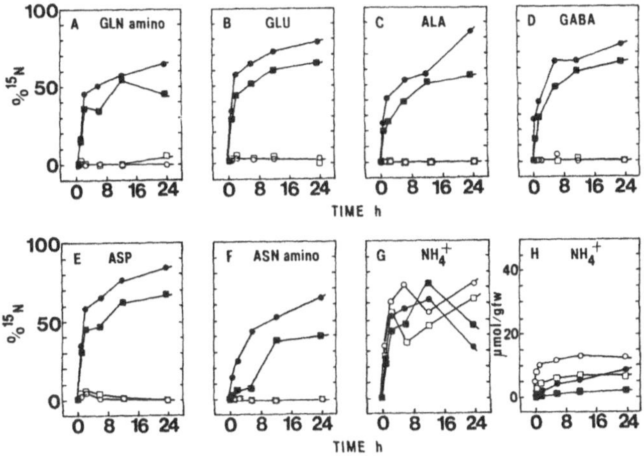

Fig. 3. Labeling behavior of ammonia and amino acids of root tissue of GDH wild-type and mutant of Zea mays in the presence and absence of a 2 h preincubation with 1 mM MSX (Magalhães and Rhodes, unpublished). 5 mM 99% $(^{15}NH_4)_2SO_4$ was supplied to the aerated root medium at t = 0 h in the absence of MSX (closed symbols), or following a 2 h preincubation with 1 mM MSX (open symbols) in minus N medium, as described in Figure 2 legend. Circles, wild-type; squares, GDH-deficient mutant. Panels A - G show changes in ^{15}N abundance of glutamine-amino N (A), glutamate (B), alanine (C), GABA (D), aspartate (E), asparagine-amino N (F) and free NH_4^+ (G). Panel H shows the changes in free NH_4^+ pool size (μmol/g f.w.).

and 4) in the wild-type and mutant in the absence of MSX, strongly suggest that appreciable flux of $^{15}NH_4^+$ directly into glutamic acid, catalyzed via GDH, occurs in maize (perhaps up to 30% of total $^{15}NH_4^+$ assimilation of the plant).

However, MSX completely inhibits incorporation of ^{15}N into all amino acids in both genotypes (Figs. 3 and 4); these results are in direct conflict with prediction (iii) above. If MSX is assumed to have no secondary effects on carbon economy of the root and potential flux via the GDH pathway [a key qualifying component of prediction (iii)], then these results provide no support

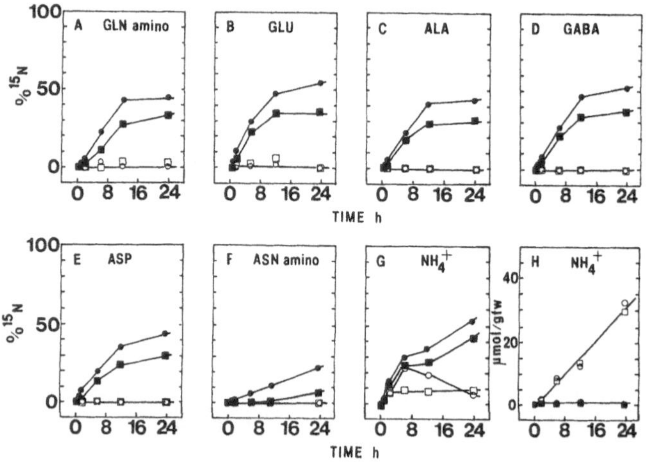

Fig. 4. Labeling behavior of ammonia and amino acids of shoot tissue of GDH wild-type and mutant of Zea mays in the presence and absence of a 2 h preincubation with 1 mM MSX (Magalhães and Rhodes, unpublished). 5 mM 99% $(^{15}NH_4)_2SO_4$ was supplied to the aerated root medium at t = 0 h in the absence of MSX (closed symbols), or following a 2 h preincubation with 1 mM MSX (open symbols), as described in Figure 2 legend. Circles, wild-type; squares, GDH-deficient mutant. Panels A - G show changes in ^{15}N abundance of glutamine-amino N (A), glutamate (B), alanine (C), GABA (D), aspartate (E), asparagine-amino N (F) and free NH_4^+ (G). Panel H shows the changes in free NH_4^+ pool size (µmol/g f.w.).

for the operation of a GDH pathway in maize roots, and alternative explanations/hypotheses must be invoked to account for the observed differences in total N assimilation rate and labeling behavior of amino acids between the wild-type and mutant. To account for the lower rate of N assimilation in the mutant in the absence of MSX, one must assume that flux through the GS/GOGAT pathway is critically dependent upon sink/source relationships determined by the different shoot/root ratios of these genotypes. It is notable that the difference in ammonia assimilation rate of the wild-type and mutant plants (Fig. 2C) is proportional to the difference in their shoot/root ratios. The lower NH_4^+ assimilation rate of the mutant in the absence of MSX could be a consequence of a lower supply of photosynthetically derived carbon from the shoot to the root of the mutant which restricts flux via the GS/GOGAT pathway in the mutant. The latter interpretation thus concludes that the difference in shoot/root ratio is the cause of the difference in N assimilation characteristics rather than GDH deficiency *per se*.

MSX elicits rapid accumulation of $^{14}NH_4^+$ in the shoots of both genotypes (Fig. 4G,H), and, as in tomato (above) and *L. minor*,[27] causes dramatic accumulations of ^{14}N-amino acids such as valine, leucine, isoleucine, phenylalanine, tyrosine and proline in roots and shoots of both genotypes (results not shown), presumably due to protein turnover.[27,30] Given the inhibitory effects of MSX on photosynthesis,[17,28,29] and organic acid metabolism,[18] and the perturbations of amino acid metabolism related to protein turnover,[27] we suggest that caution should be exercised in interpreting these MSX data to indicate that GDH makes no contribution to ammonia assimilation in maize.

These studies with a GDH-deficient mutant fail to distinguish conclusively between an insignificant and a significant role of GDH in ammonia assimilation. The ambiguity of these results centers around the fact that we simply do not know enough about the effects of inhibitors such as MSX on the transport and metabolism of photosynthetic carbon, and compartmentation of 2-oxoglutarate *in vivo* in order to state definitively that MSX is without secondary effects on flux via GDH. These limitations have led us to consider alternative approaches to discriminating between the operation of the GDH and GS/GOGAT pathways which do not rely on the use of metabolic inhibitors.

INTERPRETATION OF $^{15}NH_4^+$ TRACER KINETICS IN *LEMNA MINOR* L.

As outlined by Rhodes *et al.*,[22] computer modeling of the ^{15}N labeling behavior of the amide- and amino-N moieties of glutamine and the amino-N moiety of glutamate offers an approach to distinguishing between the GS/GOGAT and GDH pathways of ammonia assimilation. In the GS/GOGAT pathway, glutamate is both the precursor and product of glutamine-amino N, and the product of glutamine-amide N;[22] thus the labeling kinetics of glutamate are highly dependent upon the labeling kinetics of both N atoms of glutamine. In the GDH pathway, the labeling kinetics of glutamate are essentially independent of glutamine.[22] We have reasoned that detailed evaluation of the labeling kinetics of glutamine-amide N and glutamine-amino N could lead to predictions concerning the labeling behavior of the pool(s) of glutamate serving as precursor(s) and product(s) of discrete glutamine pools. These predictions could then be tested against the observed labeling behavior of glutamate. Discrepancies between observed and predicted labeling kinetics might reveal alternative pathways of glutamate biosynthesis such as that catalyzed by GDH, thereby facilitating assignment of upper bounds to potential flux via such an alternative pathway relative to flux via the GS/GOGAT pathway.

Using *L. minor* as a model system,[23] we have employed the derivatization scheme described by Fortier *et al.*[34] to determine unique features of the ^{15}N labeling behavior of glutamine. This procedure yields derivatives which contain both the amide- and amino-N moieties of glutamine, facilitating mass spectral estimation not only of amide- and amino-^{15}N abundance, but also the relative percentages of unlabeled ($^{14}N,^{14}N$), single-^{15}N labeled ($^{14}N,^{15}N$ plus $^{15}N,^{14}N$), and double ^{15}N-labeled ($^{15}N,^{15}N$) glutamine species in the free pool. The changes in the ratios of these three species, as a function of time of exposure to $^{15}NH_4^+$, prove to be highly diagnostic of the compartmentation of glutamine, the isotopic abundance of its precursor pools, and the turnover of discrete glutamine pools of the tissue.[23] This is illustrated in the following simplified hypothetical example.

If a metabolic pool of NH_4^+ was labeled to 95% ^{15}N, and a metabolic pool of glutamate was labeled to 90% ^{15}N, one would expect (from simple probability calculations) the following ratios of unlabeled : single-^{15}N : double-^{15}N labeled glutamine species in the metabolic pool of glutamine derived from these precursors, namely (0.05×0.1) : $[(0.05 \times 0.9) + (0.1 \times 0.95)]$: (0.95×0.9) or 0.005 : 0.14 : 0.855, respectively. The occurence of an unlabeled, vacuolar, storage pool of glutamine (perhaps derived from protein turnover) representing 9.3% of the total glutamine pool would be revealed by an observed ratio of unlabeled : single-^{15}N labeled : double-^{15}N labeled glutamine species of 0.097 : 0.127 : 0.776, respectively. That is, the ratio of double : single labeled glutamine species in the metabolic pool would be maintained at 6.107 : 1, in accordance with the expected ratios dictated by the isotopic abundance of the glutamine precursors, but the percentage of unlabeled glutamine (composed of discrete vacuolar and metabolic pools) would be higher than the expected value calculated from the isotopic abundance of the glutamine precursors (9.7% versus 0.5%). In practice, interpretation of these labeling patterns is far more complex due to continually changing isotopic abundance of the precursor pools, and the occurrence of multiple pools of glutamine with different turnover rates. Nevertheless, computer simulation can assist in solving these problems.

Computer assisted analysis of the labeling behavior of glutamine-amide N and glutamine-amino N in L. minor supplied with 5 mM $^{15}NH_4^+$ in the absence of inhibitors has revealed 3 pools of glutamine: a small rapidly turning-over pool presumably localized in the chloroplast, a relatively large cytoplasmic pool turning-over less rapidly, and a small vacuolar pool derived from protein turnover (Rhodes, Rich and Brunk, unpublished) (Fig. 5). The chloroplastic glutamine pool is envisaged to be derived from a small rapidly turning-over glutamate pool (presumed to be synthesized in the GS/GOGAT cycle), and glutamate is presumed to be exported from that compartment to sustain glutamine synthesis in the cytosol. The latter could involve glutamate : 2-oxoglutarate exchange across the chloroplast envelope.[32] Only this model appears to adequately account for the complex labeling behavior of glutamine (Fig. 6), i.e., the amide- and amino-N labeling kinetics, the relative percentages of unlabeled : single-^{15}N : double-^{15}N labeled glutamine species in the free pool, and the observed departures from steady-state briefly outline in Figure 5 [it was necessary, for example, to envisage that the observed accumulation of glutamine during the labeling experiment was confined to the cytoplasmic compartment (Fig. 5)]. It was impossible to accommodate these labeling data assuming only 2 pools of glutamine. The model accounts

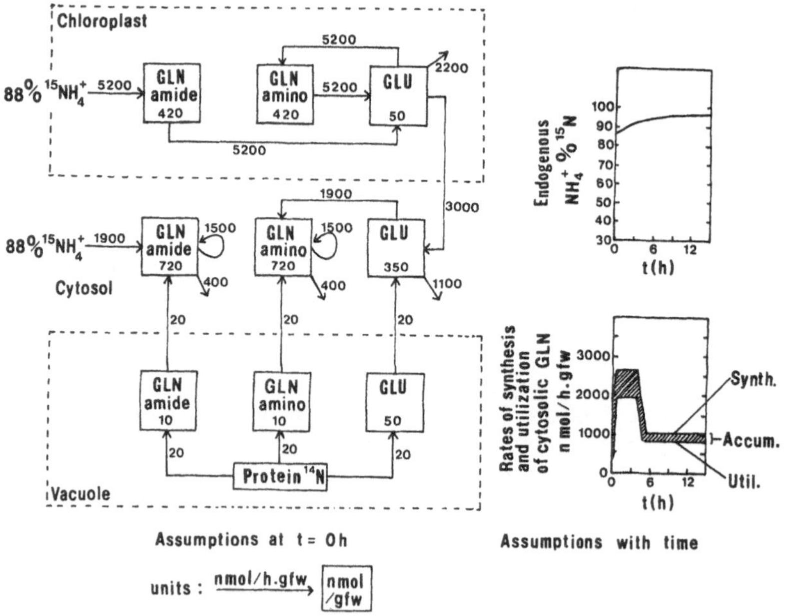

Fig. 5. Model to account for the labeling kinetics of glutamine-amide N and glutamine-amino N in Lemna minor pre-grown on 5 mM KNO_3 as sole nitrogen source and then supplied with 5 mM 99% $^{15}NH_4Cl$ plus $CaCO_3$ to prevent medium acidification due to ammonia assimilation (Rhodes, Rich and Brunk, unpublished).

for known rates of release of glutamine and glutamate into the free pool via protein turnover in L. minor, if one estimates that turnover to be about 20 nmol/h/g f.w., and assumes that it is confined to the vacuole (Fig. 5). With the exception of this vacuolar pool of glutamine, this model is similar to that proposed earlier by Rhodes et al.[22]

It is notable that the model of Figure 5, although accounting for the labeling of glutamine (Fig. 6), over-estimates the isotope abundance of glutamate (dashed line of Fig. 7). This discrepancy can be rectified by envisaging a mitochondrial pool of glutamate being synthesized at a slow rate (0.3 μmol/h g f.w.) from NH_4^+ and 2-oxoglutarate with GDH. The latter scenario is

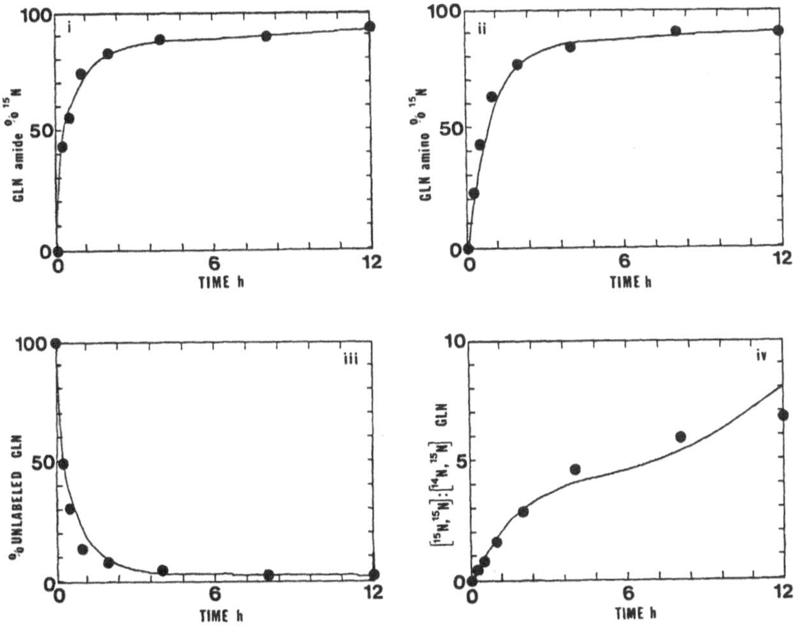

Fig. 6. Simulated and observed labeling of glutamine-amide N and glutamine-amino N in <u>Lemna minor</u> using the model parameters specified in Figure 5. Panel (i), glutamine-amide ^{15}N abundance; panel (ii), glutamine-amino ^{15}N abundance; panel (iii), percentage unlabeled $^{14}N,^{14}N$ glutamine; panel (iv), ratio of double : single labeled glutamine species in the free pool. Closed symbols represent observed values derived from Rhodes <u>et al.</u>,[23] and lines shown represent computer simulated values.

simulated in Figure 7 (solid line), and is in close agreement with observed values. We, therefore, estimate that, in <u>L. minor</u>, flux via GDH can be no more than 4% of the total N assimilation rate, which in turn implies that the GS/GOGAT cycle catalyzes at least 96% of the assimilation of NH_4^+ in this species. Although there are other possible explanations for this discrepant labeling of glutamate, relative to that predicted from the labeling behavior of its precursor and product glutamine (e.g., a peroxisomal and/or mitochondrial pool of

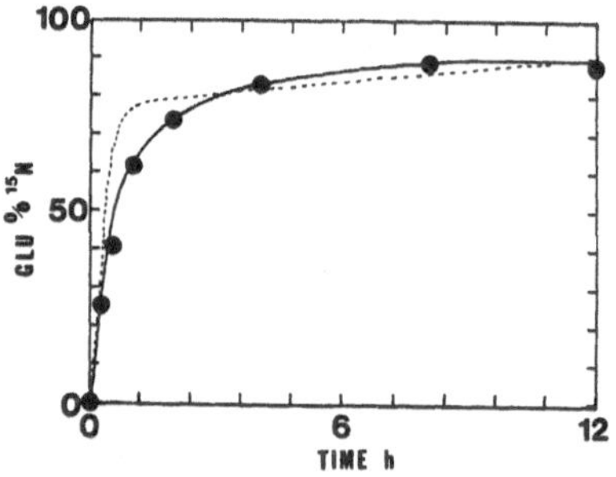

Fig. 7. Simulated and observed labeling of glutamate in *Lemna minor* using the model parameters specified in Figure 5 (dashed line through closed symbols) and the simulated labeling kinetics obtained when a fourth pool of glutamate is invoked; pool size = 0.4 µmol/g f.w., with a flux of 0.3 µmol/h g f.w. from $^{15}NH_4^+$ catalyzed by GDH (solid line through closed symbols).

glutamate becoming slowly labeled by transamination reactions, or a large vacuolar pool of glutamate becoming slowly labeled by transport exchange with the cytosolic glutamate pool), we believe that this approach to the quantitative interpretation of isotopic labeling kinetics holds promise for estimating the upper bounds of potential flux via the GDH pathway. This approach will be facilitated if additional independent kinetic criteria can be developed to test the various assumptions and improve the resolution of the model(s). One approach to increasing the resolution of these models, particularly with respect to the deduction of protein turnover rates and the size and turnover of "storage" pools of amino acids, is outlined below.

INTERPRETATION OF $^{15}NH_4^+$ TRACER KINETICS IN TOBACCO SUSPENSION CULTURES

Studies similar to those described above for *L. minor* are in progress in our laboratory with tobacco suspension cultures (*Nicotiana tabacum* L. var. Wisconsin 38). These are designed to extend the kinetic analysis of $^{15}NH_4^+$ assimilation by considering amino acid catabolism rates,

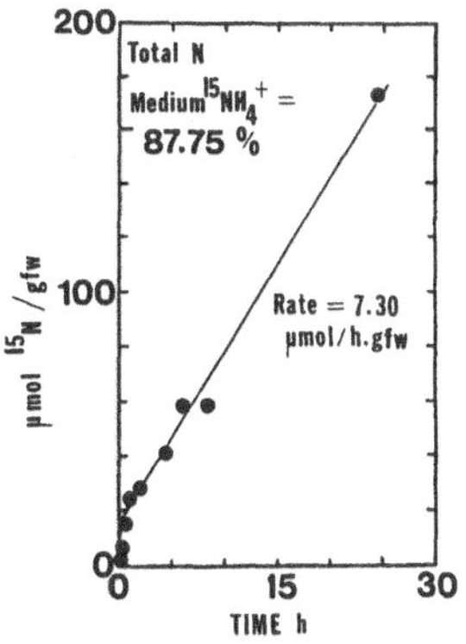

Fig. 8. Incorporation of $^{15}NH_4^+$ into total reduced N in tobacco suspension cultures (derived from Rhodes and Handa[40]).

and amino acid utilization in protein synthesis.[40] Tobacco suspension cultures in the mid-exponential growth phase typically exhibit the amino acid profile and soluble protein amino acid composition summarized in Table 2. They also assimilate ammonia at the rate of 7.3 µmol/h/g f.w. based on studies of steady-state $^{15}NH_4^+$ incorporation into total reduced N, correcting for isotope abundance of the medium (Fig. 8). With an observed doubling time of 47 h and a total N content of 492 µmol N/g f.w., the calculated rate of N assimilation required to maintain this total N pool with growth is equal to 7.26 µmol/h/g f.w.[40] This value is extremely close to the observed value shown in Figure 8. For each free amino acid pool one can derive an expansion flux (ef) defined as the rate of synthesis of the amino acid required simply to maintain the free amino acid pool with growth (Table 2):

$$ef = \frac{\text{free amino acid pool size x Ln2}}{\text{doubling time (h)}}$$

(where Ln2 = 0.693).

Similarly, for each protein bound amino acid pool, one can derive a flux (p) that is required to maintain the protein amino acid pool with growth (Table 2):

Table 2. Free and Total Soluble Protein Bound Amino Acid Pools of Tobacco Cells During Steady-state $^{15}NH_4^+$ Assimilation.

Amino Acid	Free Amino Acid Pool Size (nmol/g f.w.)	Expansion Flux (ef)[b] (nmol/h/g f.w.)	Soluble Protein Amino Acid Pool (nmol/g f.w.)	Protein Synthesis Flux (p)[b] (nmol/h/g f.w.)
Alanine	6801	100.31	4656	68.68
ACC[a]	1147	16.92	----	----
Asparagine	311	4.58	2217[c]	32.70
Aspartate	555	8.19	2217[c]	32.70
GABA[a]	4643	68.49	----	----
Glutamine	3023	44.59	2532[d]	37.35
Glutamate	2852	42.07	2532[d]	37.35
Glycine	1282	18.92	3965	58.49
Hydroxyproline	<10	<1	90	1.33
Isoleucine	71	1.05	2131	31.43
Leucine	172	2.54	4100	60.48

Lysine	<10	<1	2108	31.09
Phenylalanine	<10	<1	1820	26.85
Proline	1038	15.31	2304	33.98
Serine	991	14.62	3028	44.66
Threonine	<10	<1	1695	25.00
Valine + β-ala[a]	877	12.94	3131	46.18

[a] ACC = 1-aminocyclopropane-1-carboxylate; GABA = γ-aminobutyrate; β-ala = β-alanine which co-chromatographs with valine in the free amino acid pool.

[b] See text for definitions.

[c] Asparagine and aspartate assumed to be present in equimolar amounts in the protein hydrolysates.

[d] Glutamine and glutamate assumed to be present in equimolar amounts in the protein hydrolysates (data derived from Rhodes and Handa[40]).

$$p = \frac{\text{protein amino acid pool size x Ln2}}{\text{doubling time (h)}}$$

In an expanding steady-state, the rate of synthesis (r) of an amino acid is equal to the sum of three utilization components:

$$r = ef + p + c$$

where ef and p have been defined above, and c is the catabolism rate of the amino acid. It follows that if ef and p are known (Table 2), and r can be deduced from the labeling kinetics of the free amino acid pool, then it should be possible to determine c by difference, thus:

$$c = r - (ef + p)$$

Strictly speaking, a fifth term should be included in these equations, namely the protein turnover flux (pt) which equals the rate of release of a given amino acid back to the free amino acid pool by protein hydrolysis:

$$r + pt = ef + (p + pt) + c$$

Although pt cancels out on both sides of the equation, this variable accommodates isotope dilution of the free amino acid pool by protein turnover, and dictates that the observed rate of incorporation of the amino acid into protein (p + pt) will be greater than the theoretical growth requirement (p) whenever pt is greater than zero. If (p + pt) can be deduced from the observed labeling kinetics of total soluble protein for each amino acid, then since p is known (Table 2), it follows that pt can be deduced by difference:

$$pt = (p + pt) - p$$

An important point concerning protein synthesis and turnover is that the rates of incorporation of amino acids into protein in any one cell culture should be stoichiometrically related as follows:

$$(p + pt)_{ala} = \underline{k} \cdot p_{ala}$$

$$(p + pt)_{gly} = \underline{k} \cdot p_{ala}$$

$$(p + pt)_{pro} = \underline{k} \cdot p_{pro}$$

etc. for all protein bound amino acids, where $\underline{k}$ is a constant for all protein amino acids. Similarly, rates of release of amino acids back to the free amino acid pool via protein turnover should be stoichiometrically related as follows:

$$pt_{ala} = (\underline{k}-1) \cdot p_{ala}$$

$$pt_{gly} = (\underline{k}-1) \cdot p_{gly}$$

$$pt_{pro} = (\underline{k}-1) \cdot p_{pro}$$

etc. for all protein amino acids. When $\underline{k} = 1$, protein turnover becomes equal to zero.

From the above theoretical considerations, it is clear that any model(s) of ^{15}N flux developed to account for the observed labeling of both free and protein bound amino acid pools should not only account for the observed $^{15}NH_4^+$ assimilation rate (Fig. 8) and the associated parameters describing the interdependent labeling behavior of glutamine and glutamate as outlined for *Lemna minor* in the previous section, but also the rate of synthesis (r) of an amino acid should always be sufficient to sustain pool maintenance with growth (ef plus p) (Table 2) and any catabolism assumed in the model(s). Moreover, the model(s) should also yield a single value of $\underline{k}$ which is applicable to all protein amino acids in the culture. As will be shown below, these criteria collectively become powerful tools in assessing the compartmentation of free amino acids between metabolically active (M) (cytosolic) and metabolically inactive or storage (S) (vacuolar) pools, where only the metabolically active (M) (cytosolic) pool is assumed to serve as a precursor for protein.

Figure 9 shows a preliminary model developed to account for the labeling kinetics of the free glutamine-amide N, glutamine-amino N and glutamate pools and the glutamic acid residues of total soluble protein of

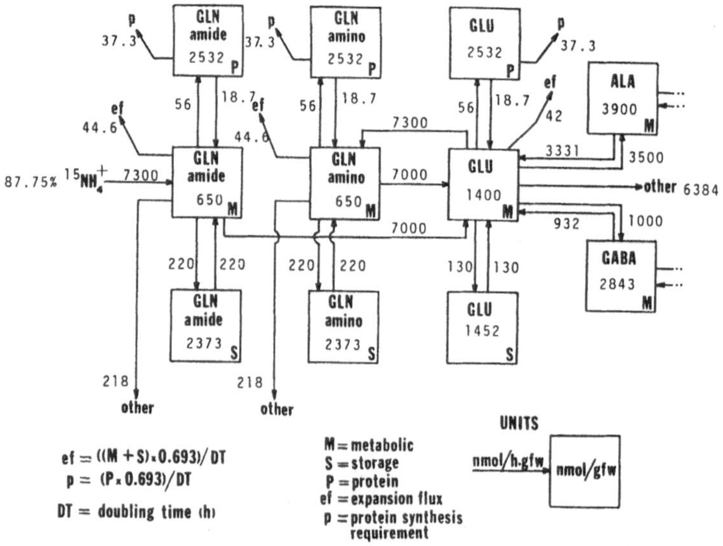

Fig. 9. Model of $^{15}NH_4^+$ assimilation into glutamine and glutamate in tobacco suspension cultures (derived from Rhodes and Handa[40]). The model assumes assimilation of ammonia exclusively via the GS/GOGAT pathway.

tobacco cells in steady-state, where the glutamic acid residues of protein are assumed to be comprised of an equimolar mixture of glutamine- and glutamate-amino N groups. This initial working model assumes exclusive assimilation of ammonia via the GS/GOGAT pathway, with both glutamine and glutamate being compartmentalized between metabolic (M) and storage (S) pools, such that the total pools yield the average (A) simulated labeling kinetics shown in Fig. 10. Only the metabolic pools (M) of glutamine-amino N and glutamate are envisaged to serve as precursors of protein glutamic acid residues (P) (Fig. 10iii). For this specific steady-state experiment, a rate of $^{15}NH_4^+$ assimilation of 7.3 μmol/h/g f.w. is accommodated, although it is notable that the simulated isotope abundance of glutamine-amide, glutamine-amino N, and glutamate slightly overestimate observed values (Fig. 10). In order to account for the labeling of glutamic acid residues of protein (Fig. 10iii), a $\underline{k}$ value of 1.5 is required (Fig. 9); that is, the observed

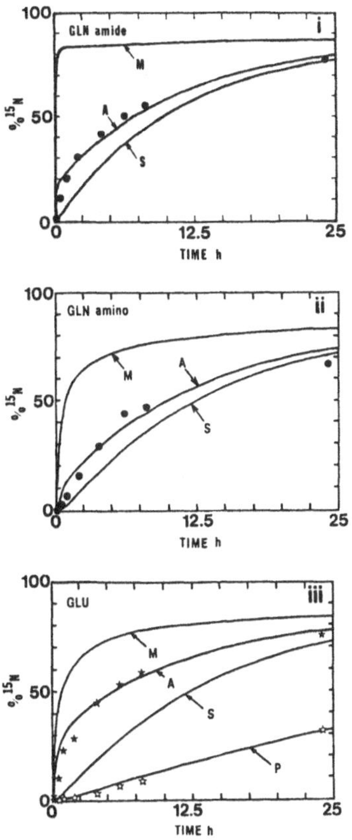

Fig. 10. Simulated and observed labeling kinetics of the free pools of glutamine-amide N (i), glutamine-amino N (ii), and glutamate (iii), and the glutamic acid residues of soluble protein (iii) of tobacco cells using the model parameters specified in Figure 9. Data of panels (ii) and (iii) are derived from Rhodes and Handa[40] using the method of Rhodes _et al._[33] for determination of amino-^{15}N abundance; data of panel (i) are unpublished results of Brunk and Rhodes, using the method of Fortier _et al._,[34] as adapted by Rhodes _et al._[23] for determination of amide-^{15}N abundance of glutamine. M, simulated labeling kinetics of the metabolic pool of the amino acid; S, simulated labeling kinetics of the storage pool of the amino acid; A, simulated labeling kinetics of the total free pool of the amino acid (M + S) (fitted to observed values); and P, simulated labeling kinetics of the protein-bound amino acid pool (fitted to observed values).

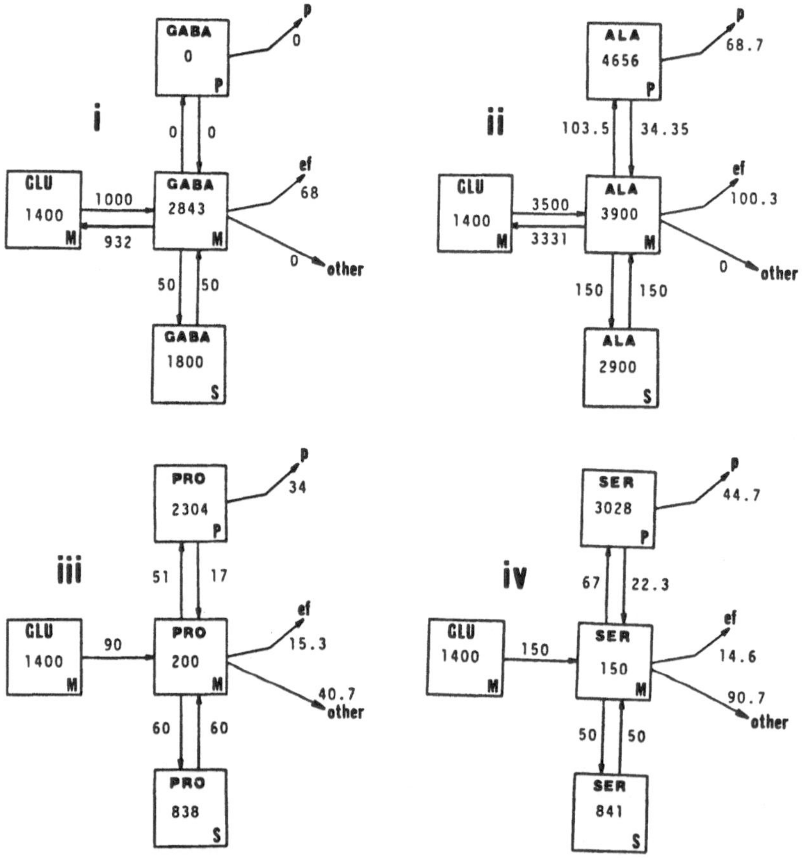

Fig. 11. Model of flux from the metabolic pool of glutamate to GABA, alanine, proline and serine in tobacco cells, derived from Rhodes and Handa.[40]

rate of incorporation of ^{15}N into glutamic acid residues of protein is 1.5 times the growth requirement (p).

Assuming that the metabolic pool (M) of glutamate serves as a precursor for alanine, γ-aminobutyrate (GABA), serine and proline (Fig. 11) the labeling kinetics of the free and protein amino acid pools (with the obvious exception of GABA which is a non-protein amino acid) can be adequately accommodated assuming the same _k_ value

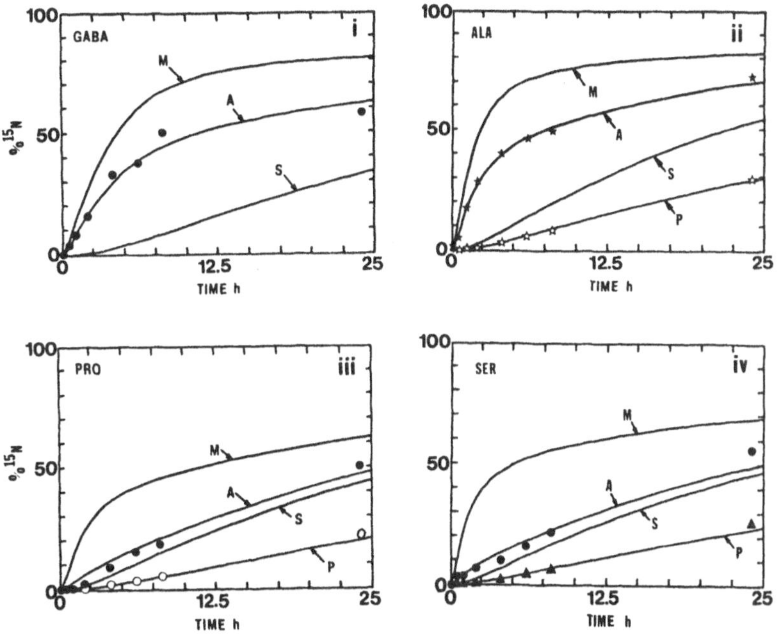

Fig. 12. Simulated and observed labeling kinetics of GABA (i), alanine (ii), proline (iii) and serine (iv) of tobacco cells using the model parameters specified in Figure 11. Data are derived from Rhodes and Handa.[40] Definitions of M, S, A and P are as described in Figure 10 legend.

(1.5) (Fig. 12), as demanded by the theoretical considerations above. This is possible only because small metabolically active pools of glutamate and glutamine are invoked (Fig. 9). Thus, it would not be possible to account for the heavy labeling of alanine by a model in which glutamate and glutamine existed as single pools. First, such a model would not account for the observed $^{15}NH_4^+$ assimilation rate; second, exceptionally high $\underline{k}$ values for glutamic acid residues of protein would have to be invoked ($\underline{k}$ >3); and third, one would have to assume direct assimilation of ammonia into alanine via an alanine dehydrogenase. Using the basic model of Figure 9 we have been able to account for the labeling of several other amino acid residues of protein assuming a *k* value of 1.5.[40] In essence, the labeling kinetics of protein amino acids provides an accurate fix on the size and turnover of the metabolic pools of amino acids serving as precursors for protein amino acid residues, and hence the size and rate of labeling of the storage pools.[40]

Collectively these results present strong preliminary evidence for the operation of the GS/GOGAT cycle in tobacco cells as the principal route of ammonia assimilation, as also concluded by Skokut *et al.*[16] using $^{13}NH_4^+$. But closer examination of the labeling behavior of glutamine-amide N, glutamine-amino N, and glutamate strongly suggests that caution should be excercised in assuming that this is the exclusive pathway of ammonia assimilation. The overestimation of the labeling of the pools of glutamine-amide, glutamine-amino N and glutamate by this model (Fig. 10), together with subtle discrepancies between the observed and predicted changes in ratios of unlabeled : single labeled : double labeled glutamine species (results not shown) (predictions dictated by the model of Fig. 9), suggests that actual flux via the GS/GOGAT cycle may be less than assumed. When a second pathway of glutamate synthesis, such as that catalyzed by GDH, is introduced into the model, allowing this second pathway of glutamate synthesis to operate at variable percentages of the total $^{15}NH_4^+$ assimilation rate, simultaneously decreasing the assumed flux through the GS/GOGAT cycle, but maintaining the same total N assimilation rate (7.3 μmol/h/g f.w.), we find that it is possible to substantially improve the goodness-of-fit between simulated and observed values. Our preliminary calculations suggest that a model which invokes a 5% contribution of the GDH pathway to total ammonia assimilation rate in tobacco cells can correct the deviations between simulated and observed values without markedly affecting the numerous additional interdependent parameters of the system, including the labeling behavior of other free amino acid pools and protein amino acid residues (results not shown). Thus, a model which reduces the assumed flux via the GS/GOGAT pathway from 7.3 to 6.9 μmol/h/g f.w. and increases the contribution of GDH from 0 to 0.4 μmol/h/g f.w. of direct ammonia assimilation into glutamate is not incompatible with the observed data, and indeed provides a more accurate simulation of the observed data.

It is notable that Skokut *et al.*,[16] likewise could not conclusively discriminate between the simultaneous operation of the GS/GOGAT and GDH pathways, and an exclusive role for the GS/GOGAT pathway in $^{13}NH_4^+$ assimilation in tobacco cells. These limitations of isotopic tracer methodologies are not always recognized. As noted by Oaks and Hirel,[11] labeling experiments which demonstrate that label first appears in the amide-N of glutamine and subsequently in glutamate, glutamine-amino N, alanine, serine, glycine, aspartate, etc.[19-21] prove only that GS is a fast reaction and not necessarily that glutamine is the *only* primary product in the assimilation of NH_4^+.

There are clearly significant technical problems associated with resolving alternative pathways in highly compartmentalized plant systems. The labeling behavior of the free amino acid pools of tobacco cells is clearly largely dominated by the storage pool components which considerably distort the labeling kinetics and make it difficult to distinguish conclusively between the simultaneous operation of alternative pathways.

Extensions of these models to whole-plant systems, particularly the GDH wild-type and mutant of maize described earlier, now seem warranted, but may prove exceedingly challenging due to the added complexity introduced by root and shoot nitrogenous metabolite exchange via the phloem and xylem transport streams. Alternatively, it may be possible to derive tissue cultures from the wild-type and GDH mutant to address these questions using the tobacco cell culture modeling approach described above. Fentem *et al.*[35] have developed a 3-compartment model to account for $^{15}NH_4^+$ assimilation in barley roots via the GS/GOGAT cycle. They noted that attempts were made to fit models in which ammonia is assimilated exclusively by GDH, but that it proved impossible to derive models capable of fitting the labeling data for both glutamate and glutamine-amide N, while accommodating the necessary rates of ammonia assimilation. This is also true of tomato cells[41] and the tobacco cell experiment described above; ammonia assimilation solely by GDH cannot account fully for observed labeling patterns and $^{15}NH_4^+$ assimilation rates. However, as noted above, such results do not necessarily preclude simultaneous operation of the GS/GOGAT and GDH pathways at a relative flux ratio of 20 : 1.

CONCLUSIONS

The controversy surrounding the precise role of GDH in ammonia assimilation in higher plants clearly persists. In the authors' opinion this controversy centers around the perturbation of carbon metabolism by inhibitors of the GS/GOGAT pathway, specifically the unknown effects of such compounds on the supply of carbon skeletons from the shoot to the root, and the supply of 2-oxoglutarate to discrete metabolic compartments, such as chloroplasts (or plastids), peroxisomes and mitochondria. The data presented for the labeling kinetics of amino acids in the maize GDH mutant and wild-type illustrate this dilemma; zero flux via the GDH pathway is suggested by the results obtained in the presence of MSX, yet the data obtained in the absence of MSX suggest a significant contribution of GDH to net ammonia assimilation. It is hoped that this chapter will encourage further studies of the effects of inhibitors of the GS/GOGAT cycle on carbon metabolism (cf.[17,18]), and the consequences of these perturbations on flux via GDH *in vivo*. The MSX inhibition experiments described here currently set lower limits of potential flux via GDH of 0% of the total N assimilation rate in maize roots, and 1% of the total N assimilation rate in tomato roots. At the same time, we acknowledge that these values may possibly be underestimates of the true flux *in vivo*.

Amino acid metabolism in higher plants is highly compartmentalized. This compartmentation poses serious difficulties for the unambiguous interpretation of $^{15}NH_4^+$ labeling kinetics of amino acids in plant tissues, in terms of flux and precursor-product relationships, in the absence of metabolic inhibitors. Models developed to account for observed labeling behavior of amino acids in both *L. minor* and tobacco cells reveal multiple pools of glutamine and glutamate whose labeling kinetics are complicated by factors such as protein turnover and metabolite exchange between compartments. The interpretation of these complex kinetic data is challenging. Computer simulation analyses will likely continue to play an important role in the development of our understanding of the labeling patterns, and in defining limitations of methods and needs for additional experiments to improve the resolution of the models. Our preliminary simulation studies currently place upper limits on the potential flux via GDH of about 4 to 5% of total NH_4^+ assimilation in L. minor and tobacco cells. We recognize that these rates may be overestimates of true flux via GDH in vivo if alternative metabolic pathways are responsible for the deviations between predicted and observed labeling kinetics. We wish to encourage further development and application of these quantitative kinetic approaches to elucidation of pathway fluxes in vivo as alternatives to metabolic inhibitor experiments; unfortunately neither approach seems devoid of inherent interpretive problems.

It would seem particularly important to try to extend such simulation models to detailed kinetic evaluation of the labeling of amino acids in the maize GDH wild-type and mutant, either at the whole-plant level or in tissue cultures derived therefrom. It is anticipated that this may permit much more precise estimates of the contribution of GDH to net NH_4^+ assimilation than have hitherto been possible. It would further be of interest to apply such models to consideration of $^{15}NH_4^+$ assimilation kinetics of GS and GOGAT mutants of higher plants.[8] Joy[15] notes that while most studies confirm the predominance of the GS/GOGAT cycle as the major assimilatory pathway, they do not rule out the possibility of some very small contribution from GDH. In GOGAT mutants in air, a small amount (a few percent) of ammonia assimilation continues.[15] These mutants can still grow (i.e., assimilate ammonia) as long as they are not exposed to the stress of photorespiration.[15] Kinetic analysis of ammonia assimilation in such mutants might distinguish between activity of GDH versus continued low flux via the GS/GOGAT cycle. Morris et al.[42] argue that, in ferredoxin-dependent GOGAT mutants of Arabidopsis thaliana, the level of NADH-dependent GOGAT is sufficient to account for the assimilation of NH_4^+ by a GS/NADH-GOGAT cycle. Demonstration of sufficient activity to account for assimilation of NH_4^+ is, however, insufficient proof of function in vivo, as has become abundantly clear for GDH.[7]

ACKNOWLEDGMENTS

This work was in part supported by a grant from the USDA (contract #85-CRCR-1-1652), and by Purdue University Agricultural Experiment Station funds. We thank C.N.P.H.-Brazil for a scholarship awarded to J.R. Magalhães. We further express our thanks to Kelly Stringham and Patrick Rich for skilled technical assistance in mass spectrometry, and Drs. Ann Oaks, Ken Joy, Tony Sims, Ray Bressan and Joe Cherry for their critical evaluation of this manuscript and many helpful suggestions. This manuscript represents Purdue University Agricultural Experiment Station Journal Article No. 11,661.

REFERENCES

1. DOUGALL, D.K. 1974. Evidence for the presence of glutamate synthase in extracts of carrot cell cultures. Biochem. Biophys. Res. Commun. 58: 639-646.
2. LEA, P.J., B.J. MIFLIN. 1974. Alternative route for nitrogen assimilation in higher plants. Nature 251: 614-616.
3. FOWLER, M.W., W. JESSUP, G.S. SARKISSIAN. 1974. Glutamate synthetase type activity in higher plants. FEBS Lett. 46: 340-342.
4. SIMS, A.P., B.F. FOLKES. 1964. A kinetic study of the assimilation of ^{15}N-ammonia and the synthesis of amino acids in an exponentially growing culture of *Candida utilis*. Proc. R. Soc. Lond. *B*. Biol. Sci. 159: 479-502.
5. FOLKES, B.F., A.P. SIMS. 1974. The significance of amino acid inhibition of NADP-linked glutamate dehydrogenase in the physiological control of glutamate synthesis in *Candida utilis*. J. Gen. Microbiol. 82: 77-95.
6. MIFLIN, B.J., P.J. LEA. 1977. Amino acid metabolism. Annu. Rev. Plant Physiol. 28: 299-329.
7. MIFLIN, B.J., P.J. LEA. 1980. Ammonia assimilation. *In* The Biochemistry of Plants: A Comprehensive Treatise. (B.J. Miflin, ed.), Vol. 5, Academic Press, New York, pp. 169-202.
8. LEA, P.J. 1988. The use of mutants lacking glutamine synthetase and glutamate synthase to study their role in plant nitrogen assimilation. *In* Plant

Nitrogen Metabolism. (E.E. Conn, ed.), Plenum Press, New York (in press).
9. LEWIS, O.A.M., D.M. JAMES, E.J. HEWITT. 1982. Nitrogen assimilation in barley (Hordeum vulgare L. cv. Mazurka) in response to nitrate and ammonium nutrition. Ann. Bot. 49: 39-49.
10. STEWART, G.R., A.F. MANN, P.A. FENTEM. 1980. Enzymes of glutamate formation: glutamate dehydrogenase, glutamine synthetase and glutamate synthase. Op. cit. Reference 7, pp. 271-327.
11. OAKS, A., B. HIREL. 1985. Nitrogen metabolism in roots. Annu. Rev. Plant Physiol. 36: 345-365.
12. OAKS, A. 1986. Biochemical aspects of nitrogen metabolism in a whole plant context. In Fundamental, Ecological and Agricultural Aspects of Nitrogen Metabolism in Higher Plants. (H. Lambers, J.J. Neetson, I. Stulen, eds.), Martinus Nijhoff, Dordrecht, pp. 133-151.
13. SRIVASTAVA, H.S., R.P. SINGH. 1987. Role and regulation of L-glutamate dehydrogenase activity in higher plants. Phytochemistry 26: 597-610.
14. YAMAYA, T., A. OAKS. 1987. Synthesis of glutamate by mitochondria - an anaplerotic function for glutamate dehydrogenase. Physiol. Plant. 70: 749-756.
15. JOY, K.W. 1988. Ammonia, glutamine and asparagine: a carbon - nitrogen interface. Can. J. Bot. (in press).
16. SKOKUT, T.A., C.P. WOLK, J. THOMAS, J.C. MEEKS, P.W. SHAFFER. 1978. Initial organic products of assimilation of ^{13}N-ammonium and ^{13}N-nitrate by tobacco cells cultured on different sources of nitrogen. Plant Physiol. 62: 299-304.
17. WALKER, K.A., A.J. KEYS, C.V. GIVAN. 1984. Effects of L-methionine sulphoximine on the products of photosynthesis in wheat (Triticum aestivum) leaves. J. Exp. Bot. 35: 1800-1810.
18. WALKER, K.A., C.V. GIVEN, A.J. KEYS. 1984. Glutamic acid metabolism and the photorespiratory nitrogen cycle in wheat leaves. Metabolic consequences of elevated ammonia concentrations and of blocking ammonia assimilation. Plant Physiol. 75: 60-66.
19. ARIMA, Y., K. KUMAZAWA. 1977. Evidence of ammonium assimilation via the glutamine synthetase

- glutamate synthase system in rice seedling roots. Plant Cell Physiol. 18: 1121-1129.
20. ARIMA Y. 1979. ^{15}N-Nitrate assimilation in association with glutamine synthesis in rice seedling roots. Soil Sci. Plant Nutr. 25: 311-322.
21. YONEYAMA, T., K. KUMAZAWA. 1974. A kinetic study of the assimilation of ^{15}N-labelled ammonium in rice seedling roots. Plant Cell Physiol. 15: 655-661.
22. RHODES, D., A.P. SIMS, B.F. FOLKES. 1980. Pathway of ammonia assimilation in illuminated *Lemna minor*. Phytochemistry 19: 357-365.
23. RHODES, D., P.J. RICH, D.G. BRUNK. 1989. Amino acid metabolism of *Lemna minor* L. IV. ^{15}N Labeling kinetics of the amide and amino groups of glutamine and asparagine. Plant Physiol. (in press).
24. STEWART, G.R., D. RHODES. 1976. Evidence for the assimilation of ammonia via the glutamine pathway in nitrate-grown *Lemna minor* L. FEBS Lett. 64: 296-299.
25. BERGER, M.G., H.P. FOCK. 1983. Effects of methionine sulfoximine and glycine on nitrogen metabolism of maize leaves in the light. Aust. J. Plant Physiol. 10: 187-194.
26. LEWIS, O.A.M., S. CHADWICK, J. WITHERS. 1983. The assimilation of ammonium by barley roots. Planta 159: 483-486.
27. RHODES, D., L. DEAL, P. HAWORTH, G.C. JAMIESON, C.C. REUTER, M.C. ERICSON. 1986. Amino acid metabolism of *Lemna minor* L. I. Responses to methionine sulfoximine. Plant Physiol. 82: 1057-1062.
28. ACHHIREDDY, N.R., D.R. VANN, J.S. FLETCHER, L. BEEVERS. 1983. The influence of methionine sulfoximine on photosynthesis and nitrogen metabolism in excised pepper (*Capsicum annuum* L.) leaves. Plant Sci. Lett. 32: 73-78.
29. JOHANSSON, L., C.-M. LARSSON. 1986. Relationship between inhibition of CO_2 fixation and glutamine synthetase inactivation in *Lemna gibba* L. treated with L-methionine-D,L-sulphoximine (MSO). J. Exp. Bot. 37: 221-229.
30. BRUNK, D.G., D. RHODES. 1988. Amino acid metabolism of *Lemna minor* L. III. Responses to amino-

oxyacetate. Plant Physiol. 87: 447-453.

31. YAMAYA, T., A. OAKS, D. RHODES, H. MATSUMOTO. 1986. Synthesis of N-glutamate from NH $^+$ and N-glycine by mitochondria isolated from pea and corn shoots. Plant Physiol. 81: 754-757.

32. WOO, K.C., U.I. FLUGGE, H.W. HELDT. 1987. A two-translocator model for the transport of 2-oxoglutarate and glutamate in chloroplasts during ammonia assimilation in the light. Plant Physiol. 84: 624-632.

33. RHODES, D., A.C. MYERS, G. JAMIESON. 1981. Gas chromatography-mass spectrometry of N-heptafluorobutyryl isobutyl esters of amino acids in the analysis of the kinetics of $^{15}NH_4^+$ assimilation in *Lemna minor* L. Plant Physiol. 68: 1197-1205.

34. FORTIER, G., D. TENASCHUK, S.L. MacKENZIE. 1986. Capillary gas chromatography micro-assay for pyroglutamic, glutamic and aspartic acids, and glutamine and asparagine. J. Chromatogr. 361: 253-261.

35. FENTEM, P.A., P.J. LEA, G.R. STEWART. 1983. Ammonia assimilation in the roots of nitrate- and ammonia-grown *Hordeum vulgare* (cv Golden Promise). Plant Physiol. 71: 496-501.

36. FENTEM, P.A., P.J. LEA, G.R. STEWART. 1983. Action of inhibitors of ammonia assimilation on amino acid metabolism in *Hordeum vulgare* L. (cv Golden Promise). Plant Physiol. 71: 502-506.

37. RAI, A.N., P. ROWELL, W.D.P. STEWART. 1981. $^{15}N_2$ Incorporation and metabolism in the lichen *Peltigera aphthosa* Willd. Planta 152: 544-552.

38. HARTMANN, T., A. EHMKE. 1980. Role of mitochondrial glutamate dehydrogenase in the reassimilation of ammonia produced by glycine-serine transformation. Planta 149: 207-208.

39. RHODES, D., G.A. RENDON, G.R. STEWART. 1976. The regulation of ammonia assimilating enzymes in *Lemna minor*. Planta 129: 203-210.

40. RHODES, D., S. HANDA. 1989. Amino acid metabolism in relation to osmotic adjustment in plant cells. *In* Biochemical and Physiological Mechanisms Associated with Environmental Stress Tolerance in Plants. (J.H. Cherry, ed.), Springer-Verlag, Berlin (in press).

41. RHODES, D., S. HANDA, R.A. BRESSAN. 1986. Metabolic changes associated with adaptation of plant cells to water stress. Plant Physiol. 82: 890-903.

42. MORRIS, P.F., D.B. LAYZELL, D.T. CANVIN. 1988. Ammonia production and assimilation in glutamate synthase mutants of *Arabidopsis thaliana*. Plant Physiol. 87: 148-154.

Chapter Seven

SITES OF ACTION OF HERBICIDES IN AMINO ACID METABOLISM: PRIMARY AND SECONDARY PHYSIOLOGICAL EFFECTS

DALE L. SHANER

American Cyanamid Co.
P.O. Box 400
Princeton, New Jersey 08540

INTRODUCTION

Plants can synthesize everything they need to survive, including amino acids. The interrelationships of amino acid regulation and coordination with other metabolic processes has been studied for many years. One of the obstacles in this research has been the inability to probe individual pathways. Such obstacles preclude the understanding of the integration of these various pathways. An analogous situation existed for research on photosynthetic light reactions, and major advances in photosynthesis research were made with the discovery of potent, specific inhibitors of these pathways.

Table 1. Inhibitors of amino acid biosynthesis.

INHIBITOR	AMINO ACID PATHWAY	ENZYME INHIBITED	REF.	STRUCTURE
Glyphosate	Aromatic Amino Acid	5-enolpyruvyl-shikimate-3-phosphate synthase	16	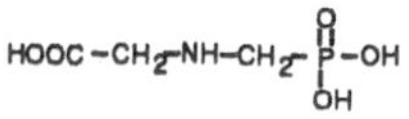
		3-deoxy-d-arabino-heptulosonate - 7-phosphate synthase	28	
Methionine Sulfoximine	Glutamine	Glutamine Synthetase	33	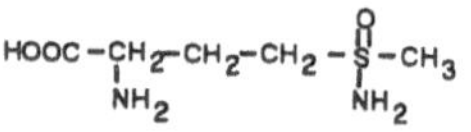
Phosphino-thricin	Glutamine	Glutamine Synthetase	33	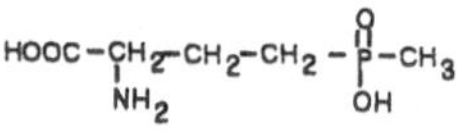
Tabtoxinine-β-lactam	Glutamine	Glutamine Synthetase	39	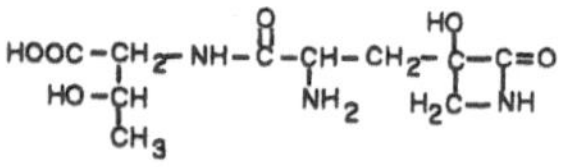
Sulfonylurea (Sulfometuron methyl)	Branched Chain Amino Acid	Acetohydroxyacid Synthase	68 74	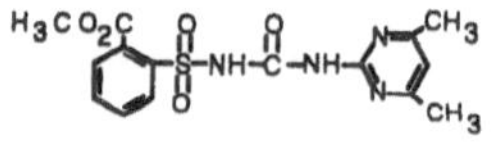
Imidazolinone (Imazapyr)	Branched Chain Amino Acid	Acetohydroxyacid Synthase	72	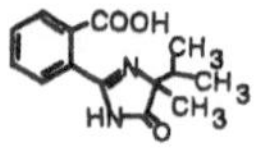
Phaseolotoxin (octicidin)	Arginine	Ornithine Carbamoyl Transferase	81	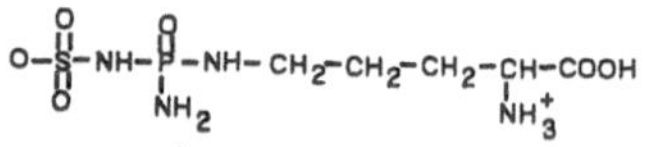

Recently, probes for amino acid metabolism have emerged in the discovery of several potent herbicides and other inhibitors of various amino acid pathways (Table 1). Through the judicious use of these inhibitors much progress has been made in understanding how these amino acid pathways work and how various amino acid pathways are integrated with the rest of plant metabolism.

Because of the agronomic importance of these herbicides, a large body of information has accumulated on the physiological effects of herbicides that inhibit amino acid biosynthesis in plants. Research effort is also being spent on the search for new inhibitors of amino acid synthesis. These types of herbicides can have extremely low mammalian toxicity and can be highly potent. Some of the new herbicides, such as the sulfonylureas and the imidazolinones, are used at a fraction of the rate of earlier herbicides and are extremely non-toxic to animals.

To aid in the search for new herbicides that act on amino acid biosynthesis, it is becoming increasingly important to be able to separate the primary effects of these herbicides from the secondary effects and to understand why the plant dies. As we gain understanding in these areas, we will be able to more quickly discern how a herbicide is killing the plant and may gain some understanding of how to improve the efficacy of such herbicides.

The purpose of this paper is to review what is known about herbicides and other inhibitors that affect plants by inhibiting amino acid biosynthesis and to attempt to separate primary effects from secondary ones. Furthermore, physiological responses of plants to these inhibitors will be compared in an attempt to understand the integration of amino acid metabolism with other metabolic pathways.

INHIBITION OF AROMATIC AMINO ACID BIOSYNTHESIS

The aromatic amino acid pathway is important in plants not only as the primary source of these amino acids but for other major end products that arise from this pathway. Such products include the phenolics, which lead to lignin precursors, flavonoids and tannins, as well as auxins and nucleotides, which are derived from tryptophan.

Aromatic amino acid biosynthesis is catalyzed by 17 different enzymes.[1] Seven of these enzymes are common to the synthesis of both phenylalanine and tryptophan, and catalyze the formation of 5-phosphoshikimate.

Table 2. Metabolic and physiological responses to glyphosate.

RESPONSE	PLANT SPECIES	TIME AFTER TREATMENT (H = hours, D = Days)	REF
Growth inhibition	Many species		26
Reduction of chlorophyll	Vigna mungo	2 D	13
	Zea mays	6 D	15
	Medicago sativa	1 D	17
	Trifolium pratense	1 D	17
Reduced respiration	Zea mays	6 H	15
Reduced photosynthesis	Phaseolus vulgaris	6 H	23
	Beta vulgaris	7.5 H	12
Increased PAL activity	Agropyron repens	2 D	10
	Triticum aestivum	2 D	10
	Glycine max	1 D	9
Increased nitrate reductase	Agropyron repens	2 D	10
Decrease in nitrate reductase	Glycine max	2-4 D	17
Decrease in PAL	Cyperus esculentum	14 D	104
Alteration of chloroplast			
Disorganization	Glycine max	3 D	20
Bursting	Lycopodium esculentum	1 D	21
Decreased protein synthesis	Xanthium sp.	2 H	
Accumulation of			
Shikimate	Many species		26
Shikimate-3-phosphate	Fagopyrum esculentum	24 H	21
Free amino acids	Many species		
Soluble carbohydrates	Xanthium sp	4 D	113
Decrease in			
Aromatic amino acids	Many species	6 H	4
Decreased translocation of photosynthate	Beta vulgaris	10 H	3 12

(continued next page)

(Table 2 continued)

Decreased absorption and translocation of ions	Glycine max	2 D	24
	Xanthium sp.	2 D	105
Decreased transpiration	Phaseolus vulgaris	6-8 H	22
	Beta vulgaris	10 H	12
Decreased anthocyanin levels	Gallium mullogo	10 D	12
	Fagopyrum escultentum	1 Day	
Increased susceptibility to pathogens	Phaseolus vulgaris	6 D	106
Reduced water potential	Medicago sativa	1-4 D	18
	Trifolium pratense	1-4 D	
Swollen root tips	Glycine max	3 D	20

Control of the phenylalanine pathway appears to be through feedback inhibition by phenylalanine and tyrosine of chorismate mutase, at the branch point in phenylalanine and tyrosine biosynthesis. Similarly, tryptophan synthesis is controlled by feedback regulation of the end product on anthranilate synthase.

Glyphosate was introduced in 1971 and has become one of the most widely used herbicides in the world.[2] It is non-selective and controls a broad spectrum of annual and perennial monocots and dicots. Glyphosate is phloem mobile and readily translocates to the active points of the plant in a source to sink direction.[3] It appears to exert its primary inhibitory action at the meristems, which cease growing soon after treatment and slowly become necrotic and die. Glyphosate is not active in the soil, and so is applied only post-emergent to the weeds.

Glyphosate treatment causes a plethora of biochemical changes within a treated plant (Table 2). These changes include a decrease in the level of phenolic compounds and their derivatives,[4] an increase in the level of free amino acids,[5-9] a decrease in soluble proteins as well as in total amino acids.[6,10] Glyphosate also causes an increase in the levels of sucrose but a decrease in the level of starch,[11,12] and a decrease in the levels of chlorophyll and other plant pigments.[8,13-17] There are also changes in the levels of extractable enzyme activities including an increase in the level of phenylalanine ammonia lyase, and a decrease in the levels of nitrate reductase, nitrite reductase and aspartate aminotransferase.[4,13,17a,18] In addition, glyphosate has been reported to cause a change in the IAA levels, possibly by increasing the levels or activity of IAA oxidase.[19]

Morphological and physiological changes occur as well. Observed morphological changes include a swelling and bursting of the chloroplasts and a loss of chloroplast organization,[20,21] a decrease in the deposition of starch granules,[20] a club-root appearance of root tips and highly disorganized nuclei.[20] Physiological changes that have been recorded include a decrease in transpiration,[12,17,22] a decrease in carbon transport from a source leaf and decreased import of carbon into a sink leaf,[3,12] decreased photosynthesis,[12,23] and decreased uptake and translocation of divalent cations.[24,25]

There is controversy about the mode of action of glyphosate. Early work by Jaworski[5] suggested that glyphosate killed plants by interfering with the synthesis of the aromatic amino acids. Jaworski found that glyphosate's inhibitory effects on *Lemna gibba* could be prevented by exogenous applications of phenylalanine. Likewise, the inhibitory effects of glyphosate on a bacterium, *Rhizobium japonicum*, could be alleviated by a combination of phenylalanine and tyrosine. Jaworski also reported that while glyphosate increased the levels of total amino acids in *Lemna*, the level of phenylalanine decreased. These results led him to propose that glyphosate kills plants by interfering with the aromatic amino acid pathway.

Following this report by Jaworski, many researchers tried to prevent the toxic effects of glyphosate by exogenous applications of phenylalanine and/or tyrosine or to correlate the endogenous levels of the aromatic amino acids with glyphosate toxicity. These experiments had mixed results.[26] In some cases applications of the aromatic amino acids could alleviate the toxic effects of glyphosate while in other cases they could not. Similarly, in some instances, glyphosate treatment caused a decrease in the levels of phenylalanine and/or tyrosine while in other cases, the levels of these amino acids increased. Because of these differences, there was speculation that glyphosate inhibited pathways other than the aromatic amino acid pathway.[26]

It was the work of Amrhein and co-workers, however, which clearly demonstrated that glyphosate inhibited the aromatic amino acid pathway *in vivo*.[8] Amrhein's group found that plants treated with glyphosate accumulated massive amounts of shikimate and shikimate-3-phosphate, important intermediates in the aromatic amino acid pathway.[8,16,21] They further showed that glyphosate was a potent inhibitor of 5-enolpyruvylshikimate-3-phosphate synthase (EPSPS). Glyphosate acts as a competitive inhibitor of phosphoenolpyruvate (PEP) and a mixed inhibitor of shikimate-3-phosphate.[27]

Glyphosate inhibits another enzyme in the aromatic amino acid pathway, 3-deoxy-D-arabino-heptulosonate-7-phosphate synthase [DAHP synthase].[28] This enzyme is much less sensitive to glyphosate than EPSPS. The I_{50} for inhibition of EPSPS by glyphosate is 5 to 7 μM while it is 10 mM for DAHP synthase.[26] This would suggest that inhibition of DAHP synthase by glyphosate is probably of secondary importance for expression of its herbicidal toxicity on plants.

Further proof that glyphosate kills plants by inhibiting the aromatic amino acid pathway is found in the work on glyphosate resistant organisms. Glyphosate resistant lines of petunia,[27] *Corydalis sempervirens*,[29] *Aerobacter aerogenes*,[30] and *Salmonella typhimurium*[31] were selected through tissue culture techniques. These lines were found to either over-express EPSPS or to have an altered EPSPS which is resistant to glyphosate. The existence of these lines of glyphosate-resistant plants and microbes provides evidence that EPSPS is the primary site of action of glyphosate. If EPSPS were not the primary site of action of glyphosate, these mutations would not provide any resistance to the herbicide.

INHIBITION OF GLUTAMINE BIOSYNTHESIS

In plants, glutamine synthetase (GS) has been shown to be one of the key enzymes involved in nitrogen metabolism; GS is the primary means by which a plant recycles ammonia from photorespiration and amino acid catabolism into anabolic pathways.[32] It is also the first step in the complex pathway of protein and nucleotides synthesis. This enzyme has been extensively studied in plants, animals and bacteria particularly through the use of specific, irreversible inhibitors, some of which are potent herbicides. The physiological responses to these various inhibitors are shown in Table 3.

Three irreversible inhibitors of GS have been studied quite extensively. These are methionine sulfoximine (MSO), phosphinothricin (PPT), and tabtoxinine-β-lactam (tabtoxin). The K_i's of MSO and PPT on GS are 0.61 mM and 0.073 mM, respectively.[33]

Methionine sulfoximine (MSO) is a methionine analog that mimics the action of a plant toxin produced by the pathogen, *Pseudomonas tabaci*.[34] MSO inhibits the growth of algae[35] and of many plant species. Tobacco leaves treated with MSO develop a necrotic spot at the site of application that is surrounded by a chlorotic halo.[36] Growth of corn, soybean and wheat seedlings watered with 2.5 ppm of MSO was extremely stunted.[37]

PPT, first isolated as a dialanyl derivative from *Streptomyces viridachromogens*, has potent herbicidal properties.[33] PPT does not translocate in plants and is not active through the soil.[38] Therefore, its registration as a herbicide will be limited to postemergent, contact use.[38] Plants treated with PPT turn yellow within 2-5 days after treatment and then wither and die.[38]

Tabtoxin is a toxic dipeptide produced by *Pseudomonas syringae* that causes chlorotic halos on susceptible plants.[39] It has been shown that tabtoxin is hydrolyzed within the plant to a lactam which is an irreversible inhibitor of GS.[39,40] The physiological responses of plants to tabtoxin are similar to their responses to MSO and PPT.

Table 3. Metabolic and physiological responses to glutamine synthetase inhibitors.

RESPONSE	PLANT SPECIES	TIME AFTER TREATMENT (HOUR,DAY)	REF.
Growth inhibition	Zea mays	14 D	37
	Glycine max		
	Triticum aestivum		
	Vigna mungo	15 D	102
Accumulation of Ammonia and Decreased Photosynthesis	Triticum aestivum	1-3 H	43
	Hordeum vulgare		
	Zea mays		
	Sorghum bicolor		
	Datura sp.		50
	Spinacia oleracea		44
	Capsicum annuum		46
	Sinapis alba		
Accumulation of free amino acids	Lemna gibba	2-28 H	41
	Pisum sativa	6 H	32
	Triticum vulgare		
	Helianthus annuum		
Decrease in Gln,Glu, Asn, Asp,Ser	Lemna minor	2-28 H	41
	Pisum sativa	6 H	32
	Triticum vulgare	6 H	
	Helianthus annuum	6 H	
	Many spp	2-5 D	38
Chlorosis In vivo loss of GS	Nicotiana tabacum	2 H	49

These GS inhibitors affect the ability of plants to assimilate ammonia. MSO has been shown to inhibit ammonia assimilation in oats,[34] Lemna minor,[41] rice,[42] Datura,[43] wheat,[43] and spinach.[44] This inhibition of ammonia assimilation resulted in a rapid increase in the levels of ammonia in the tissue and this increase appeared to be coupled with a decrease in photosynthesis. Furthermore, these effects appear to be related to photorespiration.

Platt and Anthon[44] were some of the first researchers to show that MSO rapidly inhibited photosynthesis under conditions of 20% O_2, and this inhibition of photosynthesis was correlated with a rapid increase in the levels of ammonia in spinach leaves. A 10 mM concentration of MSO caused photosynthesis to fall 50% within 55 min after treatment and caused complete inhibition by 120 min. When exposed to 2.2% O_2 the spinach leaves accumulated as much ammonia as under 20% O_2 but the effects on photosynthesis were slower. The lag time increased over 3-fold under the decreased levels of O_2.

Similar work by Ikeda _et al._[45] and Achhireddy _et al._[46] showed that MSO inhibited photosynthesis in wheat and pepper, respectively, under 20% O_2, but this effect was greatly diminished under 2% O_2. Unlike Platt and Anthon,[44] however, Achhireddy _et al._[46] found that the increase in ammonia levels were also decreased under low O_2. In fact, they found a negative correlation between the endogenous levels of ammonia and photosynthesis rates.

Work by Wild _et al._[47] showed that PPT inhibited photosynthesis in _Sinapis alba_ under normal CO_2 and O_2 conditions by 50% within 30 min after treatment, while under low O_2 (2%) and high CO_2 (1000 ppm) conditions, photosynthesis remained constant for 7.5 h after treatment. The reduction in photosynthesis under normal conditions was coupled with an accumulation of ammonia which was reduced 60% when the plants were grown under nonphotorespiratory conditions.[47] The increase in ammonia was shown to be dependent on light.

Tabtoxin caused the accumulation of ammonia in excised oat leaves under photorespiratory conditions, but this accumulation was decreased over 85% under nonphotorespiratory conditions.[40] Turner _et al._[48] found that tabtoxin rapidly inhibits ammonia assimilation into amino acids and proteins in isolated _Asparagus sprengeri_ cells. This effect was accompanied by a dramatic increase in the levels of ammonia within the cell and was correlated with a rapid loss in GS activity. When they examined the effects of tabtoxin on intact _Nicotiana tabacum_ plants, they found that the chlorotic spots contained high levels of ammonia and low levels of GS under normal atmospheric conditions.[49] Under high CO_2

(1%) levels, the effects of tabtoxin on chlorophyll and ammonia levels was much less than under normal CO_2 concentrations although there was no difference in GS inhibition.

These reports suggest that the increase in ammonia caused by these GS inhibitors is related to nitrogen metabolism associated with photorespiration.[32] Furthermore, several researchers proposed that the decrease in the rate of photosynthesis was due to ammonia toxicity, possibly through an uncoupling of photophosphorylation.[34,44,46,50]

It was the work by Ikeda et al.[45] and Sauer et al.,[51] however, which showed that the effects of MSO and PPT on photosynthesis were not directly connected to the increase in ammonia levels. Sauer et al.[51] showed that exogenous applications of ammonia could not mimic the effects of PPT on photosynthesis. Furthermore, Ikeda et al.[45] and Sauer et al.[51] found that the inhibitory effects of MSO and PPT on photosynthesis could be relieved by adding glutamine simultaneously with the inhibitors, even though the levels of ammonia within the plant were even higher in plants treated with glutamine plus the inhibitors than in plants treated with the inhibitors alone.

Ikeda et al.[45] suggested that inhibition of GS was the primary site of action of MSO and that the reduced rate of photosynthesis associated with MSO could be due to the depletion of intermediates in the Calvin cycle or to the accumulation of toxic intermediates such as glyoxylate, which is a reversible inhibitor of ribulose bisphosphate carboxylase.[52] Sauer et al.[51] suggested that the inhibition of photosynthesis by PPT is due to the disturbance of amino acid metabolism. This aspect will be discussed below.

In addition to the direct effect on GS and indirect effects on photosynthesis, these inhibitors also cause other changes in the plant. MSO causes a rapid change in the levels of free amino acids in *L. gibba* as shown by Rhodes et al.[41] Certain amino acids which decreased rapidly in concentration included glutamine, glutamate, aspartate, asparagine, alanine, and serine. Other amino acids which increased in concentration were proline, valine, leucine, isoleucine, threonine, lysine, phenylalanine, tyrosine, histidine, and methionine. Lea et al.[53] found that PPT caused a rapid decrease in the concentrations of glutamine, glutamate, glycine, alanine, and asparagine in Pisum sativum, Triticum vulgare, and Helianthus annuus.

The increase in these free amino acids appears to be due to increase turnover of proteins;[41] this aspect of the GS inhibitors will be discussed later in the paper.

MSO also causes a change in the flow of carbon from glutamate. Walker et al.[54] found that when glutamate alone was fed to excised wheat leaves, most of the glutamate was converted to glutamine via the GOGAT cycle. However, when the plants were treated with MSO along with the glutamate, the levels of glutamine rapidly dropped and the glutamate was incorporated into organic acids, sugar phosphates and CO_2.

INHIBITION OF BRANCHED CHAIN AMINO ACID BIOSYNTHESIS

The branched chain amino acids include valine, leucine and isoleucine. The pathway producing these amino acids consists of nine enzymes, four of which are shared by all three amino acids.[55] The control points of this pathway are threonine dehydratase (TD), acetohydroxyacid synthase (AHAS) and isopropylmalate synthase.[55] TD is feedback regulated by isoleucine, isopropylmalate synthase is feedback regulated by leucine, and AHAS is feedback regulated by valine, leucine and isoleucine. There are, at the present time, two classes of commercially available herbicides that inhibit branched chain amino acid biosynthesis. These are the imidazolinones and the sulfonylureas.

The sulfonylureas represent a new class of highly potent herbicides which can be used in a wide variety of crops, including small grains, soybeans, corn, rice, and canola.[56] This is the first class of herbicides to be used at rates as low as 5 to 50 g/ha. A review of the sulfonyl ureas was recently written by Blair and Martin.[56] Some of the metabolic and physiological effects of sulfonylureas are shown in Table 4.

Table 4. Metabolic and Physiological Responses to Sulfonylureas.

RESPONSE	PLANT SPECIES	TIME AFTER TREATMENT (HOUR,DAY)	REF.
Growth Inhibiton	*Lemna minor*	1 D	60
	Zea mays	6 H	64
	Pisum sativum	4 D	69
Mitosis inhibition	*Pisum sativum*	4-8 H	70
	Vicia faba	1 D	64
Inhibition of DNA Synthesis	*Zea mays*	2-6 H	64
Increased Anthocyanin	*Glycine max*	6-10 D	62
Increased PAL	*Glycine max*	3-4 D	62
Increased Ethylene	*Glycine max*	3-4 D	62
	Abutilon theophrasti	1.5 D	103
Increased Cellulase	*Abutilon theophrasti*	1.5 D	103
Accelerated Leaf Abscission	*Abutilon theophrasti*	3 D	103
Inhibition of Ca Absorption and Translocation	*Zea mays*	14 D	104
Increased Levels of Free Amino Acids	*Lemna minor*	1 D	60

The imidazolinones are a new class of potent, broad spectrum herbicides that are active on annual and perennial monocots and dicots. Imidazolinones can be used in soybeans and other leguminous crops, small grains, forestry, plantation crops, railroads, and industrial sites. These herbicides are applied at rates from 63 to 1000 g/ha. Some of the metabolic and physiological effects of imidazolinones are summarized in Table 5.

Both classes of herbicides are absorbed by the foliage and roots of plants and appear to be ambimobile.[56,57] More herbicide translocates throughout the plant when a sulfonylurea is absorbed by the roots

Table 5. Metabolic and Physiological Responses to Imidazolinones.

RESPONSE	PLANT SPECIES	TIME AFTER TREATMENT (HOUR,DAY)	REF.
Growth Inhibition	Zea mays	2-6 H	61
	Imperata cylindrica	2-6 H	59
	Avena fatua	1 Day	65
	Alopecurus myosuroides	1 Day	
Increase Sugar Levels	Zea mays	1 Day	61
Increased Free Amino Acids	Zea mays	1 Day	61
	Zea mays	2 Day	71
Decreased Soluble Proteins Levels	Zea mays	1 Day	61
			71
Inhibition of DNA Synthesis	Zea mays	6 H	61
	Avena fatua	6 H	65
Inhibition of Mitosis	Avena fatua	24 H	65
Decreased Val/Leu Levels	Zea mays	24 H	71

rather than the shoots of plants. In one experiment with Cirsium arvense, Peterson and Swisher[58] found that only 26% of the chlorsulfuron absorbed by the leaves translocated to the rest of the plant while over 60% of the herbicide absorbed by the roots translocated to other parts of the plant.

The imidazolinones move well in both the xylem and the phloem. Imazapyr is extremely mobile in the plant when applied to the leaves; over 75% of the imazapyr absorbed by an Imperata cylindrica leaf translocated to the rest of the plant,[59] while 90% of the imazaquin absorbed by roots of Xanthium strumarium translocated to the shoot within 24 hours after application.[57]

Although the imidazolinones and sulfonylureas do not inhibit germination of seeds, they do prevent the first true leaves from emerging. When these herbicides are applied to older plants, death slowly becomes evident with symptoms that include a cessation of growth, followed by chlorosis and necrosis of the growing points,[56] as well as anthocyanin expression, particularly in the veinal regions. Sublethal doses cause a cessation of growth, followed by chlorosis and necrosis of the growing points,[56] as well as anthocyanin expression, particularly in the veinal regions. Sublethal doses cause a temporary stunting of the plant by a shortening of the internodes that were developing at the time of application. Other physiological effects of imidazolinones and sulfonylureas include an increase in the levels of free amino acids,[60,61] an increase in PAL activity,[62] and an inhibition of carbohydrate transport from source to sink leaves.[63]

The first physiological response of plants treated with either herbicide is the cessation of growth. Ray[64] and Shaner and Reider[61] found that corn leaves stopped growing within 6 hours after application of chlorsulfuron and imazapyr, respectively. Pillmoor and Caseley found a similar inhibitory effect of AC 222,293 on growth of wild oats and blackgrass.[65] This inhibitory effect was not due to an inhibition of photosynthesis, respiration, hormone-regulated cell elongation, protein synthesis or RNA synthesis.[61 64 65] However, DNA synthesis was severely inhibited and cell division was stopped soon after application of either an imidazolinone or a sulfonylurea.

Further work by Ray[66] showed that the inhibition of DNA synthesis, as measured by ^{14}C-thymidine incorporation was not the result of any direct effect of chlorsulfuron on DNA polymerase or thymidine kinase. Rost[67] found that, in excised pea roots, chlorsulfuron arrested the dividing cells in the G_2 to mitosis stage and secondarily from G_1 to DNA synthesis. Rost did not find that either the DNA synthesis stage or the mitosis stage were inhibited by chlorsulfuron. Both Rost and Ray speculated that the effects of chlorsulfuron on DNA synthesis was a secondary response to its herbicidal action.

The imidazolinones and sulfonylureas kill plants by interfering with the biosynthesis of the branched chain amino acids. The discovery of this site of action for the two herbicides followed two different paths. The mode of action of the sulfonylureas was discovered through their inhibition of microbial growth. La Rossa and Schloss[68] found that sulfometuron methyl, another potent sulfonylurea, inhibited the growth of Salmonella typhimurium in the presence of valine, and that this inhibition could be prevented with the addition of isoleucine. These results suggested that the sulfonylureas interfered with the branched chain amino acid pathway. Further work showed that sulfometuron methyl was a slow-binding inhibitor of acetohydroxyacid synthase (also called acetolactate synthase), the first enzyme in the biosynthetic pathway for the branched chain amino acids,[68] as were other sulfonylureas on AHAS from a large number of species.[69]

This site of action for sulfonylureas was confirmed in plants by work in which Ray[69] showed that the inhibitory effects of chlorsulfuron on excised pea root growth could be prevented by supplementing the plants with 100 μM valine and isoleucine. Rost and Reynolds[70] found that valine and isoleucine completely prevented the inhibition of mitosis in pea roots caused by chlorsulfuron.

The discovery of the site of action of the imidazolinones started with the observation that the imidazolinones caused an increase in free amino acid levels and a decrease in the soluble protein levels. This led Anderson and Hibberd[71] to measure the effects of imazapyr on the individual amino acids in corn suspension cells. They found that while the levels of most of the amino acids increased, the levels of valine and leucine decreased. They then supplemented the corn cells with the branched chain amino acids and found that these supplements could prevent the inhibitory action of imazapyr. They further observed that it took all three of the branched chain amino acids to maximize the safening effects.

Shaner and Reider[61] then demonstrated that these three amino acids could prevent the inhibitory activity of imazapyr on both whole plant growth and DNA synthesis in corn. Likewise, Pillmoor and Caseley[65] showed that the herbicidal effect of AC 222,293 could be prevented by supplementing susceptible plants with these three amino acids.

Shaner et al.[72] reported that the imidazolinones were potent inhibitors of AHAS. The imidazolinones appear to be uncompetitive, slow-binding inhibitors of AHAS extracted from a wide variety of plants.[73]

Proof that AHAS is the only site of action of the imidazolinones and sulfonylureas was provided by the selection of plant lines that were resistant to these herbicides. Imidazolinone and sulfonylurea resistant plant lines have been selected in tobacco,[74] corn,[75] flax,[76] *Arabidopsis*,[77] *Datura innoxia*,[78] *Escherichia coli*,[79] and yeast.[79] All of these resistant lines contain an altered AHAS that is much more resistant to the sulfonylureas and/or the imidazolinones than AHAS from wild type lines. Genetic studies have shown that this herbicide resistant trait is as inherited as a single, semidominant trait.[74-79]

INHIBITION OF ARGININE BIOSYNTHESIS

Although arginine is synthesized by both animals and plants, it is rapidly metabolized in animals by arginase and therefore is considered an essential amino acid. In plants there are nine enzymes required for arginine biosynthesis.[80] Arginine is used for protein synthesis and is one of the main storage forms in seed proteins. Nitrogen release from arginine is also thought to occur during leaf senescence.[80]

Phaseolotoxin is a tripeptide toxin produced by *P. syringae* pv *phaseolicola* and is the causal agent of the halo blight disease of *Phaseolus vulgaris*.[81] This toxin is not host specific and so causes stunting of many types of plants as well as inhibition of the growth of plant cell cultures and bacteria.[82] A summary of some of the metabolic and physiological effects of phaseolotoxin on plants is shown in Table 6.

Turner and Mitchell[83] and Turner[84] reported that plants treated with this toxin show an increase in ornithine levels and a decrease in arginine levels within 4 hours after treatment. The other amino acids increase over time except for arginine. Chlorotic spots developed on newly expanding leaves but not on older leaves where the chlorophyll concentration had reached a maximum. Phaseolotoxin also decreased the levels of soluble protein in the chlorotic spots.

Table 6. Metabolic and Physiological Responses to Phaseolotoxin.

RESPONSE	PLANT SPECIES	TIME AFTER TREATMENT (HOUR,DAY)	REF.
Growth Inhibition	*Daucus carota*	30 D	105
Chlorosis	*Nicotiana tabacum*	3 D	84
	Phaseolus vulgaris	2 D	106
Decreased Soluble Protein	*Nicotiana tabacum*	3 D	84
Increased Free Amino Acids	*Nicotiana tabacum*	15 H	83
	Phaseolus vulgaris	12 D	107
Decreased Arginine Levels	*Daucus carota*	30 D	105
	Phaseolus vulgaris	12 D	107
	Nicotiana tabacum	3 DAYS	83

The mode of action of phaseolotoxin appears to be as an irreversible inhibitor of ornithine carbamoyl-transferase.[85] The toxic effects of phaseolotoxin in bacteria and plants can be reversed by the exogenous application of arginine.[83,84]

INHIBITION OF HISTIDINE BIOSYNTHESIS

Histidine biosynthesis has not been well studied in plants. In bacteria the biosynthetic pathway for histidine involves a series of 8 reactions.[86] The first product in the pathway (phosphoribosyl pyrophosphate) is also the starting point for purine and pyrimidine biosynthesis.

Amitrole is an effective herbicide on annual and perennial monocots and dicots.[87] It rapidly translocates throughout the plant via the phloem and is probably one of the most mobile herbicides known.

There is controversy about the mode of action of amitrole in plants. The most obvious response of plants treated with amitrole is albinism of the emerging leaves.[87] However, amitrole causes cessation of growth before these leaves emerge. Amitrole has also been shown to decrease the levels of soluble proteins and increase the free amino acid levels.[88,89] In addition, this herbicide causes a rapid decrease in the growth of roots, inhibits the incorporation of radiolabelled precursors into RNA, DNA, and acid-soluble nucleotides, and disrupts the development of chloroplasts.[87]

Amitrole interferes with histidine biosynthesis in yeast and Salmonella by inhibiting the enzyme imidazoleglycerol phosphate (IGP) dehydratase which leads to the accumulation of imidazoleglycerol (IG) in the growth medium.[90] The toxicity of amitrole to these organisms can be reversed through exogenous applications of histidine.

The site of action of amitrole in plants has not been clearly established. IGP dehydratase from plants have been reported to be sensitive to amitrole,[91] and Davies[91] has shown that IG accumulated in Paul's Scarlet Rose suspension cells treated with the herbicide. Furthermore he found that he could prevent the toxicity of amitrole by exogenous applications of histidine. Work done in our labs has shown that the toxicity of amitrole to suspension cultures of Black Mexican sweet corn could be prevented by exogenous applications of histidine (unpublished data).

In whole plant studies, however, the herbicidal effects of amitrole could not be reversed by exogenous applications of histidine.[87] It was also found that, in young maize seedlings, the endogenous levels of histidine were higher in amitrole treated plants than in control plants and that amitrole inhibited growth of these seedlings during this period of high histidine levels.[87] Based on these observations it appears that amitrole may have sites of action in addition to histidine biosynthesis. Since it would be difficult to separate the effects of various sites of action from one another, amitrole will not be considered any further in this review.

COMPARISON OF INHIBITORS OF AMINO ACID SYNTHESIS

Table 7 compares the physiological responses of plants to four different types of amino acid synthesis inhibitors. Some of the common characteristics of plant

Table 7. Comparison of Amino Acid Inhibitors.

INHIBITOR	INCREASED AMINO ACIDS LEVELS	DECREASED AMINO ACIDS LEVELS	PRECURSOR ACCUMULATED
Glyphosate	Pro,Val,Leu Ile,Ser, Thr	Phe, Tyr	Shikimate Shikimate-3-phosphate
MSO	Pro,Val,Leu Ile, Thr Lys, Phe, Tyr, His Met	Gln, Glu, Asn, Asp, Ser, Ala	Ammonia
Chlorsulfuron	Asn/Asp, Gln/Glu, Ser	Leu, Val	2-oxybutyrate α-aminobutyrate
Imazapyr	Asn/Asp, Gln/Glu, Ser, Thr	Leu, Val	α-aminobutyrate
Phaseolotoxin	His, Met, Asn/Gln, β-ala, Lys, Gly, Ser	Arg, Phe, Tyr	Ornithine

responses to these inhibitors are a reduction in chlorophyll levels in young leaves, inhibition of apical growth, relative slow death of the plant, decreased soluble protein levels, and a change in the carbohydrate level of the plant.

Changes in Free Amino Acid Levels

Soon after treatment with an inhibitor of amino acid biosynthesis there is a rapid change in the free amino acid levels with the specific changes depending on the inhibitor used. Following glyphosate application there is a rapid loss in phenylalanine and/or tyrosine; while following the application of GS inhibitors there is a rapid loss in glutamine and other amino acids related to ammonia metabolism and photorespiration. Sulfonylureas and imidazolinones cause rapid losses in the levels of leucine and valine while phaseolotoxin causes a decrease in the level of arginine. Most of these changes are accompanied by increases in the levels of many of the other amino acids, either as a result of increased synthesis of the amino acids, increased turnover of existing proteins or a decreased rate of protein synthesis.

Rhodes *et al.*[41,60] showed that the increase in the free amino acid levels in *L. minor* treated with either MSO or chlorfulfuron was due to increased protein turnover. They used ^{15}N tracer studies to show that chlorsulfuron rapidly inhibited the incorporation of nitrogen into valine, leucine and isoleucine and that this inhibition was accompanied by a 2 to 4-fold increase in the total free amino acid levels. However, a base level of the three branched chain amino acids was maintained in chlorsulfuron-treated plants by protein turnover.

Similarly, MSO prevented nitrogen incorporation into glutamine. In MSO-treated *Lemna*, there was a rapid increase in the levels of valine, leucine, isoleucine, proline, and threonine. Rhodes *et al.*[41] showed that these amino acids were derived from the catabolism of proteins. They suggested that these amino acids increased because they may be sequestered in the vacuole and so would be unavailable for catabolic reactions.

In *Lemna* treated with both MSO and chlorsulfuron, the latter herbicide did not inhibit the MSO-induced accumulation in valine, leucine and isoleucine. Rhodes *et al.*[60] concluded that this interaction between MSO and chlorsulfuron showed that the increase in valine, leucine and isoleucine after MSO treatment is due to recycling of amino acids from pre-existing proteins and not from stimulation of synthesis.

Changes in amino acid metabolism caused by these inhibitors could be related to nitrogen starvation. Davies and Humphrey[92] found that the rate of turnover of proteins in *L. minor* increased if the plants were starved for nitrogen; the half life of proteins in normal plants was 143 hours, but was 28 hours in plants starved for nitrate.

Amino acid inhibitors would create a shortage of certain amino acids which could, in turn, cause an increase in the rate of turnover of proteins. The consequence of this action would be increased free amino acid levels, particularly of those amino acids that are either sequestered in the vacuole, or are poorly catabolyzed.[41] A concomitant decrease in the soluble protein levels would also be observed. Both of these changes have been observed for all of the inhibitors of amino acid biosynthesis.

If this theory is true, then the lack of correlation between the proposed sites of action of these herbicides and the amino acid levels, as observed by some researchers, can be explained. Many people have tried to correlate the disappearance of certain amino acids with the toxicity of an herbicide. For example, Haderlie _et al._[7] reported that glyphosate treatment did not change the internal concentration of the aromatic amino acids in carrot cells although exogenous applications of these amino acids could protect these cells from the inhibitory effects of the herbicide. This same lack of correlation between the levels of phenylalanine and tyrosine and the toxic effects of glyphosate have also been reported for wheat, soybeans, and tobacco.[26] Similarly, in corn[71] and _Lemna_[60] treated with an imidazolinone or sulfonylurea the levels of isoleucine are actually higher in the treated plants as compared to untreated plants. Based on these kinds of measurements, researchers have erroneously concluded that these herbicides must not be interfering with amino acid biosynthesis because there are adequate amounts of the amino acids available within the plant tissue.

However, measurements of the levels of the free amino acids after herbicide treatment may be meaningless unless it can be determined what percentage of these free amino acids are available for protein synthesis. Holleman and Key[93] and Oaks[94] showed that, in soybean hypocotyls and corn roots respectively, there are two pools of leucine and valine, a large storage pool and a small protein precursor pool. Most measurements of the free amino acid levels in plant tissue do not separate these two pools from one another. If these amino acids are arising from the turnover of pre-existing proteins and if these amino acids are sequestered in sites within the cell, they may not be readily available to the plant for protein synthesis.

Measurements of changes in free amino acid levels in meristematic tissue may be more indicative of the herbicidal activity of an amino acid inhibitor than the same measurements made on fully differentiated tissue. Meristematic tissue has a high demand for amino acids and does not have highly developed vacuoles, so the protein precursor pool should be a larger percentage of the total free amino acid level in this tissue as compared to mature tissue. Changes in the precursor amino acid pool should, thus, be more easily measured in this kind of tissue.

Accumulation of Precursors

Another response of plants treated with these different inhibitors is the buildup of the precursors of the inhibited enzyme (Table 7). PPT-treated plants accumulate high levels of ammonia.[48] Glyphosate treatment causes a rapid rise in the levels of shikimate-3-phosphate and its derivative shikimate.[21] Sulfonylureas and imidazolinones cause an increase in a new amino acid, α-amino-n-butyrate, which is the transamination product of 2-oxybutyrate.[61] In *Salmonella*, the sulfonylureas cause an increase in the levels of 2-oxybutyrate.[95] Phaseolotoxin causes a rapid increase in the levels of ornithine.[84]

The role of the increased levels of these precursors in the modes of action of the various inhibitors is not clear. Many researchers have speculated that the primary toxicity of PPT is due to the rapid rise in ammonia, which is known to have toxic effects on plants. However, Sauer *et al.*[51] were not able to mimic the toxic effects of PPT by exogenous applications of ammonia at levels equivalent to the measured endogenous levels following PPT treatment. Furthermore, the toxic effects of PPT could be prevented by supplying glutamine to the plants even though the levels of ammonia were even higher in these plants than in plants treated with PPT alone. A similar phenomenon was observed for the inhibitory effects of MSO on plants.[45]

The effect of glyphosate on chloroplasts may be related to its effects on shikimate-3-phosphate which increases due to the inhibition of EPSPS. Mollenhauer *et al.*[21] have found that the disruption of chloroplasts by glyphosate is correlated with increased concentrations of shikimate-3-phosphate and shikimate in tomato. However, the levels of shikimate-3-phosphate in these chloroplasts was only 3 mM, which did not seem high enough to account for the swelling and bursting of the chloroplasts. The relationship between shikimate and shikimate-3-phosphate with glyphosate toxicity has yet to be determined.

LaRossa *et al.*[95] suggested that the toxic action of the sulfonylureas and imidazolinones is due to an increase in the levels of 2-oxybutyrate. Primaerano and Burns[96] showed that 2-oxybutyrate is toxic to microorganisms by interfering with methionine biosynthesis. 2-Oxybutyrate competes with the substrate of the first enzyme in the pantothenate pathway which results in a decrease in the levels of succinyl CoA, an obligate precursor for methionine biosynthesis.

2-Oxybutyrate arises from the deamination of threonine, and TD is regulated *in vivo* by isoleucine.[55] In the presence of sulfonylureas and imidazolinones, the levels of isoleucine fall and thus TD is no longer inhibited, so 2-oxybutyrate concentrations rise. LaRossa *et al.*[95] found that the levels of 2-oxybutyrate rose rapidly in *Salmonella* treated with sulfometuron methyl and that this increase, along with the inhibitory effects of sulfometuron methyl, could be prevented by exogenous applications of isoleucine. They also found that a mutant *Salmonella* which contained TD that was not feedback regulated by isoleucine, but that was hypersensitive to sulfometuron methyl, could not be protected from the herbicide by isoleucine applications.

The significance of the accumulation of 2-oxybutyrate in plants is not clear. Rhodes *et al.*[61] found that chlorsulfuron and imazapyr increased the levels of α-amino-n-butyrate in *Lemna*. Since α-aminobutyrate is the transamination product of 2-oxybutyrate, these results show that these inhibitors increase the levels of AHAS precursors in plants. However, no one has been successful in protecting plants from the inhibitory effects of these herbicides by adding either isoleucine alone or methionine alone to the plant; saening requires both valine and isoleucine. If increases in the levels of 2-oxybutyrate were playing a significant role in the toxicity of these herbicides, then there should be some safening by isoleucine and/or methionine. Thus, the relationship between 2-oxybutyrate accumulation and herbicide toxicity has yet to be determined.

Primary and Secondary Effects

The primary effect of inhibitors of amino acid synthesis is the inhibition of a specific enzyme reaction within a biosynthetic pathway. The secondary consequences of this inhibition are varied, depending on which pathway is inhibited.

The first event that occurs from enzyme inhibition is a rapid increase in precursors of the inhibited enzyme and a decrease in the enzyme product which results in a decrease in the level(s) of a particular amino acid(s). This decrease in amino acid levels appears to trigger a change in protein metabolism so that there is a decrease in the levels of soluble proteins. The disruption of other metabolic process is probably due either to these changes in protein synthesis or to other effects of amino acids and enzyme precursors on plants.

For example, the decrease in phenolic levels in glyphosate-treated plants is due to the decrease in the levels of phenylalanine and tyrosine.[26] The rapid decrease in photosynthesis caused by PPT and MSO is related, in some way, to the decrease in glutamine levels and to the disruption of photorespiration.[45 51] A third example is the inhibition of cell division and DNA synthesis caused by the sulfonylureas and imidazolinones. This inhibition may be directly related to the levels of valine, leucine and isoleucine in the plant. Rost and Reynolds[70] and Shaner and Reider[61] found that cell division and DNA synthesis began to recover almost immediately in plants pretreated with a sulfonylurea or an imidazolinone when branched chain amino acids were exogenously supplied to the plants 24 hours after the herbicide treatment.

Although it has not been shown in plants, a relationship between isoleucine and cell division has been found in Chinese hamster ovary (CHO) cells by Tobey and Ley[97] and Tobey.[98] They found that CHO cells grown on isoleucine-deficient media were arrested in the G_1 stage of cell division. Rost found that pea cells were arrested in a similar stage after chlorsulfuron treatment. In CHO cells isoleucine seems to be playing some role in the initiation of DNA synthesis. A similar phenomenon may be occurring in plants. This could explain the relatively rapid inhibition of cell division and DNA synthesis that occurs after treatment with imidazolinones or sulfonylureas.

Another secondary effect caused by the disruption of protein synthesis and inhibition of growth, particularly of the meristematic tissue, will be a change in the hormonal balance of the plant. One of the effects of sublethal doses of amino acid inhibitors is a loss of apical dominance with multiple lateral bud breaks, indicating that hormonal levels within the plant have changed. Another possible result from this change in hormonal levels could be a triggering of premature senescence in the plant, so that the plant dies, not from the direct effect of the herbicide, but from the change in hormonal signals.

The ramifications of the inhibition of amino acid biosynthesis for the rest of plant metabolism still remains to be determined. The availability of these different types of inhibitors and the intensive search for new inhibitors promises to open up new areas of research which will help us understand how amino acid metabolism is regulated and integrated with the rest of plant metabolism.

REFERENCES

1. GILCHRIST, D.G., T. KOSUGE. 1980. Aromatic amino acid biosynthesis and its regulation. In The Biochemistry of Plants: A Comprehensive Treatise. (B.J. Miflin, P.K. Stumpf, E.E. Conn, eds.), Academic Press, New York, Vol. 5, pp. 507-532.
2. BAIRD, D.D., R.P. UPCHURCH, W.B. HOMESLEY, J.E. FRANZ. 1971. Introduction of a new broad spectrum post emergence herbicide class with utility of herbaceous perennial weed control. Proc. 26th North Central Weed Control Conf. pp. 64-68.
3. GOUGLER, J.A., D.R. GEIGER. 1984. Carbon partitioning and herbicide transport in glyphosate-

treated sugarbeet (Beta vulgaris). Weed Sci. 32: 546-551.
4. DUKE, S.O., HOAGLAND, R.E. 1985. Effects of glyphosate on metabolism of phenolic compounds. In The Herbicide Glyphosate. (E. Grossbard, D. Atkinson, eds.), Butterworth, London, pp. 75-91.
5. JAWORSKI, E.G. 1972. The mode of action of N-phosphonomethylglycine. Inhibition of aromatic amino acid biosynthesis. J. Agric. Food Chem. 20: 1195-1198.
6. NILSSON, G. 1977. Effects of glyphosate on the amino acid content in spring wheat plants. Swed. J. Agric. Res. 7: 153-157.
7. HADERLIE, L.C., J.M. WIDHOLM, F.W. SLIFE. 1977. Effect of glyphosate on carrot and tobacco cells. Plant Physiol. 60: 40-43.
8. HOLLANDER, H., N. AMRHEIN. 1980. The site of inhibition of the shikimate pathway by glyphosate I. Inhibition by glyphosate of phenylpropanoid synthesis in buckwheat (Fagopyrum esculentum Moench.). Plant Physiol. 66: 823-829.
9. HOAGLAND, R.E., S.O. DUKE, D. ELMORE. 1979. Effects of glyphosate on metabolism of phenolic compounds III. Phenylalanine ammonia lyase, free amino acids, and soluble protein levels in dark-grown maize roots. Physiol. Plant. 46: 357-366.
10. COLE, D.J., A.D. DODGE, J.C. CASELEY. 1980. Some biochemical effects of glyphosate on plant meristems. J. Expt. Bot. 31: 1665-1674.
11. OSGOOD, R.V., T. TESHIMA. 1980. The effect of several growth regulators on dry matter production and partitioning in sugarcane cv. H59-3775. Proc. 7th Plant Growth Regulator Meet., Dallas, pp. 150-153.
12. GEIGER, D.R., S.W. KAPITAN, M.A. TUCCI. 1986. Glyphosate inhibits photosynthesis and allocation of carbon to starch in sugar beet leaves. Plant Physiol. 82: 468-472.
13. COLE, D.J., J.C. CASELEY, A.D. DODGE. 1983. Influence of glyphosate on selected plant processes. Weed Res. 23: 173-183.
14. KITCHEN, L.M., W.W. WITT, C.E. RIECK. 1981. Inhibition of chlorophyll accumulation by glyphosate. Weed Sci. 29: 513-516.

15. ALI, A., R.A. FLETCHER. 1978. Phytotoxic action of glyphosate and amitrole on corn seedlings. Can. J. Bot. 56: 2196-2202.
16. AMRHEIN, N., J. SCHAB, H.C. STEINRUCKEN. 1980. The mode of action of the herbicide glyphosate. Naturwissenschaften 67: 356-357.
17. MUNOZ-RUEDA, A., C. GONZALEZ-MURUA, J.M. BECERRIL, M.F. SANCHEZ-DIAZ. 1986. Effects of glyphosate [N-(phosphonomethyl)glycine] on photosynthetic pigments, stomatal response and photosynthetic electron transport in _Medicago sativa_ and _Trifolium pratense_. Physiol. Plant. 66: 63-68.
17a. HOAGLAND, R.E. 1985. Influence of glyphosate on nitrate reductase activity in soybean (_Glycine max_). Plant Cell Physiol. 26: 565-570.
18. MUNOZ-RUEDA, A., C. GONZALEZ-MURUA, F.L. HERNANDO, M. SANCHEZ-DIAZ. 1986. Effects of glyphosate N-(phosphonomethyl)-glycine on water potential, and activities of nitrate and nitrite reductase and aspartate aminotransferase in lucerne and clover. J. Plant Physiol. 123: 107-115.
19. LEE, T.T. 1982. Mode of action of glyphosate in relation to metabolism of indole-3-acetic acid. Physiol. Plant. 54: 289-294.
20. VAUGHN, K.E., S.O. DUKE. 1986. Ultrastructural effects of glyphosate on _Glycine max_ seedlings. Pest. Biochem. Physiol. 26: 56-65.
21. MOLLENHAUER, C., C.C. SMART, N. AMRHEIN. 1987. Glyphosate toxicity in the shoot apical region of the tomato plant I. Plastid swelling is the initial ultrastructural feature following _in vivo_ inhibition of 5-enolpyruvylshikimic acid 3-phosphate synthase. Pestic. Biochem. Physiol. 29: 55-65.
22. SHANER, D. 1978. Effects of glyphosate on transpiration. Weed Sci. 26: 513-515.
23. SHANER, D.L., J.L. LYON. 1979. Stomatal cycling in _Phaseolus vulgaris_ L. in response to glyphosate. Plant Sci. Lett. 15: 83-87.
24. DUKE, S.O., R.D. WAUCHOPE, R.E. HOAGLAND, G.D. WILLS. 1983. Influence of glyphosate on uptake and translocation of calcium ion in soybean seedlings. Weed Res. 23: 133-139.
25. NILSSON, G. 1985. Interactions between glyphosate and metals essential for plant growth. _Op. cit._ Reference 4, pp. 35-47.

26. COLE, D.J. 1985. Mode of action of glyphosate-a literature analysis. Op. cit. Reference 4, pp. 48-74.
27. STEINRUCKEN, H.C., A. SCHULZ, N. AMRHEIN, C.A. PORTER, R.T. FRALEY. 1986. Overproduction of 5-enolpyruvylshikimate-3-phosphate synthase in a glyphosate-tolerant Petunia hybrida cell line. Arch. Biochem. Biophys. 244: 169-178.
28. RUBIN, J.L., C.G. GAINES, R.A. JENSEN. 1982. Enzymological basis for herbicidal action of glyphosate. Plant Physiol. 70: 833-839.
29. SMART, C.C., D. JOHANNING, G. MULLER, N. AMRHEIN. 1985. Selective overproduction of 5-enol-pyruvylshikimic acid 3-phosphate synthase in a plant cell culture which tolerates high doses of the herbicide glyphosate. J. Biol. Chem. 260: 16338-16346.
30. SOST, D., A. SCHULZ, N. AMRHEIN. 1984. Characterization of a glyphosate-insensitive 5-enolpyruvyl-shikimic acid-3-phosphate synthase. FEBS Lett. 173: 238-242.
31. STALKER, D.M., W.R. HIATT, L. COMAI. 1985. A single amino acid substitution in the enzyme 5-enolpyruvylshikimate-3-phosphate synthase confers resistance to the herbicide glyphosate. J. Biol. Chem. 260: 4724-4728.
32. LEA, P.J., R.M. WALLSGROVE, B.J. MIFLIN. 1978. Photorespiratory nitrogen cycle. Nature 275: 741-743.
33. LEASON, M., D. CUNLIFFE, D. PARKIN, P.J. LEA, B.J. MIFLIN. 1982. Inhibition of pea leaf glutamine synthetase by methionine sulphoximine, phosphinothricin, and other glutamate analogues. Phytochemistry 21: 855-857.
34. FRANTZ, T.A., D.M. PETERSON, R.D. DURBIN. 1982. Sources of ammonium in oat leaves treated with tabtoxin or methionine sulfoximine. Plant Physiol. 69: 345-348.
35. SINGH, H.N., J.K. LADHA, H.D. KUMAR. 1977. Genetic control of heterocyst formation in the blue-green algae Nostoc muscorum and Nostoc linckia. Arch. Microbiol. 144: 155-159.
36. CARLSON, P.S. 1973. Methionine sulfoximine-resistant mutants of tobacco. Science 180: 1366-1368.

37. SINGH, M., J.M. WIDHOLM. 1975. Inhibition of corn, soybean and wheat seedling growth by amino acid analogs. Crop Sci. 15: 79-81.
38. BASTA Technical Bulletin. Hoechst Aktiengesellschaft, Verkauf Landwirtschaft, Zentrales Marketing, D-6230 Frankfurt am Main 80.
39. STEWART, W.W. 1972. Isolation and proof of structure of wildfire toxin. Nature 229: 174-178.
40. LANGSTON-UNKEFER, P.L., P.A. MACY, R.D. DURBIN. 1984. Inactivation of glutamine synthetase by tabtoxinine-β-lactam: Effects of substrates and pH. Plant Physiol. 76: 71-75.
41. RHODES, D., L. DEAL, P. HAWORTH, G.C. JAMIESON, C.C. REUTER, M.C. ERICSON. 1986. Amino acid metabolism of Lemna minor L. I. Responses to methionine sulfoximine. Plant Physiol. 82: 1057-1062.
42. ARIMA, Y., K. KUMAZAWA. 1977. Evidence of ammonium assimilation via the glutamine synthetase-glutamate synthase system in rice seedling roots. Plant Cell Physiol. 18: 1121-1129.
43. MARTIN, F., M.J. WINSPEAR, J.D. MacFARLANE, A. OAKS. 1983. Effect of methionine sulfoximine on the accumulation of ammonia in C_3 and C_4 leaves: The relationships between NH_3 accumulation and photorespiratory activity. Plant Physiol. 71: 177-181.
44. PLATT, S.G., G.E. ANTHON. 1981. Ammonia accumulation and inhibition of photosynthesis in methionine sulfoximine-treated spinach. Plant Physiol. 67: 509-513.
45. IKEDA, M., W.L. OGREN, R.H. HAGEMAN. 1984. Effect of methionine sulfoximine on photosynthetic carbon metabolism in wheat leaves. Plant Cell Physiol. 25: 447-452.
46. ACHHIREDDY, N.R., D.R. VANN, J.S. FLETCHER, L. BEEVERS. 1983. The influence of methionine sulfoximine on photosynthesis and nitrogen metabolism in excised pepper (Capsicum annuum L.) leaves. Plant Sci. Lett. 32: 73-78.
47. WILD, A., H. SAUER, W. RUHLE. 1987. The effect of phosphinothricin (glufosinate) on photosynthesis I. Inhibition of photosynthesis and accumulation of ammonia. Z. Naturforsch. 42: 263-269.
48. TURNER, J.G., R.R. TAHA, J. DEBBAGE. 1986. Effects of tabtoxin on nitrogen metabolism. Physiol. Plant. 67: 649-653.

49. TURNER, J.G. 1981. Tabtoxin, produced by Pseudomonas tabaci, decreases Nicotiana tabacum glutamine synthetase in vivo and causes accumulation of ammonia. Physiol. Plant Path. 19: 57-67.

50. PLATT, S.G., L. RAND. 1982. Methionine sulfoximine effects on C_4 plant leaf discs: Comparison with C_3 species. Plant Cell Physiol. 23: 917-921.

51. SAUER, H., A. WILD, W. RUHLE. 1987. The effects of phosphinothricin (glufosinate) on photosynthesis II. The causes of inhibition of photosynthesis. Z. Naturforsch. 42: 270-277.

52. COOK, C.M., N.E. TOLBERT. 1982. Inhibition of spinach ribulose bisphosphate carboxylase/ oxygenase by glyoxylate. Plant Physiol. 69: S-290.

53. LEA, P.J., K.W. JOY, J.L. RAMOS, M.G. GUERRERO. 1984. The action of 2-amino-4-(methylphosphinyl)-butanoic acid (phosphinothricin) and its 2-oxo-derivative on the metabolism of cyanobacteria and higher plants. Phytochemistry 23: 1-6.

54. WALKER, K.E., C.V. GIVAN, A.J. KEYS. 1984. Glutamic acid metabolism and the photorespiratory nitrogen cycle in wheat leaves: Metabolic consequences of elevated ammonia concentrations and of blocking ammonia assimilation. Plant Physiol. 75: 60-66.

55. BRYAN, J.K. 1985. Synthesis of the aspartate family and branched-chain amino acids. Op. cit. Reference 2, pp. 403-453.

56. BLAIR, A.M., T.D. MARTIN. 1988. A review of the activity, fate, and mode of action of sulfonylurea herbicides. Pestic. Sci. 22: 195-219.

57. SHANER, D.L., P.A. ROBSON. 1985. Absorption, translocation and metabolism of AC 252,214 in soybean (Glycine max), common cocklebur (Xanthium strumarium) and velvetleaf (Abutilon theophrasti). Weed Sci. 33: 469-471.

58. PETERSON, P.J., B.A. SWISHER. 1985. Absorption, translocation and metabolism of ^{14}C-chlorsulfuron in Canada thistle (Cirsium arvense). Weed. Sci. 33: 7-11.

59. SHANER, D.L. 1988. Absorption and translocation of imazapyr in Imperata cylindrica (L.) Raeuschel and effects on growth and water usage. Trop. Pest Manag., in press.

60. RHODES, D., A.L. HOGAN, L. DEAL, G.C. JAMIESON, P. HAWORTH. 1987. Amino acid metabolism of Lemna minor L. I. Responses to chlorsulfuron. Plant Physiol. 84: 775-780.

61. SHANEL, D.L., M.L. REIDER. 1986. Physiological responses of corn (Zea mays) to AC 243,997 in combination with valine, leucine, and isoleucine. Pestic. Biochem. Physiol. 25: 248-257.

62. SUTTLE, J.C., D.R. SCHREINER. 1982. Effects of DPX-4189 (2-chloro-N-((4-methoxy-6-methyl-1,3,5-triazin-2-yl)aminocarboxyl)benzenesulfonamide) on antocyanin synthesis, phenylalanine ammonia lyase activity, and ethylene production in soybean hypocotyls. Can. J. Bot. 60: 741-745.

63. BESTMAN, H.D., M.D. DEVINVE, W.H. VANDENBORN. 1987. Chlorsulfuron reduces assimilate translocation out of treated leaves of field pennycress (Thlaspi arvense L.) seedlings. WSSA Abstracts 1987 Meeting #189.

64. RAY, T.B. 1982. The mode of action of chlorsulfuron: A new herbicide for cereals. Pestic. Biochem. Physiol. 17: 10-17.

65. PILLMOOR, J.B., J.C. CASELEY. 1987. The biochemical and physiological effects and mode of action of AC 222,293 against Alopecurus myosuroides Huds. and Avena fatua L. Pestic. Biochem. Physiol. 27: 340-349.

66. RAY, T.B. 1982. The mode of action of chlorsulfuron: The lack of direct inhibition of plant DNA synthesis. Pestic. Biochem. Physiol. 18: 262-266.

67. ROST, T.L. 1984. The comparative cell cycle and metabolic effects of chemical treatments on root tip meristems. III. Chlorsulfuron. J. Plant Growth Regul. 3: 51-63.

68. LAROSSA, R.A., J.V. SCHLOSS. 1984. The sulfonylurea herbicide sulfometuron methyl is an extremely potent and selective inhibitor of acetolactate synthase in Salmonella typhimurium. J. Biol. Chem. 259: 8753-8757.

69. RAY, T.B. 1984. Site of action of chlorsulfuron: Inhibition of valine and isoleucine biosynthesis in plants. Plant Physiol. 75: 827-831.

70. ROST, T.L., T. REYNOLDS. 1985. Reversal of chlorsulfuron-induced inhibition of mitotic entry by isoleucine and valine. Plant Physiol. 77: 481-482.

71. ANDERSON, P.C., K.A. HIBBERD. 1985. Evidence for the interaction of an imidazolinone herbicide with leucine, valine and isoleucine metabolism. Weed Sci. 33: 479-483.
72. SHANER, D.L., P.C. ANDERSON, M.A. STIDHAM. 1984. Imidazolinones: Potent inhibitors of acetohydroxyacid synthase. Plant Physiol. 76: 545-546.
73. SHANER, D., K. NEWHOUSE, B. SINGH, M. STIDHAM. 1988. Acetohydroxyacid synthase from resistant maize lines. Plant Physiol. 86: S386.
74. CHALEFF, R.S., C.J. MAUVIS. 1984. Acetolactate synthase is the site of action of two sulfonylurea herbicides in higher plants. Science 224: 1443-1444.
75. ANDERSON, P.C., M.A. GEORGESON, K.A. HIBBERD. 1984. Cell culture selection of herbicide tolerant corn. Agron. Abst. 76: 56.
76. JORDAN, M.C., A. MCHUGHEN. 1987. Selection of chlorsulfuron resistance in flax (*Linum usitatissimum*) cell cultures. J. Plant Physiol. 131: 333-338.
77. HAUGHN, G.W., C.S. SOMERVILLE. 1986. Sulfonylurea resistant mutants of *Arabidopsis thaliana*. Mol. Gen. Genet. 204: 430-434.
78. SAXENA, P.K., J. KING. 1988. Herbicide resistance in *Datura innoxia*: Cross resistance of sulfonylurea-resistant cell lines to imidazolinones. Plant Physiol. 86: 863-867.
79. YADAV, N., R.E. McDEVITT, S. BENARD, S.C. FALCO. 1986. Single amino acid substitutions in the enzyme acetolactate synthase confer resistance to the herbicide sulfometuron methyl. Proc. Natl. Acad. Sci. 83: 4418-4422.
80. THOMPSON, J.F. 1985. Arginine synthesis, proline synthesis, and related processes. *Op. cit.* Reference 2, pp. 375-402.
81. MITCHELL, R.E. 1976. Isolation and structure of a chlorosis-inducing toxin of *Pseudomonas phaseolicola*. Phytochemistry 15: 1941-1947.
82. STASKAWICZ, B.J., N.J. PANOPOULOS. 1979. A rapid and sensitive microbiological assay for phaseolotoxin. Phytopathology 69: 663-666.
83. TURNER, J.G., R.E. MITCHELL. 1985. Association between symptom development and inhibition of ornithine carbamoyltransferase in bean leaves

treated with phaseolotoxin. Plant Physiol. 79: 468-473.
84. TURNER, J.G. 1986. Effect of phaseolotoxin on the synthesis of arginine and protein. Plant Physiol. 80: 760-765.
85. MOORE, R.E., W.P. NIEMCZURA, O.C.H. KWOK, S.S. PATIL. 1984. Inhibitors of ornithine carbamoyltransferase from Pseudomonas syringae pv. phaseolicola: Revised structure of phaseolotoxin. Tetrahedron Lett. 25: 3931-3934.
86. MIFLIN, B.J. 1985. Histidine biosynthesis. Op. cit. Reference 2, pp. 533-541.
87. ASHTON, F.M., A.S. CRAFTS. 1981. Mode of Action of Herbicides. John Wiley & Sons, New York, 525 p.
88. McWHORTER, C.G. 1963. Effects of 3-amino-1,2,4-triazole on some chemical constituents of Zea mays. Physiol. Plant. 16: 31-39.
89. BARTELS, P.G., F.R. WOLF. 1965. The effect of amitrole upon nucleic acid and protein metabolism of wheat seedlings. Physiol. Plant. 18: 805-812.
90. HILTON, J.L. P.C. KEARNEY, B.N. AMES. 1965. The mode of action of herbicide 3-amino-1,2,4-triazole (amitrole) inhibition of an enzyme in histidine biosynthesis. Arch. Biochem. Biophys. 112: 544-547.
91. DAVIES, M.E. 1971. Regulation of histidine biosynthesis in cultured plant cells: Evidence from studies on amitrole toxicity. Phytochemistry 10: 783-788.
92. DAVIES, D.D., T.J. HUMPHREY. 1978. Amino acid recycling in relation to protein turnover. Plant Physiol. 61: 54-58.
93. HOLLEMAN, J.M., J.L. KEY. 1967. Inactive and protein precursor pools of amino acids in the soybean hypocotyl. Plant Physiol. 42: 29-36.
94. OAKS, A. 1965. The soluble leucine pool in maize root tips. Plant Physiol. 40: 142-149.
95. LAROSSA, R.A., T.K. VAN CYK, D.R. SMULSKI. 1987. Toxic accumulation of α-ketobutyrate caused by inhibition of the branched-chain amino acid biosynthetic enzyme acetolactate synthase in Salmonella typhimurium. J. Bact. 169: 1372-1378.
96. PRIMAERANO, D.A., R.O. BURNS. 1982. Metabolic basis for the isoleucine, pantothenate or

methionine requirement of ilvG strains of Salmonella typhimurium. J. Bact. 150: 1202-1211.
97. TOBY, R.A., K.D. LEY. 1971. Isoleucine-mediated regulation of genome replication in various mammalian cell lines. Cancer Res. 31: 46-51.
98. TOBY, R.A. 1973. Production and characterization of mammalian cells reversibly arrested in G1 by growth in isoleucine-deficient medium. Methods Cell Biol. 6: 67-112.
99. CANAL, M.J., R.T. SANCHEZ, B. FERNANDEZ. 1987. Effects of glyphosate on phenolic metabolism in yellow nutsedge leaves. Physiol. Plant. 69: 627-632.
100. NAFZIGER, E.D., F.W. SLIFE. 1983. Physiological response of common cocklebur (Xanthium pensylvanicum) to glyphosate. Weed Sci. 31: 874-878.
101. JOHAL, G.S., J.E. RAHE. 1984. Effect of soilborne plant-pathogenic fungi on the herbicidal action of glyphosate on bean seedlings. Phytopathology 74: 950-955.
102. DESSAUER, D.W., L.C. HANNAH. 1978. Inhibition of cowpea seedling growth by methionine analogs. Crop Sci. 18: 593-597.
103. HAGEMAN, L.H., R. BEHRENS. 1984. Chlorsulfuron induction of leaf abscission in velvetleaf (Abutilon theophrasti). Weed Sci. 32: 132-137.
104. CROWLEY, J., G.N. PRENDEVILLE. 1985. Effect of chlorsulfuron and 1,8 naphthalic anhydride on uptake of ^{44}Ca in maize. Weed Res. 25: 341-345.
105. JACQUES, S., Z.R. SUNG. 1981. Regulation of pyrimidine and arginine biosynthesis investigated by the use of phaseolotoxin and 5-fluorouracil. Plant Physiol. 67: 287-291.
106. MITCHELL, R.E., R.L. BIELESKI. 1977. Involvement of phaseolotoxin in halo blight of beans: Transport and conversion to functional toxin. Plant Physiol. 60: 723-729.
107. PATELL, P.N., J.C. WALKER. 1963. Changes in free amino acid and amide content of resistant and susceptible beans after infection with the halo blight organism. Phytopathology 53: 522-528.
108. FOLEY, M.E., E.D. NAFZIGER, F.W. SLIFE, L.M. WAX. 1983. Effect of glyphosate on protein, nucleic acid synthesis and ATP levels in common cocklebur (Xanthium pensylvanicum) root tissue. Weed Sci. 31: 76-80.

Chapter Eight

METABOLISM OF 1-AMINOCYCLOPROPANE-1-CARBOXYLIC ACID IN RELATION TO ETHYLENE BIOSYNTHESIS

SHANG FA YANG

Vegetable Crops Department
Mann Laboratory
University of California
Davis, California 95616

INTRODUCTION

1-Aminocyclopropane-1-carboxylic acid (ACC) was first isolated in 1957 from ripe cider apples and perry pears by Burroughs[1] and from ripe cowberries by Vahatalo and Virtanen.[2] Although Burroughs[3] observed that the amount of ACC in perry pears increased during storage and speculated that ACC might be related in some way to fruit ripening, he could not detect ACC in other varieties of apple or pear. Undoubtedly this was because of the insensitive method employed at that time. Interest in this unusual, non-protein amino acid was sparked when it was recognized as the immediate precursor of ethylene in 1979. Adams and Yang[4] unravelled this through tracer studies of methionine metabolism, while Lurssen *et al.*[5] showed that ACC greatly stimulated ethylene production in a number of plant tissues.

Ethylene was unwittingly used as a plant growth regulator for a long time. The early use of smoke in hastening fruit ripening and to induce pineapple flowering can be attributed to its ethylene content. The rate of ethylene production by plant tissue is normally low. As part of the normal life of a plant, an increase in ethylene production is induced during certain stages of growth such as seed germination, fruit ripening, flower fading, leaf senescence and abscission. Ethylene production can also be induced by auxin application or environmental stresses such as physical wounding and cutting, chilling, drought, and water flooding. It has been recognized that this increased ethylene production, in turn, brings about many important physiological responses.[6]

In plant tissues, ACC is biosynthesized from S-adenosyl-L-methionine (SAM) by ACC synthase and is metabolized to ethylene by ACC oxidase or to N-malonyl-ACC (MACC) by ACC N-malonyltransferase. This chapter deals with these processes as they relate to ethylene production.

ACC SYNTHASE

The role of methionine as an ethylene precursor in plant tissues was first demonstrated by Lieberman *et al.*[7] who found that methionine labeled at C-3,4 positions was efficiently incorporated into labelled ethylene in apple fruit tissue. These findings were subsequently confirmed in other plant tissues. Later, Adams and Yang[8] demonstrated that, during the conversion of methionine to ethylene, the methylthio group of methionine is released as 5'-methylthioadenosine (MTA) and its hydrolysis product, 5-methylthioribose (MTR). These observations indicated that methionine must be converted into SAM before ethylene is released.

It has long been known that a nitrogen atmosphere causes a cessation in ethylene production by apples and that a surge of ethylene production occurs when the tissue is returned to air. These observations were interpreted to indicate that an intermediate accumulates during anaerobic incubation and is subsequently converted to ethylene upon exposure to oxygen. Adams and Yang[4] compared the metabolism of methionine in air and in a mitrogen atmosphere. In nitrogen methionine was not metabolized to ethylene but to MTR and a compound later identified as ACC. In the presence of air, ACC was rapidly converted to ethylene, indicating that ACC is an intermediate and that the conversion of ACC to ethylene requires oxygen. These data indicate the following sequence for the pathway of ethylene biosynthesis in apple tissue.

Methionine --> SAM --> ACC --> Ethylene

Aminoethoxyvinylglycine (AVG), a known inhibitor of pyridoxal phosphate-mediated enzyme reactions, strongly inhibited the conversion of methionine to ACC, but it did not block the conversion of methionine to SAM or the conversion of ACC to ethylene. These results indicate that AVG inhibits the conversion of SAM to ACC, and that this conversion may be mediated by a pyridoxal enzyme.[4] Sonn after the pathway for ethylene biosynthesis became known, Boller _et al._[9] reported a cell-free extract prepared from tomato fruit which catalyzed the conversion of SAM to ACC. They found that the enzyme was soluble and was strongly inhibited by AVG, as predicted by Adams and Yang.[4] The enzyme utilized SAM specifically as the substrate. Employing labeled SAM, Yu _et al._[10] demonstrated that SAM was converted to ACC and MTA by the tomato enzyme preparation. Moreover, they demonstrated that ACC synthase was activated by low concentrations of pyridoxal phosphate and strongly inhibited by 2-aminooxyacetic acid, another well-known inhibitor of pyridoxal enzymes. These observations further support the view that ACC synthase is a pyridoxal enzyme. ACC synthase activity has since been demonstrated in many other plant tissues. The conversion of SAM to a cyclopropane amino acid is a γ-elimination (1,3-elimination) reaction. In organic reactions γ-elimination can proceed readily through an intramolecular nucleophilic attack by a carbanion on the α-position as depicted below:

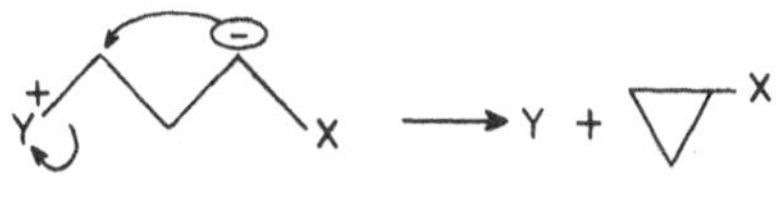

It is well known that a pyridoxal enzyme facilitates elimination of the proton from the α-carbon of an amino acid, yielding a carbanion. The positive sulfonium ion further facilitates the intramolecular nucleophilic displacement reaction by the carbanion, yielding ACC and MTA, as depicted in Figure 1.[4,10] This reaction mechanism

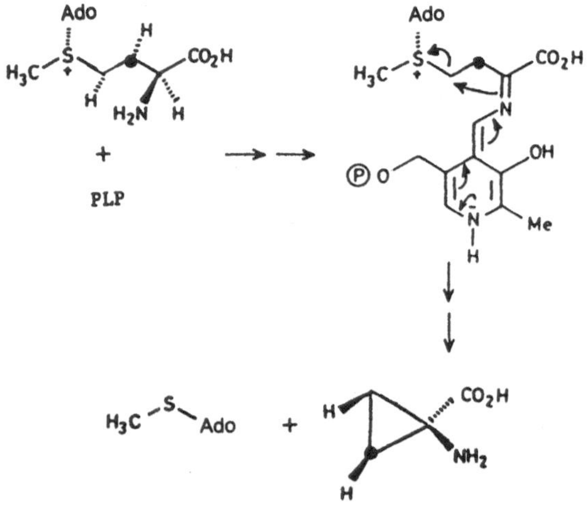

Fig. 1. Reaction mechanism for the conversion of SAM to ACC catalyzed by the pyridoxal phosphate (PLP) dependent ACC synthase. PLP facilitates the elimination of the α-hydrogen of the methionine moiety of SAM, and the resulting carbanion equivalent undergoes γ-displacement leading to the formation of ACC and MTA.[4,10,11] Shown also is the stereochemical course of the reaction. C-3 of the L(S)-methionine portion of SAM as represented by • becomes the pro-(R) methylene group of ACC, indicating that the reaction involves an inversion at the α-carbon center of the methionine moiety of SAM.[13] The reaction also involves an inversion at the γ-carbon center of the methionine moiety because the two hydrogens attached to C-3,4 of the methionine moiety in the cis-configuration yield cis-ACC.[11]

was supported by the experiment of Ramalingam et al.[11] who showed that the reaction proceeds via a direct γ-elimination (1,3-elimination) without the participation of β-elimination (1,2-elimination).

The stereochemical courses of the reaction catalyzed by ACC synthase have been studied in several laboratories. There are two diasteriomers of SAM on the sulfonium center. Although the biologically synthesized SAM is in the form of (-)-SAM (S-configuration on the sulfonium center), it can undergo epimerization spontaneously into (+)-SAM

(R-configuration). Khani-Oskouee et al.[12] have reported that only the natural (-)-SAM, but not the epimerized (+)-SAM, serves as the ACC synthase substrate. Ramalingam et al.[11] synthesized (±)-S-adenosyl[3,4-$^2H_2$]methionine with deuteriums specifically labeled in either cis or trans configuration, and demonstrated that trans-SAM gave trans-ACC and cis-SAM gave cis-ACC by the action of ACC synthase. Thus, the enzymatic ring formation involves an inversion at the C_4 center (γ-position) or methionine moiety of SAM. On the other hand, Wiesendanger et al.[13] synthesized S-adenosyl-L-[4-$^2H_2$]methionine as the substrate of ACC synthase, and determined that the deuterium was on the pro-(S) methylene group of ACC. Since the absolute configuration of the α-carbon of methionine in SMA is (S), and that of ACC is (R), they concluded that the enzymatic conversion of SAM to ACC involves an inversion at the α-carbon of the methionine moiety. These findings are depicted in Figure 1.

Since ACC synthase exists in very low concentration in plant tissues and is unstable, progress in the purification of this enzyme has been slow. Using a combination of various purification procedures, Bleecker et al.[14] have purified the enzyme 6500-fold from wounded tomato pericarp. Two-dimensional gel electrophoresis revealed a molecular weight of 50 kD. They prepared monoclonal antibodies against ACC synthase and demonstrated that these antibodies recognized the native enzyme. When the antibody was linked to Sepharose-4B gel, this matrix was shown to be effective in isolating ACC synthase from a crude enzyme preparation. The enzyme eluted from the immunoaffinity gel was, however, denatured and contaminated with IgG proteins derived from the antibodies. They have estimated the specific activity of the pure enzyme at about 7 μmol/min-mg protein. It is to be noted that these monoclonal antibodies could not detect ACC synthase bound to nitrocellulose. Nakajima et al.[15] in Japan have purified the enzyme to homogeneity from sliced and aged mesocarp of winter squash fruits. The enzyme had a molecular weight of 45-50 kD by SDS polyacrylamide gel electrophoresis; the specific activity of the pure enzyme was estimated to be 2.4 μmol/min-mg protein. On the other hand Tsai et al.[62] purified ACC synthase from hormone-treated etiolated mungbean hypocotyls. The molecular weight of the native enzyme was 125 kD and consisted of two subunits of 65 kD. The specific activity was 0.35

μmol/min-mg protein, a value much lower than that reported for the tomato enzyme. Thus, the enzymes from these tissues differed in final specific activity and in molecular weight.

ACC synthase plays a major role in regulating ethylene biosynthesis. Although all plant tissues are capable of producing ethylene, their rates of production of ethylene are normally low. As part of the normal life of the plant, ethylene production is induced during certain stages of development such as germination, fruit ripening, abscission, and flower senescence. Ethylene production can also be induced by a variety of external factors such as mechanical wounding, various environmental stresses (drought, water logging, chilling, and biotic infestation), and certain chemicals including auxin and other plant growth regulators. It has been recognized that this increased ethylene production can in turn bring about many important physiological consequences.[6,16] In all these cases, it has been shown that the higher levels of ethylene are accompanied by an increased ACC production.[16] The induction of ACC synthase is inferred since cycloheximide, an inhibitor of protein synthesis, effectively blocks the increased production of ethylene as do inhibitors of RNA synthesis.[17,18] Moreover, tomato pericarp slices incubated with deuterium oxide (^2H_2O) develop ACC synthase with a greater buoyant density than that in control slices incubated with H_2O, showing that ACC synthase is synthesized de novo after wounding.[18]

Another important characteristic of ACC synthase in plant tissue is its lability and short half-life. It has been well recognized that the ethylene production rate in IAA-treated tissues declines rapidly when IAA is withdrawn from the incubation medium.[17] Similarly, the ethylene production rates in mechanically wounded or water-stressed tissues undergo a rapid decline even when these stimuli are relieved.[19] Two mechanisms are responsible for this rapid decline in ethylene production: one is the conjugation of ACC into N-malonyl-ACC (MACC) catalyzed by ACC malonyltransferase resulting in the reduced level of ACC in the tissue, and the other is the inactivation of ACC synthase. The apparent half-life of ACC synthase in wounded green tomato pericarp and in IAA-treated mungbean hypocotyls has been estimated to be about 40 min, based on the decay kinetics of enzyme activity extracted from the induced tissues in the presence of cycloheximide which blocks new synthesis of the enzyme protein.[17,19] Since ACC synthase is rapidly turned over _in vivo_, the level of ACC synthase in plant tissues is determined not only by the synthesis of the enzyme but also by its inactivation.

Satoh and Esashi[20] observed that ACC synthase isolated from mungbean hypocotyls was inactivated _in vitro_ by its substrate SAM during its catalytic reaction. Since the half-life of ACC synthase _in vitro_ was similar to that reported in the tissue, they suggested that the SAM-induced inactivation is responsible for the rapid inactivation of the enzyme found in the tissue. Recently, Satoh and Yang[21] demonstrated that when a partially purified ACC synthase preparation isolated from wounded tomato pericarp tissue was incubated with S-adenosyl-L-[3,4-^{14}C]methionine and the resulting proteins were analyzed by sodium dodecylsulfate-polyacrylamide gel electrophoresis (SDS-PAGE), only one radioactive protein band was observed. This protein was judged to be ACC synthase based on the observations that its molecular weight was 50 kD and that it was specifically bound to a monoclonal antibody against ACC synthase prepared by Bleecker _et al._[14] S-Adenosyl-L-[carboxyl-^{14}C] methionine, but not S-adenosyl-L-[methyl-^{14}C]methionine, also radiolabeled ACC synthase (S. Satoh, unpublished results). These results suggest that the SAM-induced inactivation of ACC synthase involves a covalent linkage of a fragment of the SAM molecule, probably the 2-aminobutyrate moiety, into the active site of ACC synthase.

ACC OXIDASE

The final step in the biosynthesis of ethylene is catalyzed by ACC oxidase. Application of ACC to many plant tissues results in a marked increase in ethylene production,[5,22] indicating that ACC oxidase in these tissues is constitutive and not rate-limiting. In other tissues such as unripe fruit or young petals of carnation flower, ACC oxidase activity is low, but increases markedly as the tissues undergo ripening or the senescence process.[16]

Although the ACC molecule possesses two enantiotopic methylene groups, they are not geometrically equivalent. Ethyl substitution of each of the 4 methylene hydrogens results in 4 stereoisomers of 1-amino-2-ethylcyclopropane-1-carboxylic acid (AEC). If conversion of ACC to ethylene by plant tissues proceeds by a stereospecific enzyme, Hoffman et al.[23] reasoned that these four stereoisomers of AEC might be converted into 1-butene with unequal efficiency. Indeed, in apples and etiolated mungbean hypocotyls, (1R, 2S)-AEC was preferentially converted to 1-butene. ACC and AEC appear to be degraded by the same enzyme since both reactions are inhibited to the same extent by a nitrogen atmosphere or by Co^{2+}. Furthermore, when both substrates are present simultaneously, each acts as an inhibitor with respect to the other. These observations indicate that ACC oxidase exhibits a high degree of stereoselectivity.[23]

Although ACC-dependent ethylene production can be demonstrated readily in intact tissues, in vitro ACC oxidase activity has proved to be peculiarly elusive. A number of cell-free preparations derived from a variety of plant tissues have been reported to be capable of oxidizing ACC to ethylene. While plant tissues displayed strict stereospecificity toward the stereoisomers of AEC as mentioned above and exhibited a high affinity for ACC, cell-free preparations did not.[24-26] It is likely that these reported enzyme preparations catalyze the formation of one or more "active oxygen" species such as hydrogen peroxide and lipid hydroperoxide, which in turn react non-enzymatically with ACC to produce ethylene.[26,27] Guy and Kende[28] demonstrated that vacuoles isolated from pea and bean leaves retained the characteristics of tissue ACC oxidase activity, as they differentiated between AEC isomers and had a high affinity for ACC. They noted that the enzyme activity ceased following lysis or rupture of the vacuoles. Thus, intact vacuoles are the smallest biological unit which has so far been shown to retain properties characteristic of native ACC oxidase. Recently, Porter et al.[29] compared the retention of ACC oxidase activity in isolated cells, protoplasts and vacuoles derived from leaves of pea or bean treated with enzymes that degrade cell walls. They observed that the vacuoles retained much of the ACC oxidase activity of the protoplasts. However, isolated cells and protoplasts retained less than 5% of the activity associated with the parent tissue. These authors proposed that full ACC oxidase activity requires tissue integrity in addition to a previously noted requirement for cell membrane integrity.

Although ACC oxidase still has to be isolated and characterized, some important information about ACC oxidase has been generated at the tissue level. The dependence of ethylene production on the concentration of ACC and oxygen has been studied in a number of plant tissues. The K_m value of O_2 varied greatly depending on the internal ACC content. When ACC levels in the tissue were low (below its K_m value), the concentration of O_2 giving half-maximal ethylene production rate ($[S]_{0.5}$) ranged between 5 and 7%, and was similar among different tissues.[30] As the concentration of ACC was increased above its K_m value, the $[S]_{0.5}$ for O_2 decreased markedly. In contrast, the K_m value for ACC was not much dependent on O_2 concentration, but varied greatly among different plant tissues, ranging from 8 μM in apple tissue to 120 μM in etiolated wheat leaf. Such a great variation was thought to be due to differences in the compartmentation of ACC within the cells in different tissues.[30] Since ACC oxidase is a bi-substrate enzyme, the $[S]_{0.5}$ values for O_2 and for ACC can vary as the concentration of the other substrate changes. The above kinetic data are consistent with the view that ACC oxidase follows an ordered binding mechanism in which ACC oxidase binds oxygen first and then ACC.

Studies of the reaction products of ACC oxidase have provided us with valuable information about the enzyme reaction catalyzed by ACC oxidase. During our early studies on ACC we attempted to destroy the amino acids extracted from plant tissues with NaOCl. Unexpectedly, Lizada and Yang[31] found that ACC was degraded by NaOCl into ethylene with a high yield in the presence of mercuric chloride. Based on this finding, a rapid, simple and sensitive method for the determination of ACC in plant extracts was developed.[31] Hiyama _et al._[32] had previously reported that 1-phenylcyclopropylamine is oxidized by NaOCl or lead tetraacetate to ethylene and benzonitrile via the intermediacy of the nitrenium ion:

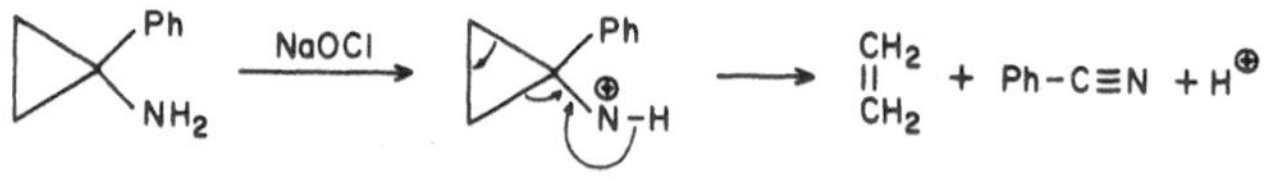

It is reasonable to assume that ACC is similarly oxidized by NaOCl into ethylene and cyanoformic acid, the latter being further degraded spontaneously into HCN and CO_2.[33] Support for this reaction scheme is derived from the observation that NaOCl oxidizes ACC, liberating equal amounts of CO_2, ethylene, and HCN from the carboxyl, C-1 and C-2,3 atoms of ACC, respectively.[34,35]

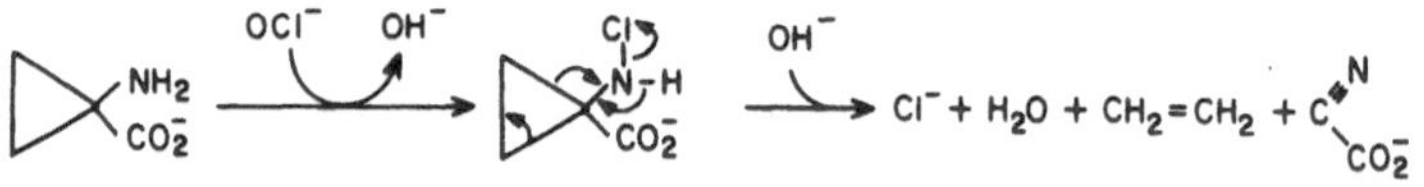

A mechanism analogous to that of chemical oxidation has been proposed,[33] where ACC is oxidized by a hydroxylase to form N-hydroxy-ACC, a nitrenium equivalent, or to form the nitrenium ion, which is then fragmented into ethylene, HCN and CO_2.

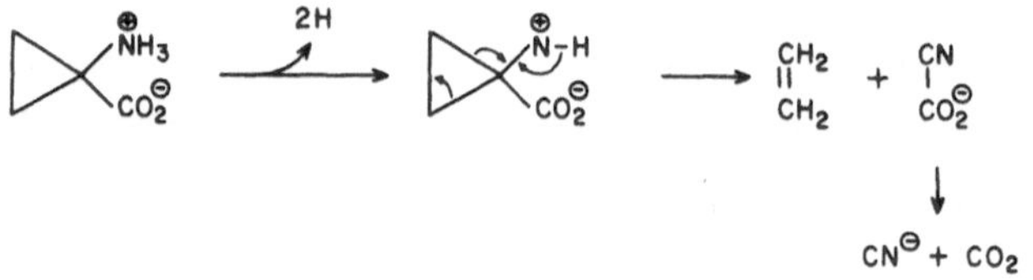

The above scheme predicts that the carboxyl and C-1 atoms of ACC are converted to CO_2 and HCN, respectively, during the enzymic degradation of ACC to ethylene. Support for this was provided by Peiser <u>et al.</u>,[34] who prepared [carboxyl-^{14}C]ACC and [1-^{14}C]ACC, and showed that the carboxyl carbon of ACC is liberated as CO_2, whereas the C-1 of ACC is recovered not as free HCN but as HCN conjugates, in an amount equivalent to that of the ethylene produced. Their data indicated that ACC is degraded into ethylene, CO_2 and HCN, with HCN being rapidly metabolized to yield β-cyanoalanine. The β-cyanoalanine is then hydrated to asparagine, but in some plants such as *Vicia sativa*, γ-glutamyl-β-cyanoalanine is the main product.[34] This sequence of reactions is illustrated below:

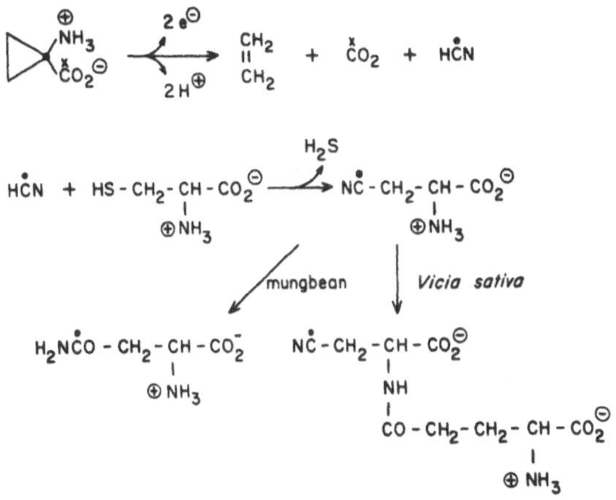

Since no free HCN was detected even in plant tissues which produced ethylene at very high rates, Peiset et al.[34] assumed that plants must have ample capacity to metabolize the HCN originating from ACC oxidase. In higher plants the key enzyme to detoxify HCN is β-cyanoalanine synthase (EC 4.4.1.9) which catalyzes the conversion of cysteine and HCN to β-cyanoalanine and H_2S. β-Cyanoalanine synthase is widely distributed in both cyanogenic and non-cyanogenic plants.[36] Since HCN is known to inhibit cytochrome oxidase and triggers cyanide-resistant respiration, it is important to know the steady-state concentration of HCN in plant tissues. By employing an isotope dilution method, Yip and Yang[37] estimated that the steady-state concentration of HCN was below 0.2 μM in ripening fruits and in auxin-treated mungbean hypocotyls which produce ethylene at high rates. The concentration of HCN which results in 50% inhibition of the cytochrome-mediated respiration in plant tissues has been estimated to be 10-20 μM.[38] Based on the early hypothesis of Solomos and Laties[39] that ethylene may induce cyanide-resistant respiration and that this cyanide-resistant respiration may operate in ripening fruits, Pirrung and Brauman[40] hypothesized that the increased biosynthesis of ethylene during fruit ripening may result in increased HCN levels, which in turn inhibit cytochrome oxidase and trigger cyanide-resistant respiration. However, the experimental data of Yip and Yang[37] indicate that the HCN level in ripening fruits is too low to cause any significant inhibition of cytochrome oxidase, and support the previous notion of Peiser *et al.*[34] that plant tissues have ample capacity to detoxify HCN formed during ethylene biosynthesis.

Recent experimental results of Theologis and Laties[38] revealed that respiration in either preclimacteric or climacteric avocado fruit is mediated by the cytochrome respiratory path and that there is no involvement of the cyanide-resistant path in either tissue. Assuming that a steady-state level of HCN in the plant tissue is maintained between the rate of HCN generation by ACC oxidase and the rate of HCN metabolism by β-cyanoalanine synthase, Yip and Yang[37] calculated that for the plant tissue to maintain HCN level below the safe level of 1 μM, the tissue β-cyanoalanine synthase activity at saturating HCN concentration should be at least 500 times higher than the tissue rate for producing ethylene. In tissues which have been examined, this relationship appeared to hold.

It should be noted that β-cyanoalanine synthase exists widely throughout the plant kingdom. In ripening fruits and senescing carnation flowers, ethylene production increases several hundred-fold, but β-cyanoalanine synthase activity increases only 1- to 2-fold.[37,41] This indicates that the basal level of β-cyanoalanine synthase in these tissues was already very high before the climacteric increase in ethylene production. The relationship between ethylene production rates, β-cyanoalanine synthase activities and tissue HCN levels in ripening apple fruits are shown in Table 1.

Recently, Adlington *et al.*[42] have prepared [*cis*-2,3-$^{2}H_2$]ACC and observed that the chemical oxidation of *cis*-dideutero-ACC with NaOCl results in complete retention of that configuration in ethylene. However, when this substrate was applied to apple slices, equal amounts of *cis*- and *trans*-dideuteroethylene were obtained, as shown below. These data indicate that NaOCl oxidation of ACC to ethylene may proceed by a concerted elimination mechanism which retains the stereochemistry, whereas the biological reaction proceeds instead by a stepwise mechanism involving an intermediate that allows scrambling of the ring hydrogens.

Table 1. Changes in ethylene production rate, extractable β-cyanoalanine synthase activity, and tissue cyanide content in ripening apple fruit.

Stage	C_2H_4 (v_1)	Cyanoalanine Synthase (V)	v_1/V	[HCN] Calculated	[HCN] Observed
	nmol/g-h	nmol/g-h		μM	μM
Unripe	0.01	630	1:63000	0.01	<0.1
Ripe	3.3	1650	1: 500	1.0	0.2

The values of [HCN] calculated = K_m X (v_1/V), where K_m is assumed to be 500 μM. Data are from Yip and Yang.[37]

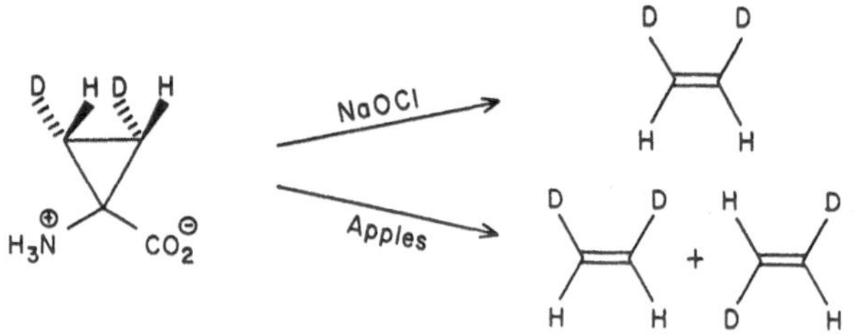

It should be noted that the conversion of ACC to CO_2, HCN and ethylene represents a two-electron oxidation.[34] Since electrolytic oxidation of cis-2,3-dideutero-ACC also yielded HCN and resulted in a scrambled label in the 1,2-dideutero-ethylene, as is observed in plant tissues, Pirrung[43] suggested that the oxidation of ACC in vivo proceeds in two sequential one-electron oxidation reactions, and that the initial oxidation yields free radical intermediates which undergo rapid ring-opening and loss of stereochemistry, as shown below.

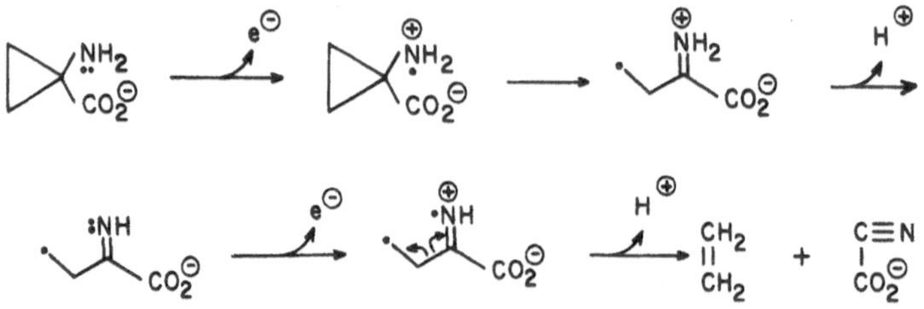

However, it should be noted that the free radical mechanism is only one of several possible mechanisms which can rationalize the loss of stereochemistry of deuterium atoms at C-2,3 of ACC during its biological oxidation to ethylene. Although ethylene cannot be biosynthesized from the ACC nitrenium ion by a concerted elimination mechanism, a stepwise elimination mechanism involving ring opening and an ionic intermediate that allows scrambling of the hydrogens on C-2,3, as illustrated below, cannot be ruled out.[44]

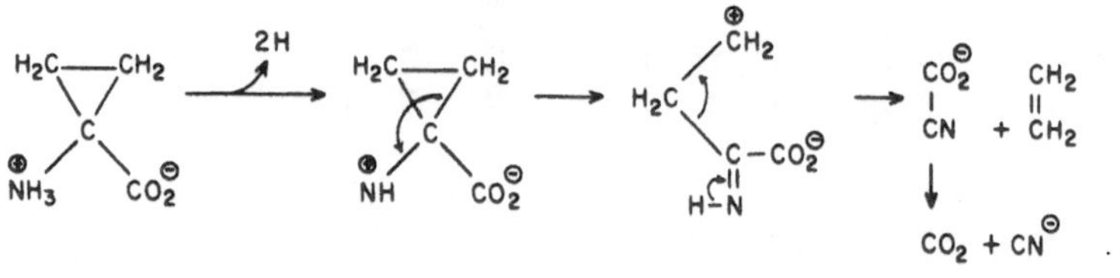

ACC N-MALONYLTRANSFERASE

Aside from its conversion to ethylene, the other metabolic fate of ACC in plant tissues is its conjugation into MACC. Because endogenous levels of ACC can increase during ent or in response to stress, it seems logical that nt would require some means to sequester ACC to prevent the overproduction of ethylene. The N-malonylation of ACC serves this purpose by reducing the tissue level of ACC and, consequently, ethylene production.

While ACC is known to be metabolized to 2-oxobutyrate and ammonia in some microorganisms,[44] Amrhein _et al_.[45] found no such deamination reaction occurring in plant tissues. Instead they discovered, initially in buckwheat seedlings, that ACC supplied exogenously was efficiently converted into a conjugate which was identified as MACC. Working on the biosynthesis of stress ethylene in wilted wheat leaves, Apelbaum and Yang[46] had previously observed that the loss of ACC, during an incubation period following the wilting treatment, was greater than the quantity of ethylene produced during the same period. This suggested that ACC must have been metabolized by some pathway other than ethylene production. Hoffman _et al_.[47] therefore examined the metabolism of exogenously supplied ACC in light-grown wheat leaves and identified MACC as the non-volatile metabolite of ACC in water-stressed as well as non-stressed tissues.

Studies on the subcellular compartmentation of ACC and MACC revealed that the synthesis of ACC and MACC occurs in the cytoplasm.[48,49] The MACC synthesized in the cytoplasm is then transported and stored in the vacuole. While MACC in the cytoplasm can be transported outside the cell, vacuolar MACC remains sequestered within the vacuole.[48] Hydrolysis of MACC to ACC was observed in some tissues when a high level of MACC was administered exogenously.[50] However, this process does not appear to play an important role under physiological conditions, because MACC is compartmentalized largely in the vauole and hence is unavailable for further metabolism.

Since MACC is a poor ethylene producer, and the conjugation of ACC to MACC is essentially irreversible, MACC is considered to be a biologically inactive end-product of ACC rather than a storage form. Support for such a conclusion

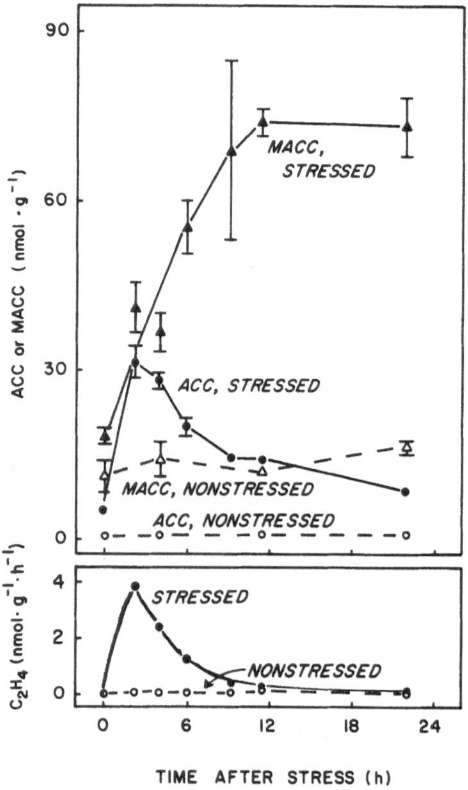

Fig. 2. Time courses of changes in ACC content, MACC content and ethylene production rate in excised wheat leaves which have been subjected to water-stress until they lost 9% of their initial fresh weight. From Hoffman et al.[52]

is found in the observation that germinating peanut seeds contained large amounts of MACC (50-100 nmol g^{-1}) but produced ethylene which was not derived from MACC and was instead synthesized de novo from SAM.[51] The time course of changes in ACC and MACC levels and of total ethylene production in water-stressed and non-stressed wheat leaves is shown in Figure 2. While there is a sharp rise followed by a decline in the ACC level and in the ethylene production rate in stressed leaves, the level of MACC increased gradually until it reached a plateau. Once formed, the MACC levels did not decrease even after the stressed tissues were rehydrated.[52] Thus, the conjugation of ACC into MACC serves as a mechanism to dissipate ACC, and thereby participates in the regulation of ethylene production. Since the enzyme responsible for the MACC formation appears to be constitutive in most plant tissues, it is expected that MACC will be formed as long as ACC synthesis is induced. Hence, the level of MACC in the plant tissue provides some information on the history of ACC synthesis in that tissue.

N-Malonylation of D-amino acids commonly occurs in higher plants.[53] It is thought that the physiological significance of N-malonylation is to inactivate foreign and potentially toxic substances such as D-amino acids or herbicides. Since ACC has no asymmetric carbon, it can be recognized by an enzyme as a D- as well as an L-amino acid. Conceivably, the malonylations of ACC and of D-amino acids are interrelated. Indeed, various D-amino acids (D-Phe, D-Met, D-Ala) inhibit the malonylation of exogenously administered ACC in segments of mungbean hypocotyl resulting in increased concentration of free ACC and in an increased rate of ethylene production. L-Enantiomers are, however, ineffective.[54] Reciprocally, ACC or D-phenylalanine greatly inhibits the formation of N-malonyl-D-methionine from exogenously administered D-methionine. These results indicate an intimate relationship between the malonylation of ACC and D-amino acids, and further suggest that both reactions are catalyzed by the same enzyme. Such a conclusion is in agreement with the data obtained with a cell-free system, as will be discussed in the next section. Previously, Satoh and Esashi[55] reported that D-amino acids stimulate ethylene production and increase ACC content in cocklebur cotyledons. Their observations can now be explained by the fact that D-amino acids inhibit malonylation of ACC resulting in higher ACC levels and thereby higher rates of ethylene production. These data clearly indicate that modification of ACC malonylation results in changes in the endogenous ACC level and thus the rate of ethylene production.

ACC N-malonyltransferase, which catalyzes the following reaction, has been isolated and partially purified from mungbean hypocotyls.[49,56-58]

$$\text{ACC + Malonyl-CoA} \dashrightarrow \text{MACC + CoA}$$

Of particular interest is the intimate relationship between ACC malonyltransferase and D-amino acid malonyl-transferase activities. Based upon the following observations, D-amino acid malonyltransferase and ACC malonyltransferase are thought to be the same enzyme: (a) the enzyme preparations malonylate both D-amino acids and ACC; (b) the K_m values of those amino acids serving as substrates for malonyltransferase agree with their corresponding K_i values when the same amino acids act as competitive inhibitors of ACC malonyltransferase; (c) among four stereoisomers of AEC, those isomers which have a D-amino acid configuration are more effective substrates and inhibitors of malonyltransferase than the L-isomers.[58] The proof that a single enzyme carries out both malonylation reactions awaits purification of the enzymes. If both reactions are catalyzed by the same enzyme, it is interesting to ask whether the primary evolutionary role of the enzyme is to regulate ethylene production by reducing the ACC level or to detoxify D-amino acids.

Although the transferase is present in a wide range of plant tissues, further expression of this enzyme activity by ethylene treatment has been recognized in certain tissues. When excised citrus flavedo tissue,[59] tobacco leaf discs,[60] or intact green tomato fruits[61] are treated with ethylene, the ability of the tissue to malonylate ACC is promoted; this ability is accompanied by an increase in extractable ACC/D-amino acid malonyl-transferase activity. In these tissues, ethylene treatment results in autoinhibition of ethylene production. In these cases ethylene promotes gene expression of ACC malonyltransferase resulting in a reduced level of ACC and thereby a reduced production of ethylene.

CONCLUSION

Ethylene has long been recognized as a plant hormone. However, its biosynthetic pathway remained elusive until the key intermediate, ACC, was shown to be the immediate precursor of the hormone. ACC synthase, which catalyzes the conversion of SAM to ACC and MTA, is a pyridoxal phosphate-dependent lyase. In plant tissues ACC synthase is, in general, the rate-controlling enzyme. The synthesis of ACC synthase can be induced by various developmental and environmental factors resulting in a dramatic increase in ACC level and hence in ethylene production. Once ACC synthase is induced, it is also rapidly inactivated in plant tissues. Using isolated ACC synthase it has been shown that the substrate SAM acts as an enzyme-activated inactivator of ACC synthase. This process involves the covalent linkage of a fragment of the SAM molecule to the active site of ACC synthase resulting in the inactivation of the enzyme. The final step in ethylene biosynthesis is catalyzed by an oxidative enzyme, which is referred to as "ACC oxidase". Since ACC oxidase activity has not been demonstrated independent of intact cellular material, it is thought that ACC oxidase is highly structural and requires membrane/tissue integrity. In spite of this, the reaction products have been identified and some important insights into the reaction mechanisms have been gained using plant tissues. Tracer studies established that ACC is aerobically oxidized in plant tissues to CO_2, HCN, and ethylene. HCN thus formed is efficiently metabolized by the action of β-cyanoalanine synthase. Since the tissue's capability to detoxify HCN is so high, the level of HCN in plant tissues is maintained at a level too low either to inhibit cytochrome oxidase or to induce the cyanide-resistant respiratory pathway. In addition to its conversion to ethylene, ACC is metabolized to a conjugate, MACC. Since MACC is inactive as an ethylene producer and the conjugation of ACC to MACC is essentially irreversible, MACC is thought to be a biologically inactive end-product of ACC rather than a storage form. ACC N-malonyltransferase, which catalyzes the malonylation of ACC with malonyl-CoA, occurs widely in plant tissues. This enzyme participates in the regulation of ethylene production by sequestering ACC thereby reducing ethylene production. Since malonylation of ACC is intimately related to the malonylation of D-amino acids, both *in vivo* and *in vitro*, it is assumed that the same enzyme catalyzes both reactions.

ACKNOWLEDGMENTS

Our work cited herein has been supported by research grants from the National Science Foundation (PCM-8114933 and PCM-8414971).

REFERENCES

1. BURROUGHS, L.F. 1957. 1-Aminocyclopropane-1-carboxylic acid. A new amino acid in perry pears and cider apples. Nature 179: 360-361.
2. VAHATALO, M.L., A.I. VIRTANEN. 1957. A new cyclic α-aminocarboxylic acid in berries of cowberry. Acta Chem. Scand. 11: 741-743.
3. BURROUGHS, L.F. 1960. The free amino acids of certain British fruits. J. Sci. Food Agric. 11: 14-18.
4. ADAMS, D.O., S.F. YANG. 1979. Ethylene biosynthesis: identification of 1-aminocyclopropane-1-carboxylic acid as an intermediate in the conversion of methionine to ethylene. Proc. Natl. Acad. Sci. USA 76: 170-174.
5. LURSSEN, K., K. NAUMANN, R. SCHRODER. 1979. 1-Aminocyclopropane-1-carboxylic acid - an intermediate of the ethylene biosynthesis in higher plants. Z. Pflanzenphysiol. 92: 285-294.
6. ABELES, F.B. 1973. Ethylene in Plant Biology. Academic Press, New York, 302 pp.
7. LIEBERMAN, M., A KUNISHI, L.W. MAPSON, D.A. WARDALE. 1966. Stimulation of ethylene production in apple tissue slices by methionine. Plant Physiol. 41: 376-382.
8. ADAMS, D.O., S.F. YANG. 1977. Methionine metabolism in apple tissue. Implication of S-adenosylmethionine as an intermediate in the conversion of methionine to ethylene. Plant Physiol. 60: 892-896.
9. BOLLER, T., R.C. HERNER, H. KENDE. 1979. Assay for and enzymatic formation of an ethylene precursor, 1-aminocyclopropane-1-carboxylic acid. Planta 145: 293-303.
10. YU, Y.B., D.O. ADAMS, S.F. YANG. 1979. 1-Aminocyclopropane-carboxylate synthase, a key enzyme in ethylene biosynthesis. Arch. Biochem. Biophys. 198: 280-286.

11. RAMALINGAM, K., K. LEE, R.W. WOODARD, A.B. BLEECKER, H. KENDE. 1985. Stereochemical course of the reaction catalyzed by the pyridoxal phosphate-dependent enzyme 1-aminocyclopropane-1-carboxylate synthase. Proc. Natl. Acad. Sci. USA 82: 7820-7824.
12. KHANI-OSKOUEE, S., J.P. JONES, R.W. WOODARD. 1984. Stereochemical course of the biosynthesis of 1-aminocyclopropane-1-carboxylic acid. I. Role of the asymmetric sulfonium pole and the α-amino acid center. Biochem. Biophys. Res. Comm. 121: 181-187.
13. WIESENDANGER, R., B. MARTINONI, T. BOLLER AND D. ARIGONI. 1986. Biosynthesis of 1-aminocyclopropanecarboxylic acid: steric course of the reaction at the α-position. J. Chem. Soc. Chem. Commun. 1986: 238-239.
14. BLEECKER, A.B., W.H. KENYOU, S.C. SOMERVILLE, H. KENDE. 1986. Use of monoclonal antibodies in purification and characterization of 1-amino-cyclopropane-1-carboxylate synthase, an enzyme in ethylene biosynthesis. Proc. Natl. Acad. Sci. USA 83: 7755-7759.
15. NAKAJIMA, N., N. NAKAGAWA, H. IMASEKI. 1988. Purification and properties of 1-aminocyclopropane-1-carboxylate synthase of mesocarp of Cucurbita maxima Duch. fruit. Plant Cell Physiol. 29: 989-998.
16. YANG, S.F., N.E. HOFFMAN. 1984. Ethylene biosynthesis and its regulation in higher plants. Annu. Rev. Plant Physiol. 35: 155-189.
17. YOSHII, H., IMASEKI, H. 1982. Regulation of auxin-induced ethylene biosynthesis. Repression of inductive formation of 1-aminocyclopropane-1-carboxylate synthase by ethylene. Plant Cell Physiol. 23: 639-649.
18. ACASTER, M.A., H. KENDE. 1983. Properties and partial purification of 1-aminocyclopropane-1-carboxylate synthase. Plant Physiol. 72: 139-145.
19. KENDE, H., T. BOLLER. 1981. Wound ethylene and 1-aminocyclopropane-1-carboxylate synthase in ripening tomato fruit. Planta 151: 476-481.
20. SATOH, S., Y. ESASHI. 1986. Inactivation of 1-aminocyclopropane-1-carboxylic acid synthase of etiolated mung bean hypocotyl segments by its

substrate, S-adenosyl-L-methionine. Plant Cell Physiol. 27: 285-291.

21. SATOH, S., S.F. YANG. 1988. S-Adenosylmethionine-dependent inactivation and radiolabeling of 1-aminocyclopropane-1-carboxylate synthase isolated from tomato fruits. Plant Physiol. 88: 109-114.

22. CAMERON, A.C., C.A.L. FENTON, Y.B. YU, D.O. ADAMS, S.F. YANG. 1979. Increased production of ethylene by plant tissues treated with 1-aminocyclopropane-1-carboxylic acid. Hortscience 14: 178-180.

23. HOFFMAN, N.E., S.F. YANG, A. ICHIHARA, A. SAKAMURA. 1982. Stereospecific conversion of 1-amino-cyclorpopanecarboxylic acid to ethylene by plant tissues. Conversion of stereoisomers of 1-amino-2-ethylcyclopropanecarboxylic acid to 1-butene. Plant Physiol. 70: 195-199.

24. McKEON, T.A., S.F. YANG. 1984. A comparison of the conversion of 1-amino-2-ethylcyclopropane-1-carboxylic acid stereoisomers to 1-butene by pea epicotyls and by a cell-free system. Planta 160: 84-87.

25. VENIS, M.A. 1984. Cell-free ethylene-forming systems lack stereochemical fidelity. Planta 162: 85-88.

26. WANG, T., S.F. YANG. 1987. The physiological role of lipoxygenase in ethylene formation from 1-aminocyclopropane-1-carboxylic acid in oat leaves. Planta 170: 190-196.

27. STEGINK, S.J., J.N. SIEDOW. 1986. Ethylene production from 1-aminocyclopropane-1-carboxylic acid _in vitro_: a mechanism for explaining ethylene production by a cell-free preparation from pea epicotyls. Physiol. Plant. 66: 625-631.

28. GUY, M., H. KENDE. 1984. Conversion of 1-amino-cyclopropane-1-carboxylic acid to ethylene by isolated vacuoles of _Pisum sativum_. Planta 160: 281-287.

29. PORTER, A.J., J.T. BORLAKOGLU, P. JOHN. 1986. Activity of the ethylene-forming enzyme in relation to plant cell structure and organization. J. Plant Physiol. 125: 207-216.

30. YIP, W.K., X.Z. JIAO, S.F. YANG. 1988. Dependence of _in vivo_ ethylene production rate on 1-amino-

cyclopropane-1-carboxylic acid content and oxygen concentration. Plant Physiol. 88: 553-558.

31. LIZADA, M.C.C., S.F. YANG. 1979. A simple and sensitive assay for 1-aminocyclopropane-1-carboxylic acid. Anal. Biochem. 100: 140-145.
32. HIYAMA, T., H. KOIDE, H. NOZAKI. 1975. Oxidation of cyclopropylamines and aziridines with lead tetraacetate. Bull. Chem. Soc. Japan 48: 2918-2921.
33. YANG, S.F. 1981. Biosynthesis of ethylene and its regulation. In Recent Advances in the Biochemistry of Fruit and Vegetables. (J. Friend, M.J.C. Rhodes, eds.), Academic Press, London, pp. 89-106.
34. PEISER, G.D., WANG, T.-T., N.E. HOFFMAN, S.F. YANG, H.-W. LIU, C.T. WALSH. 1984. Formation of cyanide from carbon 1 of 1-aminocyclopropane-1-carboxylic acid during its conversion to ethylene. Proc. Natl. Acad. Sci. USA 81: 3059-3063.
35. LEETE, E., G.J. MORRIS, H.S.P. RAO. 1985. Biosynthesis of ethylene. Revista Latinoamer Quim.
36. MILLER, J.M., E.E. CONN. 1980. Metabolism of hydrogen cyanide by higher plants. Plant Physiol. 65: 1199-1202.
37. YIP, W.K., S.F. YANG. 1988. Cyanide metabolism in relation to ethylene production in plant tissues. Plant Physiol. 88: 473-476.
38. THEOLOGIS, A., G.G. LATIES. 1978. Relative contribution of cytochrome-mediated and cyanide-resistant electron transport in fresh and aged potato slices. Plant Physiol. 62: 232-237.
39. SOLOMOS, T., G.G. LATIES. 1976. Induction by ethylene of cyanide-resistant respiration. Biochem. Biophys. Res. Commun. 70: 663-671.
40. PIRRUNG, M.C., J.I, BRAUMAN. 1987. Involvement of cyanide in the regulation of ethylene biosynthesis. Plant Physiol. Biochem. 25: 55-61.
41. MANNING, K. 1986. Ethylene production and β-cyanoalanine synthase activity in carnation flowers. Planta 168: 61-66.
42. ADLINGTON, R.M., J.E. BALDWIN, B.J. RAWLINGS. 1983. On the stereochemistry of ethylene biosynthesis.

J. Chem. Soc. Chem. Commun. 290-292.
43. PIRRUNG, M.C. 1983. Ethylene biosynthesis. 2. Stereochemistry of ripening, stress, and model reactions. J. Am. Chem. Soc. 105: 7207-7209.
44. HONMA, M., T. SHIMOMURA. 1978. Metabolism of 1-aminocyclopropane-1-carboxylic acid. Agric. Biol. Chem. 42: 1825-1831.
45. AMRHEIN, N., D. SCHNEEDBECK, H. SKORUPKA, S. TOPHOF. 1981. Identification of a major metabolite of the ethylene precursor 1-aminocyclopropane-1-carboxylic acid in higher plants. Naturwissenschaften 68: 619-620.
46. APELBAUM, A., S.F. YANG. 1981. Biosynthesis of stress ethylene induced by water deficit. Plant Physiol. 68: 594-596.
47. HOFFMAN, N.E., S.F. YANG, T. McKEON. 1982. Identification of 1-(malonylamino)cyclopropane-1-carboxylic acid as a major conjugate of 1-aminocyclopropane-1-carboxylic acid, an ethylene precursor in higher plants. Biochem. Biophys. Res. Commun. 104: 765-770.
48. BOUZAYEN, M., A. LATCHÉ, G. ALIBERT, J.-C. PECH. 1988. Intracellular sites of synthesis and storage of 1-(malonylamino)cyclopropane-1-carboxylic acid in _Acer pseudoplatanus_ cells. Plant Physiol. 88: 613-617.
49. AMRHEIN, N., C. FORREITER, C. KIONKA, H. SKORUPKA, S. TOPHOF. 1987. Metabolism, and its compartmentation, of 1-aminocyclopropane-1-carboxylic acid in plant cells. _In_ Conjugated Plant Hormones. (K. Schreiber, H.R. Schutte, G. Sembdner, eds.), Institute of Plant Biochemistry, Halle, East Germany, pp. 102-108.
50. JIAO, X.Z., S. PHILOSOPH-HADAS, L.Y. SU, S.F. YANG. 1986. The conversion of 1-(malonylamino)-cyclopropane-1-carboxylic acid to 1-aminocyclopropane-1-carboxylic acid in plant tissues. Plant Physiol. 81: 637-641.
51. HOFFMAN, N.E., J. FU, S.F. YANG. 1983. Identification and metabolism of 1-(malonylamino)cyclopropane-1-carboxylic acid in germinating peanut seeds. Plant Physiol. 71: 197-199.
52. HOFFMAN, N.E., Y. LIU, S.F. YANG. 1983. Changes in 1-(malonylamino)cyclopropane-1-carboxylic acid content in wilted wheat leaves in relation to their ethylene production rates and 1-amino-

cyclopropane-1-carboxylic acid content. Planta 157: 518-523.

53. KAWASAKI, Y., T. OGAWA, K. SASAOKA. 1982. Two pathways for formation of D-amino acid conjugates in pea seedlings. Agric. Biol. Chem. 46: 1-5.

54. LIU, Y., N.E. HOFFMAN, S.F. YANG. 1983. Relationship between the malonylation of 1-aminocyclopropane-1-carboxylic acid and D-amino acids in mung-bean hypocotyls. Planta 158: 437-441.

55. SATOH, S., Y. ESASHI. 1981. D-Amino acid-stimulated ethylene production: molecular requirements for the stimulation and a possible receptor site. Phytochemistry 20: 947-949.

56. KIONKA, C., N. AMRHEIN. 1984. The enzymatic malonylation of 1-aminocyclopropane-1-carboxylic acid in homogenates of mung-bean hypocotyls. Planta 162: 226-235.

57. SU, L.-Y., Y. LIU, S.F. YANG. 1985. Relationship between 1-aminocyclopropanecarboxylate malonyltransferase and D-amino acid malonyltransferase. Phytochemistry 24: 1141-1145.

58. LIU, Y., L.-Y. SU, S.F. YANG. 1984. Stereoselectivity of 1-aminocyclopropanecarboxylate malonyltransferase toward stereoisomers of 1-amino-2-ethylcyclopropanecarboxylic acid. Arch. Biochem. Biophys. 235: 319-325.

59. LIU, Y., N.E. HOFFMAN, S.F. YANG. 1985. Ethylene-promoted malonylation of 1-aminocyclopropane-1-carboxylic acid participates in autoinhibition of ethylene synthesis in grapefruit flavedo discs. Planta 164: 565-568.

60. PHILOSOPH-HADAS, S., S. MEIR, N. AHARONI. 1985. Autoinhibition of ethylene production in tobacco leaf discs: enhancement of 1-aminocyclopropane-1-carboxylic acid conjugation. Physiol. Plant. 63: 431-437.

61. LIU, Y., L.-Y. SU, S.F. YANG. 1985. Ethylene promotes the capability to malonylate 1-aminocyclopropane-1-carboxylic acid and D-amino acids in preclimacteric tomato fruits. Plant Physiol. 77: 891-895.

62. TSAI, D., R.N. ARTECA, J.M. BACHMAN, A.T. PHILLIPS. 1988. Purification and characterization of 1-aminocyclopropane-1-carboxylate synthase from etiolated mung bean hypocotyls. Arch. Biochem. Biophys. 264: 632-640.

Chapter Nine

MOLECULAR ASPECTS OF STORAGE PROTEIN AND STARCH SYNTHESIS IN WHEAT AND RICE SEEDS

THOMAS OKITA, ARUN ARYAN,
CHRISTOPHER REEVES, WOO TAEK KIM,
DOUGLAS LEISY, JIM HNILO, AND
DAVID MORROW

Institute of Biological Chemistry
Washington State University
Pullman, Washington 99164-6340

INTRODUCTION

Seed formation in cereals is accompanied by the mobilization and transport of nitrogen from leaves and other source tissues to the developing cereal grain. Nitrogen enters the seed primarily as amino acids, of which glutamine appears to be the predominant constituent.[1] It has been estimated that about 50% of the total nitrogen available in leaf and stem tissue of wheat is catabolized and transported to the developing grains[2] where it is converted mainly into storage proteins.

Storage proteins (defined here as any protein which serves as a nitrogen store and is packaged into protein bodies) are synthesized during the mid-stages of seed development and normally constitute about 80-90% of the total protein present in seeds. These proteins are subsequently utilized by the young developing seedling as a source of nitrogen and carbon and sometimes sulfur.

Storage proteins were first studied by Osborne[3] who classified them into various groups based on their solubility properties. Other than the water-soluble albumin fraction, the major storage proteins of the cereal grains fall in three solubility classes; globulins (dilute salt solutions), prolamines (alcohol solutions), and glutelins (dilute alkali solutions). This classification of storage proteins has been questioned in recent years. Based on a number of criteria, some of which are discussed here, storage proteins fall into two broad groups, the globulins and prolamines.

Wheat and rice are used mainly as human food and they provide much of the caloric and dietary protein for the world's population. Long term interests have centered on the genetic improvement of these cereals for increased protein content and/or better balance of amino acids.[4] Wheat usually contains about 8-15% protein and is limited in several amino acids, the main one being lysine, which are essential for human growth and development.[4] Rice proteins, on the other hand, are more nutritionally balanced, but they only constitute about 5% of the dry weight of the seed.[4] Attempts by plant breeders to increase the protein quality of cereal grains have met with very little success. Although high lysine maize, barley, and sorghum mutants have been identified, the high lysine phenotype appears to arise partly because these mutants accumulate smaller proportions of the lysine-deficient prolamine storage proteins.[5] These high lysine lines also suffer from a number of pleiotrophic effects, the most serious being the partial reduction of starch synthesis resulting in decreased grain yield. This relationship between protein and starch synthesis is also manifested in starch mutants, where blockage of starch synthesis can lead to a suppression of the synthesis of storage proteins. Despite intensive breeding efforts, high lysine cereal lines which are at par with normal yielding varieties have not been developed. It is clear that a thorough understanding of the molecular, biochemical, and cellular aspects of both storage protein and starch synthesis is required to permit genetic improvement of the nutritional quality of the cereal grains.

In this article we present results of our recent efforts to understand the structure of the storage protein genes in both wheat and rice, the mechanism by which these genes are regulated at both the transcriptional and post-transcriptional level, and the cellular processing involved in the deposition of these proteins into protein bodies. In addition, studies have been initiated which will hopefully lead to the elucidation of the molecular control of starch biosynthesis in cereals which will aid in efforts to improve the accumulation of this macromolecule in seeds.

STRUCTURE OF THE WHEAT PROLAMINES AND THEIR GENES

The storage proteins of wheat consist of two prolamine subclasses, gliadins and glutenins, which can be distinguished by their solubility in the presence of reducing agents.[6,7] Gliadins are monomers while glutenins exist as large, insoluble protein complexes which dissociate in the presence of reducing agents. Both prolamine subclasses exhibit pronounced polymorphism. The gliadins can be further divided into α/β, γ, and ω electrophoretic groups and the glutenins can be resolved into high molecular weight and low molecular weight species. Almost all of the prolamine components of wheat protein can be resolved by two-dimensional gel electrophoresis (Fig. 1). Despite the heterogeneous properties of the wheat prolamines, individual components within the various prolamine subclasses share similar net charges and molecular weights, and can be grouped into clusters on the 2-D polyacrylamide gel.

The primary structures of most of the wheat prolamines have been elucidated by analysis of recombinant DNA clones. Schematic diagrams of the derived primary sequences of the α/β-gliadins,[6-13] γ-gliadins,[14,15] low molecular weight glutenins,[9] and high molecular weight

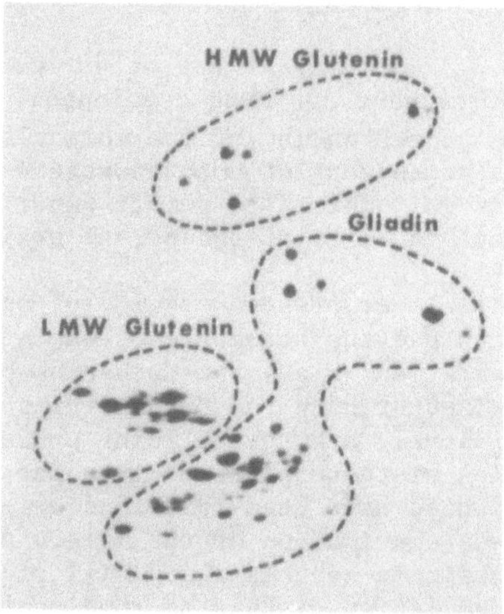

Fig. 1. Two dimensional gel electrophoresis of wheat proteins. A protein extract from the wheat cultivar "Chinese Spring" was resolved by isoelectric focusing in the first dimension followed by SDS polyacrylamide gel electrophoresis. Unpublished data kindly provided by H.P. Tao and D.D. Kasarda.

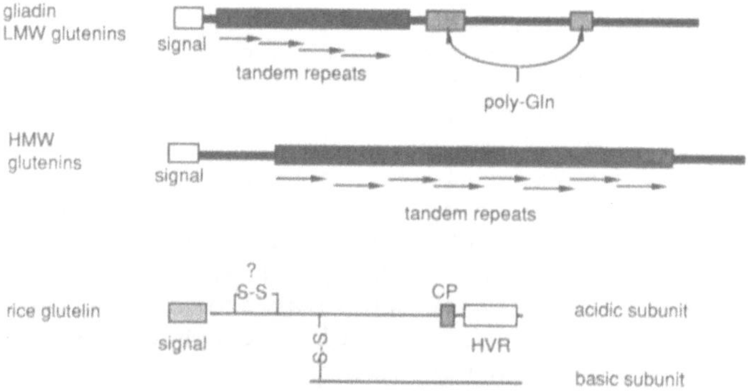

Fig. 2. Schematic representations of wheat and rice storage proteins. Only the signal peptide and regions of repetitive sequences are shown for the wheat proteins. The rice glutelin protein contains a conserved peptide (CP) evident at the N-terminus of the high molecular weight glutenins and a hypervariable region (HVR). For further details of the structural features of these proteins see References 6 and 7.

glutenins[17-19] are illustrated in Figure 2. A unique feature of most prolamine polypeptides is the presence of repetitive sequences at both the nucleotide and amino acid level. The α/β-gliadins, γ-gliadins, and low molecular weight glutenins are similar structurally and possess a N-terminal region displaying a tandem repeat of a conserved peptide. Some variation in the size of the conserved peptide (7-14 residues) is exhibited by these prolamines, but all are composed mainly of 3 amino acids, viz. glutamine, proline and an aromatic residue, phenylalanine and/or tyrosine. Two other repetitive sequences composed of consecutive stretches of polyglutamine are also evident in these proteins. On the C-terminal half of the protein, there are two peptide regions containing unique sequences which exhibit a very biased amino acid composition. Comparison of the derived primary sequences of several α/β-gliadins showed that they diverged mainly by point and DNA segmental mutations.[9] The latter mutations were clustered in the repetitive sequences, suggesting that expansion/contraction of tandem repeats accounts for the heterogeneity in size of the encoded polypeptides. The structure of the high molecular weight glutenins is different from that observed for the other wheat prolamines. Again, a repetitive domain is evident but it encompasses over 70% of the central core of the polypeptide with unique sequences at the N- and C-termini. This domain is composed of two conserved peptides that are intermixed. As recombinant clones for the ω-gliadins have yet to be isolated, their primary sequences are not known. Edman degradation of the N-terminus of several ω-gliadins indicates the presence of a tandem repeat of a small conserved peptide, enriched in glutamine and proline.[20]

Genomic clones for wheat prolamines have been studied by several laboratories. All the genes lack intervening sequences within the coding segment. The 3' flanking region contains two polyadenylation signals, both apparently utilized for poly(A) tailing of the mRNA transcript. The 5' flanking region contains a typical plant TATA box and a less conserved CCAAT motif. Kreis et al.[6] identified a conserved sequence located about -300 bp from the translational start of prolamine genes from maize, barley and wheat (Table 1). A second conserved sequence, the CACA box, is also evident in many cereal prolamine genes at about 175 bp upstream from the translational start.[13]

Table 1. Conservation of promoter elements in cereal seed genes.

A.

-300 bp conserved element	TG$\overset{T}{\underset{C}{A}}AAA^{GG}_{AT}$	Ref. 6, 21

B.

Wheat / -Gliadin	-138	GTGCCACCAAACACAAC	-122	Ref. 11-13
Wheat -Gliadin	-138	GTGCCACCAAACACAAC	-122	Ref. 16
Barley Hordein	-138	GTGCAACCTAACACAAT	-122	Ref. 74
Maize Zein (upstream promoter)	-133	TTGTCACC-AACACAAA	-118	Ref. 75
Consensus Sequence		GTGCCACC$^{A}_{T}$AACACAA$\overset{C}{\underset{A}{T}}$		

The consensus -300 bp element present in the promoter regions of cereal seed protein genes is shown in part A. A second conserved region between -140 bp to -120 from the transcriptional start is shown from several cereal storage protein genes in part B.

EXPRESSION OF WHEAT PROLAMINE GENES

Kreis *et al.*[6] suggested that the ubiquitous -300 bp motif, conserved in many cereal prolamine genes, may be involved in transcriptional control. The ability to transfer genes into several dicotyledonous hosts by *Agrobacterium*-mediated transformation presents the opportunity to identify putative regulatory sequences of cereal prolamine genes. Although differences in transcriptional control between dicot and monocot genes may exist, regulatory sequences of plant genes may be sufficiently conserved to permit the expression in heterologous host systems. Colot *et al.*[21] fused the 5'-flanking sequence of both the high and low molecular weight glutenin genes to chloramphenicol acetyltransferase (CAT) and transferred them into tobacco. Analysis for CAT activity in tissues of transgenic plants showed that both glutenin promoters were able to direct the expression of CAT in a faithful tissue- and developmental stage-specific manner. CAT activity was first detected in 14-17-day old seeds and increased steadily during development. Moreover, CAT expression was restricted to endosperm tissue; no activity was detected in tobacco seed embryos. In the 5'-end deletion analysis of the low molecular weight glutenin promoter, CAT activity was abolished when DNA between -326 bp to -160 bp from the transcriptional start was removed. This region of the low molecular weight glutenin promoter, -326 bp to -160 bp, contains two sequences of the conserved -300 bp element, suggesting its possible role in developmental expression of the glutenin genes.

Similar studies have been conducted with the gliadin promoters. Unlike results using the wheat glutenin promoters, detectable CAT expression was not directed by the gliadin 5'-flanking sequences. Detectable levels of CAT mRNA, however, were observed in tobacco seed tissues but not in leaves (Leisy and Okita, unpublished; Rafalski, personal communication). The absence of CAT activity, though this gliadin promoter-CAT construct appeared to be transcribed, may suggest that the turnover rates of the CAT enzyme were significant compared to the accumulation of its transcripts. These observations are consistent with the gliadin promoter, being functional but transcriptionally weak as compared to the promoter of the glutenins, at least in transgenic tobacco hosts.

The promoter activity of the gliadin gene was also assessed using transient expression assays in electroporated tobacco cells. The gliadin promoter alone was unable to direct detectable CAT gene expression in tobacco protoplasts but did so in rice cells. This observation indicated that one or more transcriptional and/or translational regulatory elements were not highly conserved in this monocot gene. To overcome this, the gliadin promoter was fused upstream to both a functional and nonfunctional nopaline synthase (Nos) promoter-CAT reporter cartridge (Fig. 3). The functional Nos promoter

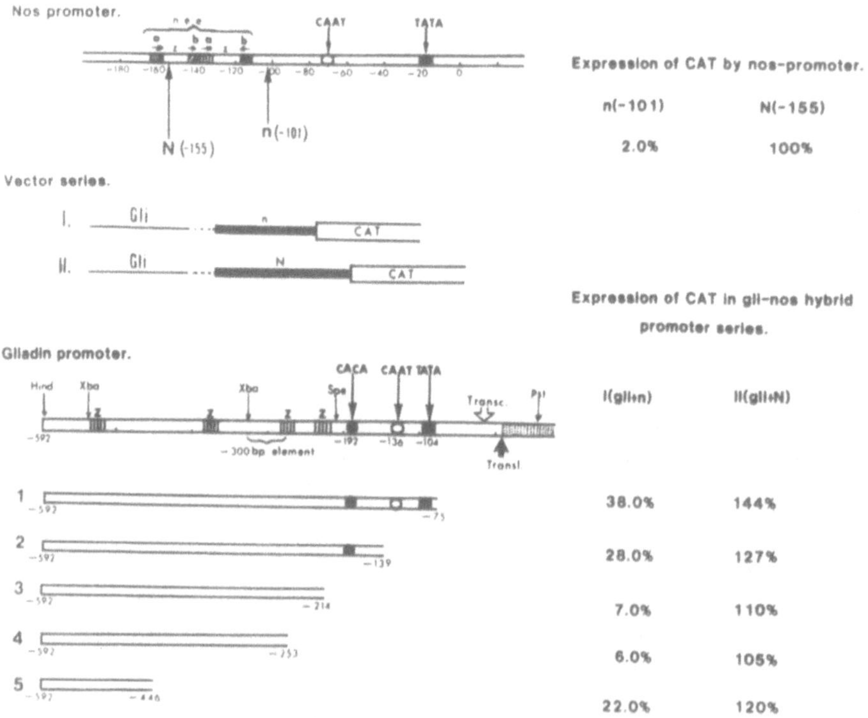

Fig. 3. Transient expression of gliadin-Nos fusion constructs in tobacco protoplasts. Gliadin promoter fragments (3' deletions) were ligated upstream to a non-functional Nos-promoter (-101, designated as "n") or a functional Nos promoter (-155, designated as "N" as described by An *et al.*[22]) which in turn were fused to the CAT reporter gene to obtain series I and series II constructs. These DNA constructs were electroporatically introduced into tobacco protoplasts and the transient expression of CAT measured after 40 h. n.e.e. = Nos enhancing elements.

contains the conserved TATA and CCAAT sequences as well as the activator sequences described by An *et al.*[22] while the inactivated Nos promoter lacks the upstream sequences essential for transcription. When these hybrid promoter constructs were introduced into tobacco protoplasts by electroporation, CAT activity was detected in both instances (Fig. 3). DNA sequences of the gliadin promoter can apparently substitute for the activating sequences of the Nos promoter and can restore the promoter function up to 40% of wild-type levels. Furthermore, the gliadin promoter sequences enhanced the functional Nos promoter when placed upstream. The level of enhancement of the functional Nos promoter by the gliadin promoter sequences was proportional to that detected for restoration of inactivated Nos promoter (Fig. 3).

To identify segments of the gliadin promoter that may serve as activator sequences for transcription, various DNA portions were successively removed at the 3' end and fused to the two Nos-CAT DNAs. When transferred into tobacco protoplasts, a decline of approximately 10% in promoter function was evident when the TATA and CCAAT boxes from the gliadin promoter were deleted. Restoration of the inactivated Nos promoter was almost completely abolished when DNA from -139 bp to -214 bp was removed. This region of the gliadin promoter contains the CACA box, and the severe reduction in Nos promoter activity suggests that this motif and/or adjacent sequences are critical for transcription. Surprisingly, removal of sequences from -75 bp to -446 bp caused an increase in Nos promoter activity. These results indicate that a second upstream sequence of the gliadin promoter can serve as an activator of the Nos promoter.

Results of expression of the wheat prolamine gene in stably integrated transgenic tobacco plants and transiently in protoplast cells suggest that there are at least two regulatory elements responsible for transcription of these genes. The presence of these regulatory sequences provides the recognition signal for binding of specific nuclear factors required for transcription.[23] Preliminary studies from several laboratories have identified such transacting factors. Maier _et al._[24] used nitrocellulose binding assays to identify a nuclear factor from developing maize endosperm that bound specifically to a DNA promoter fragment of a zein gene. The binding activity complexed at or near the -300 bp conserved element and was present only in developing seed tissue and not in leaves. Tissue specificity and sequence specific binding of this nuclear factor strongly suggested some role in transcriptional control of these genes.

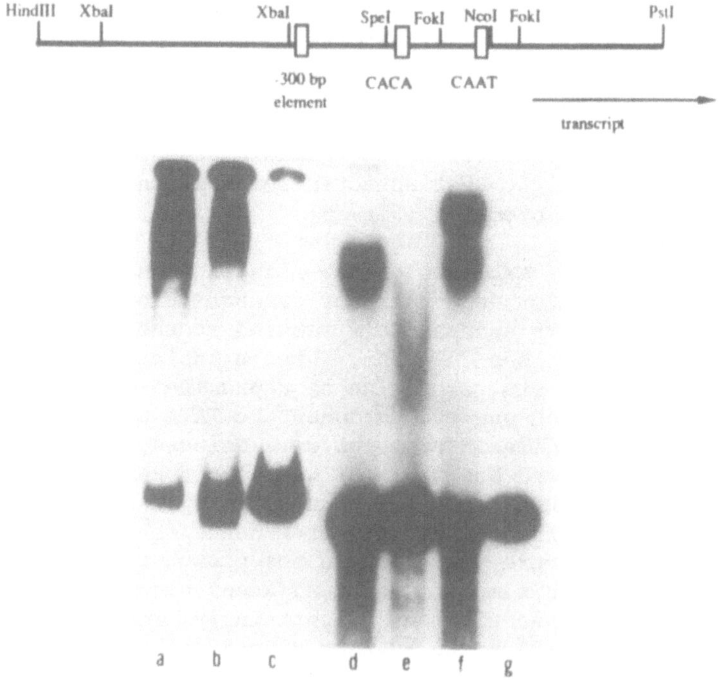

Fig. 4. Gel retardation of gliadin promoter fragments. The SpeI-NcoI (lanes a-c) and XbaI-FokI (lanes d-f) were labeled with ^{32}P, incubated with a wheat endosperm nuclear extract, and resolved by polyacrylamide gel electrophoresis. Lanes b and c depict binding reactions incubated in the presence of 25 ng of unlabeled XbaI-SpeI and SpeI-NcoI fragments, respectively. Lane e depicts a binding reaction in the presence of unlabeled XbaI-FokI, while lane g shows the free mobility of labeled XbaI-FokI.

Similar studies from our laboratory suggest a second regulatory site which mediates the binding of specific nuclear factor(s). Identification of nuclear binding activity was assessed by gel retardation (or electrophoretic mobility shift) assays. In this approach, nuclear protein extracts were incubated with a radiolabeled DNA fragment and then resolved by polyacrylamide gel electrophoresis. Specific protein binding to the DNA fragment results in retarded electrophoretic mobility of the protein-DNA complex which can be easily detected by comparison with the mobility of the free DNA fragment. Two DNA fragments, a 295 bp XbaI-XbaI segment and a 315 bp XbaI-NcoI segment containing the -300 bp element, CACA motif, and the CCAAT box were employed in the gel retardation assays (Fig. 4). Specific binding was evident for only the XbaI-NcoI fragment with nuclear extracts from developing wheat endosperm (but not coleoptile tissue). To reduce the number of possible sites of protein interaction within the gliadin promoter, smaller DNA fragments contained within the Xba-NcoI fragment were employed.

The results indicated strong binding activity exhibited by DNA sequences situated between the SpeI and FokI restriction enzyme sites of the gliadin promoter (Fig. 4). This binding activity was specific as binding to radiolabeled Spe-NcoI fragment was "chased" by the addition of unlabeled SpeI-NcoI DNA but not by the upstream fragment, XbaI-SpeI. This DNA fragment contains the CACA sequence motif. In contrast, no reproducible binding activity was evident for a XbaI-SpeI fragment which contains the -300 bp element. When binding activity was detected for the XbaI-SpeI fragment, the protein-DNA complex was not chased by addition of unlabeled XbaI-SpeI, but such binding activity was eliminated by the addition of unlabeled SpeI-NcoI. Several imperfect inverted repeats of the CACA motif are evident in the XbaI-SpeI region and may be responsible for gel retardation of this DNA segment.

Based on plant transformation and biochemical studies, two important DNA sequences which may be required for transcription of the genes for cereal storage protein have been identified. In view of the complexity of the promoters of animal genes,[23] it is likely that several other cis-acting regulatory sequences await to be identified. Some may be activator sequences,[22] a function which is suggested by the capacity of the CACA motif or nearby sequences to restore the promoter function of an inactivated Nos promoter, while others may be required specifically for tissue and developmental gene expression.[25] This is an exciting phase of research and the continuing activity of several laboratories will surely lead to a much better understanding of the regulation of cereal storage protein genes.

STRUCTURE OF THE RICE STORAGE PROTEINS AND THEIR GENES

Rice is atypical in that the major storage protein fraction accumulated in endosperm is the insoluble glutelin. The glutelin fraction composes about 80% of the total endosperm protein. Rice endosperm also stores a prolamine, however, these proteins are only a minor class (5%). Studies from several laboratories have shown that despite the insoluble nature of the glutelin, these proteins share many properties with the 11S globulins which typically accumulate in legumes.[7]

Rice glutelin is initially synthesized as a prepro-protein containing a signal peptide sequence, which mediates binding to the endoplasmic reticulum (ER) and transfer into the lumen.[26,27] The processed proprotein is then post-translationally cleaved into acidic and basic subunits.[26,28] A partial amino acid sequence of the basic subunit showed it to be homologous to the basic subunit of the pea 11S storage protein.[29] The N-terminus of the acidic subunit was blocked to Edman degradation and a direct relationship between the acidic subunits of rice and legumes could not be addressed.

Analysis of amino acid composition indicated that the basic subunit of rice glutelin was related to comparable subunits of 11S storage proteins but the acidic subunit was not.[30] Immunological studies, however, showed that antibodies raised against the rice acidic subunit cross-reacted with the acidic subunit of pea, indicating a direct homology between rice glutelins and legume 11S globulins.[31] The structural homology among these proteins has been verified by analysis of the derived primary sequences obtained from recombinant DNA clones. The primary sequence of the rice glutelin protein is about 30% identical to several 11S storage proteins in legumes.[32,33] Strong conservation to the legume 11S proteins is displayed by the signal peptide sequence, the putative post-translational proteolytic cleavage site, as well as several other peptide regions. Of the seven cysteine residues, four are conserved and are positioned at similar locations within the primary sequences of these proteins (Fig. 2). Two of these cysteine residues are known to be responsible for the interchain linkage between acidic and basic subunits of legume 11S proteins.[34] Comparable residues in rice probably have the same role. Conservation of the other pair of cysteine residues suggests that they may be involved in intrachain linkage. Nonconserved regions are found not only among different glutelin polypeptides but also between other 11S storage proteins. Rice glutelin has a divergent peptide domain which corresponds to the hypervariable region of legume 11S globulins[35] (Fig. 2). This hypervariable region tolerates large insertions of variable size suggesting that it is a likely target for genetic manipulation of these proteins.[35]

In contrast to the glutelins, very little information is known about the prolamine fraction of rice endosperm. Biochemical studies have shown that it possesses a relatively small molecular size between 14,000-17,000 daltons. The protein is rich in glutamine, leucine, and tyrosine residues and, as in other cereal prolamines, lacks lysine and sulfur containing amino acids.

Two classes of recombinant DNA clones for prolamine sequences have been isolated.[36,37] Both prolamine classes encode a polypeptide with a signal peptide of 14 residues. One class of prolamines displays a composition similar to that obtained for a 70% alcohol-soluble protein fraction. The class II prolamines are distinguished by high mole percentages of cysteine (5.9%) and methionine (3.0%). It seems likely that, due to the number of cysteine residues, the class II prolamines will be insoluble in alcohol solutions, which accounts for the presence of a 14 kD prolamine in the glutelin fraction.[27] mRNA Transcripts of class II prolamines appear to be present at 3-4 fold levels greater than class I transcripts, suggesting that the level of prolamines present in seed tissue has been underestimated by the Osborne serial extraction analysis. Unlike those from other cereals,[6,7] the rice prolamines are devoid of repetitive sequences at both the nucleic acid and amino acid levels. Direct comparison of the rice prolamine primary sequence to other cereal prolamines indicates no homology among these proteins and suggests that the rice prolamine gene evolved independently of those from wheat, barley and rye.[6,37]

A number of genomic clones have been isolated for the rice storage proteins. Takaiwa _et al_.[38] presented evidence that the glutelin gene contains three intervening sequences interspersed within the coding segments and that there are 3-4 copies of this gene in the rice genome. Results from our laboratory indicate that the organization of glutelin genes on the rice chromosomes is much more complex (Okita et al., submitted). Glutelin genes representative of three subfamilies, designated Gt1, Gt2, and Gt3, have been isolated. All glutelin genes contain three introns which interrupt the coding sequence at identical positions of the coding segments (Fig. 5). The positions of these introns are identical to those observed in the pea legumin 11S storage protein gene,[39] further supporting the hypothesis that these genes were derived from common ancestral DNA segments. Each subfamily contains about 5-8 gene copies as determined from reconstruction Southern blot analysis of rice DNA. The 11S storage proteins from soybean and pea are encoded by a small multigene family consisting of 5 and 8 copies,[7,39] respectively. The relative constant gene copy number of each glutelin subfamily suggests that the genes encoding the glutelins protein already existed as a multigene family before the duplication events that gave rise to the three gene subfamilies.

Structural studies showed that Gt1 and Gt2 are about 95% homologous in the coding sequences while Gt3 shares only about 83% identity with Gt1 and Gt2. The close relationship between Gt1 and Gt2 sequences is also evident in the 5' and 3' flanking regions and introns. Inspection of the 5' flanking sequences reveals that all genes are relatively conserved proximal to the translational start. DNA sequences located -125 bp to -1 bp from the translational start show 12% and 23% divergence for Gt1/Gt2 and Gt1/Gt3 comparisons, respectively. This region contains conserved sequences for the transcriptional initiation site, TATA and CCAAT boxes. Sequences upstream to -900 bp are well conserved in Gt1 and Gt2. The only significant differences between the promoter sequences of these genes are the presence of several insertions/deletions of DNA segments. Two of these DNA insertions are bordered by inverted repeats reminiscent of transposable elements, suggesting their involvement in the evolution of the glutelin genes. Promoter sequences of Gt3, other than the above mentioned conserved elements, diverge significantly from Gt1 and Gt2, and sequences upstream from -267 bp shared no homology (Fig. 5). The significant sequence divergence exhibited by the

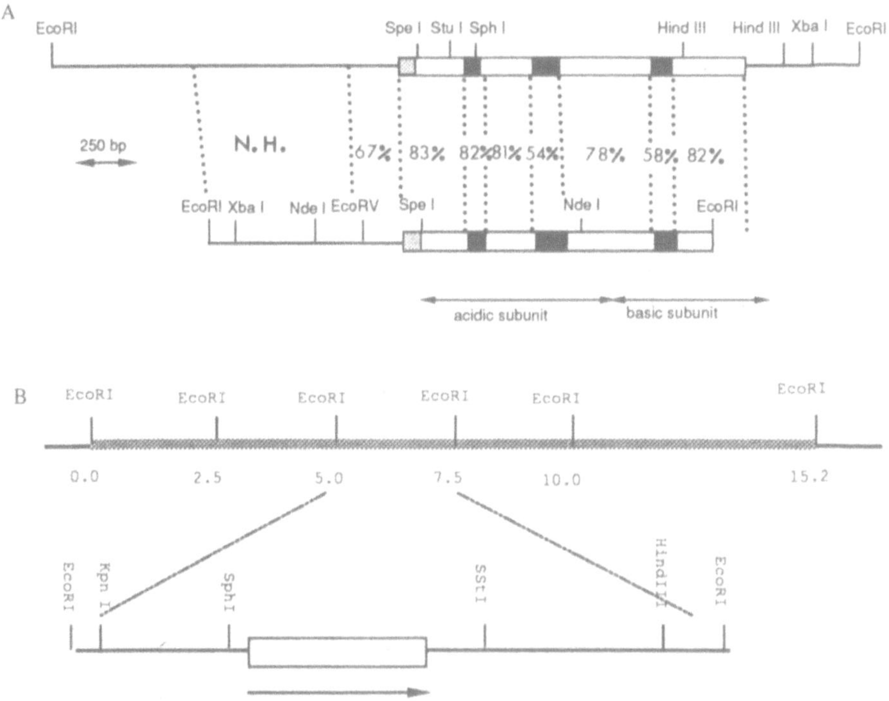

Fig. 5. Physical restriction maps of rice storage protein genes. Panel A shows the gene structures of Gt1 (top) and Gt3 (bottom). The coding segment is shown in open rectangles. The signal peptide is depicted as a dotted box while the introns are darkly shaded. The degree of homology between Gt1 and Gt3 are shown for selected regions of these genes. Panel B shows the physical map of λpro14a. An enlarged map of the tandemly repeated 2.5 kb EcoRI fragment of λpro14a is depicted. The open rectangle represents the coding region and the arrow indicates the direction of transcription.

promoter region of Gt3 upstream of the CCAAT box would suggest that genes of the Gt3 subfamily may not be coordinately regulated with other glutelin genes. This has been supported by Northern blot data which show that Gt3 mRNA transcripts accumulate to a maximum between 5 to 10 days after flowering and decline to low levels at subsequent periods of endosperm development, whereas, transcripts of Gtl increase steadily and maintain high levels at seed maturation (Okita _et al._, submitted).

The structure of the rice prolamine genes is similar to other cereal prolamine genes. As evident for wheat prolamines, the rice gene lacks introns and forms a large multigene family consisting of 75-100 copies.[37] One interesting feature of the rice prolamines is that they appear to be tightly clustered on the rice chromosomes. The genomic clone, λprol4a, contains a 2.5 kb EcoRI fragment which is duplicated in tandem four times. Each of these 2.5 kb EcoRI fragments contains a putative prolamine gene (Fig. 5). The tight clustering of the prolamine genes may reflect the relatively small genome size of rice (0.6 pg) as compared to barley or diploid wheat (4.0 pg).

POST-TRANSCRIPTIONAL CONTROL OF STORAGE PROTEIN BIOSYNTHESIS

In addition to control of biosynthesis of storage protein at the transcriptional level, there is some evidence that additional controls modulate the synthesis of these proteins during endosperm development. In wheat, gliadin mRNA transcripts are first evident at 3 days post-anthesis, but their polypeptides are not detected until about 6 days later.[40] This differential accumulation of gliadin mRNA transcripts and polypeptides contrasts with the coordinate accumulation patterns evident for ADPglucose pyrophosphorylase,[40] a key enzyme of starch biosynthesis. More substantial evidence that post-transcriptional controls may be operational in modulating seed protein synthesis is found for the rice storage proteins. Northern blot analysis of RNA isolated from 15-day old developing seeds showed that the abundance levels of glutelin and prolamine transcripts were about the same (Kim and Okita, unpublished). These abundance levels of the mRNAs, however, were inconsistent with the overall accumulation levels of these storage proteins in rice endosperm where glutelins dominate over prolamines by 4-10 fold.

To clarify this apparent disparity between mRNA transcript and protein levels, the developmental program of accumulation of these mRNA transcripts was investigated

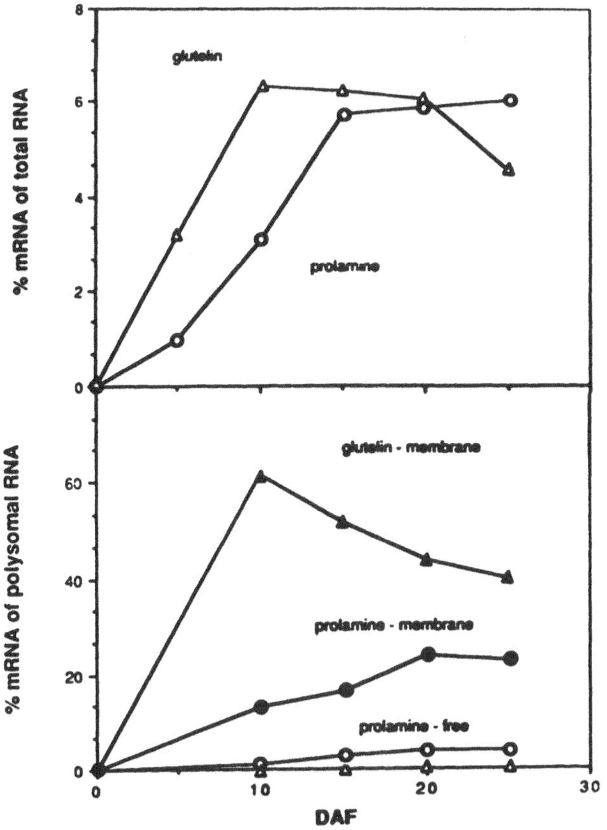

Fig. 6. The accumulation of rice storage protein mRNAs (top panel) and their recruitment into membrane and free polysomes (bottom panel) during endosperm development.

by blotting techniques and densitometry (Fig. 6). Glutelin mRNAs reach a maximum level at around 10 days post-anthesis while prolamine transcripts do not obtain the same abundance levels until 15 days post-anthesis. Glutelin mRNA transcripts are accumulated at an earlier stage of seed development than those for prolamine. This differential activation and expression of the storage protein genes therefore provides a partial explanation for the differences in the levels of these storage proteins in rice seeds. Since temporal differences in the accumulation of these mRNA transcripts alone cannot completely explain the disparate levels of their proteins, the levels of these transcripts in polysomal RNA fractions were assessed. Unlike the analysis of total RNA fractions, which only describes the translational potential of these transcripts, assessment of these species in polysome fractions should reveal their actual synthetic capabilities.

Such analysis shows that the glutelin mRNAs were readily recruited into membrane bound polysomes, consistent with their site of synthesis, whereas very little was present in the free polysome fraction. There appeared to be little, if any, temporal lag in the accumulation levels of the glutelin mRNA transcripts and their engagement in protein synthesis. In contrast, prolamine mRNA transcripts were not as efficiently recruited into membrane polysomes and a significant fraction was associated with free polysomes. Assuming that mRNA species constitute about 1% of the total RNA or polysomal RNA fractions, almost all of the glutelin transcripts were associated with membrane polysomes while only about 40% of the prolamine mRNAs were engaged in translation. Therefore, rice seeds accumulate glutelin as their major storage protein because their mRNA transcripts are accumulated at greater quantities particularly during the earlier stages of endosperm development and are more efficiently recruited into membrane polysomes as compared to the prolamine transcripts.

Although translational regulation appears to be operative in modulating rice storage protein synthesis, the molecular basis of this control is not known. The synthesis of globulins constitutes about 80% of total protein synthesis _in vivo_ in oat endosperm even though its transcripts comprise only about 30% of the total polysomal RNA.[41] Fabijanski and Altosaar[41] have observed enhanced globulin synthesis when nuclease-treated polysomes were added to _in vitro_ translation assays of oat poly(A+)-RNA. These workers suggested that protein factors associated with oat polysomes were responsible for the selective translation of oat storage proteins. If specific proteins are required for selective storage protein synthesis in rice and oat, then the control step lies either at initiation or at some other early event of translation.

THE CELLULAR PATHWAYS OF PROTEIN BODY FORMATION

The storage proteins in plants are packaged into discrete membrane-delimited organelles called protein bodies. All storage proteins are initially synthesized with a transient signal peptide which mediates binding of the nascent chain to the ER. Vectorial transport and proteolytic processing of the signal sequence results in internalization of the mature storage protein within the rough ER lumen.

With regard to their eventual packaging into protein bodies, these storage proteins can undergo two fates. The two different cellular processes in protein body formation are illustrated for rice in Figure 7. Ultrastructural analysis of rice endosperm revealed at least two different types of protein bodies, distinguishable both by morphology and staining. One type of protein body assumed a spherical shape and was electron lucent while a second type displayed an irregular shape and stained very densely (Fig. 7). Since rice endosperm accumulates glutelins (globulins) and prolamines, the questions raised were whether these proteins were packaged separately into distinct organelles or were intermixed within the same, and what were the cellular processes that mediate the formation of morphological distinct protein bodies in rice endosperm.

Using immunocytochemistry, where antigen-antibody complexes are visualized by gold particles conjugated with protein A, Krishnan _et al._[42] demonstrated that the rice glutelins and prolamines were packaged into separate organelles. Protein contents contained within the electron-lucent, spherical protein bodies reacted with the prolamine antibodies while electron-dense, irregular shaped protein bodies were labeled with the gold particles when glutelin antibodies were employed (Fig. 7). Very little cross-labeling was evident, indicating that these organelles preferentially contained only one type of storage protein. The surrounding membrane of the spherical protein bodies was studded with ribosome-like particles and in many instances a direct connection to the rough ER was observed (Fig. 7B). These observations indicate that rice prolamine is packaged by the process first described for the maize prolamine, zein, where these proteins aggregate directly within the rough ER lumen.[43]

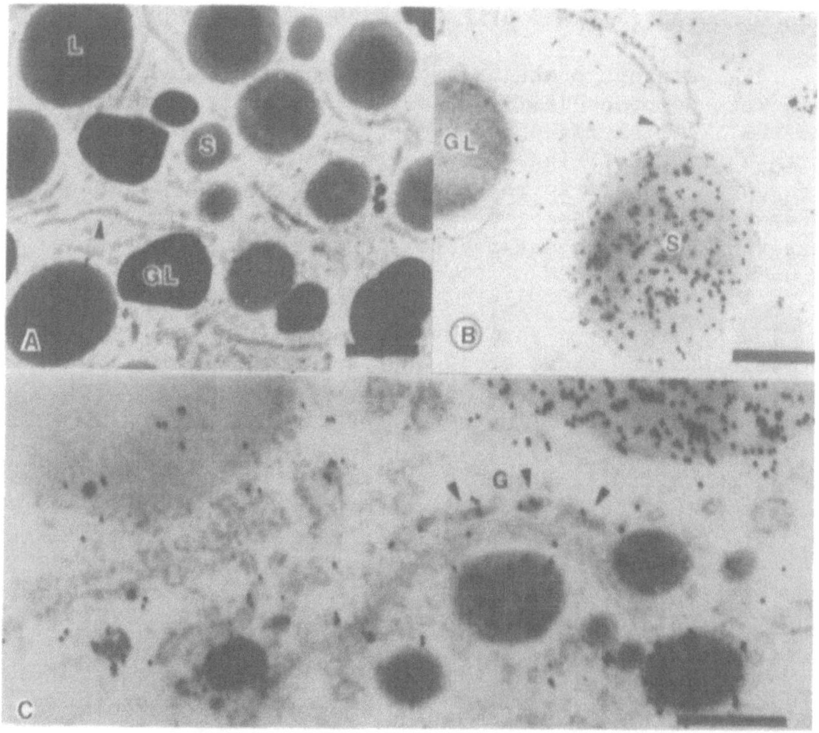

Fig. 7. Protein body formation in rice endosperm. Panel A is a low magnification view of 14-day old endosperm and shows the presence of three types of protein bodies. Extensive rough ER is routinely observed (arrowhead). L and S, large and small spherical prolamine protein bodies, respectively; GL, irregular shaped glutelin protein body. Panel B shows the anti-prolamine mediated, immunogold labeling of spherical protein body (note the direct connection to rough ER). Panel C depicts gold particle labeling of Golgi-derived vesicles after exposure to glutelin antibodies. Bar = 1 μm.

In contrast, several lines of evidence indicate that the rice glutelin is packaged by a different mechanism. The irregular-shaped protein bodies which are highly enriched for glutelin are surrounded by smooth membranes devoid of ribosomes. At higher magnification, small electron-dense particles, labeled with protein A conjugate, were commonly found in the vicinity of the Golgi complex, strongly suggesting a role for this organelle in sorting of glutelins to protein bodies (Fig. 7). Yamagata and Tanaka[44] have suggested that the irregular shaped protein bodies are formed by the discharge of protein from the dictyosomal vesicles into vacuoles. A vacuolar site of glutelin deposition was observed in subaleurone cells (the outer 2-3 layers of endosperm cells adjacent to the aleurone). Vacuoles, however, were not evident in the bulky parenchymous cells of the endosperm, indicating an alternative pathway for formation and maturation of glutelin protein bodies. Staining of endosperm sections with phosphotungstic acid, considered a selective stain for polysaccharides of the plasma membrane, revealed an elaborate infrastructure of the irregular shaped protein bodies. Distinct protein inclusions of various sizes separated by electron-dense material were evident. The distinct structural features of this organelle suggested to these workers that irregular shaped protein bodies enlarge by fusion with Golgi-derived vesicles or with smaller protein bodies.

Seed storage proteins are packaged by two different mechanisms, and rice is of notable distinction as it displays both mechanisms of protein body formation. Although rice glutelin is sorted via the Golgi complex, as demonstrated for the legume globulins, there are differences in the initiation and maturation of protein body formation between legumes and cereals. In legume cotyledons, storage proteins are deposited within the central vacuole. Subdivision of the vacuole accompanies protein deposition, resulting in the formation of numerous small protein bodies at seed maturation. Such large vacuoles are not evident in rice endosperm except in the subaleurone cells and, as explained above, these protein bodies appear to initiate and mature by fusion events. These deviations in protein body formation in legumes and cereals probably reflect the inherent differences by which these proteins are ultimately utilized. Legume protein bodies can be viewed as modified vacuoles which serve as a temporary storage organelle during seed formation and as a hydrolytic compartment during seedling growth.[45] The accumulation and subsequent hydrolysis of proteins must be carefully controlled, and storage in the vacuole provides this function. Compartmentalization of storage proteins in the vacuole of endosperm cells of cereals is probably unnecessary since it is a terminally differentiated tissue, and hydrolysis of these proteins occurs extracellularly during seedling growth and development.

The prolamines of rice, maize and sorghum are packaged into distended rough ER.[42,43,46] The packaging of the wheat prolamines has been under investigation in several laboratories but a consensus on the formation of protein body has yet to be reached. Campbell et al.[47] reported that protein bodies arose directly by enlargement of the rough ER lumen. Miflin et al.[48] favored the rough ER lumen as the site of protein deposition in wheat and barley, with the added modification that these protein aggregates were released into the cytoplasm. Parker's and Bechtel's laboratories, however, presented evidence that wheat prolamines were packaged into protein bodies via the Golgi apparatus,[48,50] Bechtel and Barnett[50] observed no direct connections between the protein bodies and rough ER. The analysis of protein body formation in these cereals, particularly wheat, has been complicated by the fact that a large protein matrix is eventually formed which is interposed with starch and lipid bodies.[51] In wheat one or few large protein matrices fills the cell's volume, unlike the cellular processes prevalent in other plants where multiple protein bodies are formed.

Protein body formation was reinvestigated by Kim et al.[52] who employed an immunocytochemical study of wheat endosperm sections. As early as 9 days after flowering, wheat endosperm cells contained abundant starch granules and numerous protein bodies up to 20 μm in diameter. Consistent with Bechtel's earlier study,[50] these workers could not demonstrate any connections between the rough ER and protein bodies, although numerous membranous structures were associated with the larger protein bodies. These larger protein bodies contained one or more spherical electron-opaque inclusions which were unreactive to anti-gliadin (Fig. 8A) or antibodies that cross-reacted with the wheat glutenins.[52] The Golgi apparatus was readily seen in endosperm tissue as early as 5 days after flowering. This organelle consisted of two or three loosely stacked cisternae layers. In 9-day old developing seeds, electron-dense vesicles were associated with the Golgi apparatus, and in many instances these vesicles were

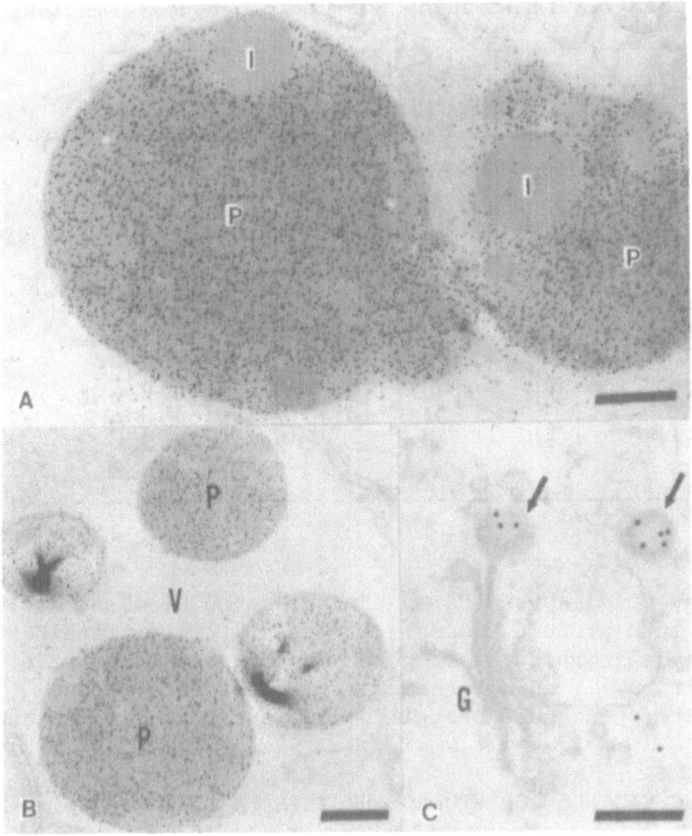

Fig. 8. Protein body formation in wheat endosperm. Panel A shows the immunogold labeling patterns of protein body after incubation with gliadin antibody. The electron-opaque inclusions are unreactive to the antibody (Bar = 2 m). Panel B shows four protein inclusions labeled with gold particles contained within a vacuole-like body (Bar = 1 m). Panel C demonstrates gold labeling of Golgi-derived vesicles (Bar = 0.5 m).

directly connected to the distal ends of the Golgi cisternae. The contents of these electron-dense vesicles were specifically labeled with gold particles when the tissue section was preincubated with gliadin antibodies (Fig. 8). Overall, these results support a specific role

Table 2. The pathways of protein body formation.

Protein	Golgi Involved?	Site of Deposition	Ref.
Globulin			
11S and 7S	Yes	Vacuole	45
Rice glutelin	Yes	Vacuole-like	42,44
Prolamine			
Maize	No	RER lumen	43
Rice	No	RER lumen	42,44
Sorghum	No	RER lumen	46
Wheat	Yes	Vacuole-like	52
Barley	Yes	Vacuole-like	--
Rye	Yes	Vacuole-like	--

for the Golgi apparatus in the packaging of wheat prolamines into protein bodies. Although it is difficult to totally eliminate the rough ER as the site of protein deposition, it is not the major mechanism by which discrete, condensed protein bodies are formed in wheat.

SORTING SIGNALS FOR STORAGE PROTEIN TARGETING

Table 2 summarizes the pathways for protein body formation in several plant species. Most of the storage proteins including the prolamines of wheat are packaged into protein bodies via the Golgi apparatus. In view of the close structural homology with wheat prolamines,[6] it is likely that storage proteins of barley and rye are also transported to protein bodies via the Golgi apparatus. The retention of storage proteins in distended rough ER appears to be confined to the prolamines of maize, sorghum, and rice.

The accumulation of these proteins within the rough ER lumen may be the result of selective retention signals displayed by the primary sequences of these protein. Munro and Pelham[53] have shown that several ER-localized proteins contain a conserved peptide, LysAspGluLeu, at their C-terminus. A comparison of the primary sequences of the 19 kD zein of maize[54] and rice prolamines[37] reveals no significant homology except at the carboxyl terminus as shown below.

rice	Thr	Leu	Gly	Gly	Val	Leu	---
maize	Ile Thr	Ile	Gly	Gly	Ala	Leu	Phe

Three to four of the six residues are identical at the C-terminus while the remaining are conserved changes that would have a minimal effect on protein conformation. This hydrophobic hexapeptide differs considerably from the highly charged tetrapeptide conserved in ER-lumen proteins of animal cells.[53] Whether this conservation displayed by the maize and rice prolamines has any biological significance remains to be elucidated. Alternatively, protein engineering experiments in animal cells showed that alterations which might affect the physico-chemical properties of the proteins can cause their normal transport to be arrested in the ER. The ER lumen is believed to be an environment of high ionic strength[55] and the retention of the prolamines in this compartment may simply be due to their insoluble character.

Hoffman et al.[56] have recently shown that a maize prolamine, a 15 kD zein, was synthesized and accumulated in transgenic tobacco. Surprisingly these workers observed this maize protein sequestered in the central crystalloid region of the protein body indicating that it was processed via the Golgi apparatus. The mis-localization of zein in transgenic tobacco would indicate that newly synthesized polypeptides maintain sufficient solubility in the ER lumen to be transported to a vacuolar compartment. It is unclear, however, whether the transport of zein was due to lack of an ER retention systen in tobacco, the capacity of the dicot targeting and transport mechanism to recognize peptide signals not used in maize endosperm, or simply the result of unusal saturation of the sorting machinery, leading to anomalous protein targeting.

The sorting of storage proteins to a vacuolar compartment is also believed to depend on specific molecular signals displayed by the primary sequence of these proteins.

Tague and Chrispeel[57] have recently demonstrated that bean seed phytohemagglutinin is correctly glycosylated and transported to the vacuole in yeast. Moreover, these workers have recently shown that only the N-terminal region of phytohemagglutinin is sufficient for protein targeting to the vacuole, indicating that the signal and sorting mechanism has been conserved between plants and yeast (Chrispeel, personal communication). Such signals and sorting mechanisms are also believed to be operational in wheat and barley endosperm cells where their prolamines are packaged into vacuole-like compartments. Although the biochemical basis for protein signalling is not yet known, a comparison of the primary sequences of these proteins may provide some interesting clues.

Kreis et al.[6] detected three localized conserved regions shared by the primary sequence of several cereal prolamines and dicot globulins suggesting that storage protein genes from both monocot and dicot plants evolved from common ancestral DNA segments. Such conserved sequences were not evident either in maize zein or in the derived primary sequences of several rice prolamine cDNAs (Okita, unpublished). Moreover, an antibody raised against the acidic subunit of the rice glutelin was found to cross-react with the high and low molecular weight glutenins of wheat.[31] A detailed comparison of the known primary sequences of the high molecular weight glutenin of wheat and rice glutelin showed a small conserved peptide with the sequence, ArgGlnLeuGlnCys, shared by both proteins. This small conserved peptide is part of the "A" domain shared by many seed proteins as discussed by Kreis et al.[6]

Although the basis for the conservation of this peptide is unknown, it is not likely to be involved in protein folding or maintenance of protein conformation as the locations of this conserved peptide are different. In the high molecular weight glutenins it lies near the N-terminus of the mature protein while it resides 54 residues from the C-terminus of the acidic subunit of rice glutelin (Fig. 2). Moreover, the high molecular weight glutenin is believed to exist predominantly as a β-coil (spring-like) structure[58] while, based on analogy to the structure of the legume 11S storage proteins, the rice glutelin is assumed to be globular with the hydrophilic acidic subunit folded around the hydrophobic basic subunit.[35]

One feature that is shared by these structurally dissimilar proteins is that they are sorted via the Golgi apparatus to their respective protein bodies.[45,52] Although purely speculative, the possible significance for the conservation of these peptides is that they may provide a signalling function or some other role in the intracellular transport of these proteins into protein bodies.

BIOCHEMICAL AND MOLECULAR ASPECTS OF STARCH BIOSYNTHESIS

The biosynthesis of starch is controlled by the activity of ADPglucose pyrophosphorylase.[59] The importance of this enzyme in starch biosynthesis is illustrated in the starch-deficient mutants of maize endosperm,[60] *shrunken*-2 and *brittle*-2. These maize mutants exhibit less than 10% of wild-type enzyme activity and accumulate only about 25% of the starch detected in normal endosperm. A starchless mutant of *Arabidopsis thaliana* has been isolated and found to lack ADPglucose pyrophosphorylase activity.[61] Levels of enzyme activity for UDPglucose pyrophosphorylase and starch phosphorylase were unaffected by the genetic lesion, indicating that these enzymes do not play a significant role in starch synthesis.[61]

ADPglucose pyrophosphorylase is subject to allosteric control by small effector molecules.[59] Synthesis of ADPglucose *in vitro* is activated by 3-P-glycerate (3-PGA) and inhibited by orthophosphate (Pi). Ample evidence has been obtained *in vivo* for a role of these effectors in modulating starch synthesis. Heldt *et al.*[62] showed that the rate of CO_2 fixation into starch corresponded directly with the ratio of 3-PGA:Pi in the stroma and, more importantly, agreed well with the *in vitro* catalytic behavior of leaf specific ADPglucose pyrophosphorylase at the different 3-PGA:Pi levels. Diurnal fluctuations of these effector molecules therefore modulate the ADPglucose pyrophosphorylase activity which in turn dictates the flow of carbon into leaf starch.

In nonphotosynthetic sink tissues such as cereal endosperm, it was previsouly thought that starch biosynthesis was not controlled by allosteric regulation of ADPglucose pyrophosphorylase. The maize endosperm pyrophosphorylase was only poorly (2-3 fold) stimulated by 3-PGA, and the concentration needed for 50% activation was quite high.[63] Moreover, one would not expect the levels of the effectors 3-PGA and Pi to fluctuate as observed in photosynthetic tissue, and hence carbon flow into starch would be dictated more by the overall enzyme levels rather than allosteric regulation. However, recent studies by Morell et al.[64] showed that the maize enzyme isolated in the presence of protease inhibitors showed greater sensitivity to allosteric activation and inhibition. Moreover, Sowokinos and Preiss[65] showed that the enzyme from potato tubers responded to 3-PGA and Pi in a fashion analogous to the leaf enzyme. ADPglucose pyrophosphorylase in nonphotosynthetic sink tissues is subject to the same allosteric controls as the leaf enzyme. Assuming the levels of these effector molecules do not fluctuate in nonphotosynthetic tissue (and there is no reason to believe they do) and are present at equimolar concentrations, then the activity of ADPglucose pyrophosphorylase is no more than 50% of maximum.

In all bacteria examined so far, ADPglucose pyrophosphorylase was found to be composed of a 200 kD tetramer with a single subunit molecular size of about 50 kD.[66] The plant enzyme appears, however, to differ in structure from the bacterial enzyme. The 200 kD holoenzyme of spinach leaf is composed of two different subunits of 51 kD and 54 kD.[67,68] The subunits are antigenically dissimilar, and exhibit different tryptic peptide patterns and N-terminal amino acid sequences. The potato tuber enzyme was purified to apparent homogeneity and only a single subunit of 50 kD was evident on SDS polyacrylamide gels.[65] Western blot analysis of leaf and seed extracts from several plant species using antibody raised against the spinach leaf enzyme also suggested that the leaf enzyme was composed of dissimilar subunits while only a single polypeptide was evident in seed tissue.[67,69]

More recent studies in Preiss's laboratory, however, suggest that the maize endosperm pyrophosphorylase is composed of two subunits of 54 kD and 60 kD which share conserved epitopes with the spinach leaf subunits. Antibodies raised against purified 51 kD and 54 kD subunits from spinach leaf reacted with maize endosperm polypeptides at 54 kD and 60 kD, respectively (Preiss, personal communication). Higher titers of spinach anti-54 kD were required to detect the cross-reactivity with the maize 60 kD polypeptide, suggesting these subunits were structurally more divergent. The presence of dissimilar pyrophosphorylase subunits would account for the two unlinked maize endosperm mutations, brittle-2 and shrunken-2.

The structure of the 54 kD subunit of rice endosperm ADPglucose pyrophosphorylase has been obtained by analysis of a cDNA clone. Figure 9 shows the derived primary sequence of the rice subunit aligned with the enzyme from *Escherichia coli*. The encoded protein from rice has a molecular size of 55.6 kD and displays about 29% homology with the *E. coli* enzyme. The apparent similarity between the plant and bacterial enzyme increases when localized regions are compared. This is readily evident between residues 48-227 and residues 297-392 where the extent of relatedness between these two proteins increases to about 42%.

The homology at the N-terminal half of the primary sequence can be readily accounted for since this region contains the putative allosteric and catalytic sites of the bacterial enzyme.[70] The allosteric site of the bacterial enzyme has been assigned to the peptide region containing the conserved lysine residue at position 68, while the conserved lysine at position 226 is believed to be part of the catalytic site. The conserved C-terminal domain (residues 297-392) may be involved in maintaining the conformation of the pyrophosphorylase subunit, particularly the allosteric site of the plant enzyme. Morell *et al.*[68] showed that a lysine at residue position 469 was reactive to [^{3}H]-pyridoxal-P which serves as a site specific probe for the allosteric site. In view of the different activators required by the bacterial and plant enzymes (fructose 1,6-P_2 and 3-PGA, respectively), structural rearrangement of ADPglucose pyrophosphorylase may have occurred during evolution resulting in modification of the specificity of the allosteric binding site.

Although the primary sequence of the large cDNA clone begins with a methionine residue, there is some

```
A Met Asn Val Leu Ala Ser Lys Ile Phe Pro Ser Arg Ser Asn Val Ala Ser Glu Gln Gln 20

  Gln Ser Lys Arg Glu Lys Ala Thr Ile Asp Asp Ala Lys Asn Ser Ser Lys Asn Lys Asn 40
                        B Met Val Ser Leu Glu Lys Asn Asp His Leu Met Leu Ala

  Leu Asp Arg Ser Val Asp Glu Ser Val Leu Gly Ile Ile Leu Gly Gly Gly Ala Gly Thr 60
  Arg Gln Leu Pro Leu Lys --- Ser Val Ala Leu Ile Leu Ala Gly Gly --- Arg Gly Thr

  Arg Leu Tyr Pro Leu Thr Lys Lys Arg Ala Lys Pro Ala Val Pro Leu Gly Ala Asn Tyr 80
  Arg Leu Lys Asp Lys Thr Asn Lys Arg Ala Leu Pro Ala Val His Pro Gly Gly Lys Phe

  Arg Leu Ile Asp Ile Pro Val Ser Asn Cys Leu Asn Ser Asn Ile Ser Leu Ile Tyr Val 100
  Arg Ile Ile Asp Phe Ala Leu Ser Asn Cys Ile Asn Ser Gly Ile Arg Arg Met Gly Val

  Leu Thr Gln Phe Asn Ser Ala Ser Pro Asn Arg His Leu Ser Arg Ala Tyr Gly Asn Asn 120
  Ile Thr Gln Tyr Gln Ser His Thr Leu Tyr Gln His Ile Gln Arg Gly Trp Ser Phe Phe

  Ile Gly Gly Tyr Leu Asn Glu Gly Phe Val Glu Val Leu Ala Ala Gln Gln Ser Pro Asp 140
  Asn Glu Glu Met --- Asn Glu --- Phe Val Asn Leu Leu Pro Ala Gln Gln Arg Met Lys

  Asn Pro Asn Trp Phe Gln Gly Thr Ala Asp Ala Val Arg Gln Tyr Leu Trp Leu Phe Glu 160
  Gly Glu Asn Trp Tyr Arg Gly Thr Ala Asp Ala Val Thr Gln Asn Leu Asp Ile Ile Arg

  Glu His Asn Val Met Glu Phe Leu Ile Leu Ala Gly Asn His Leu Tyr Arg Met Asn Tyr 180
  Arg Tyr Lys Ala Glu Tyr Val Val Ile Leu Ala Gly Asn His Ile Tyr Lys Gln Asn Tyr

  Glu Lys Phe Ile Gln Ala His Arg Glu Thr Asp Ser Asp Ile Thr Val Ala Ala Leu Pro 200
  Ser Arg Met Leu Ile Asp His Val Glu Lys Gly Val Arg Cys Thr Val Val Cys Met Pro

  Met Asp Glu Lys Arg Ala Thr Ala Phe Gly Leu Met Lys Ile Asp Glu Glu Gly Arg Ile 220
  Val Pro Ile Glu Glu Ala Ser Ala Phe Gly Val Met Ala Val Asp Glu Asn Asp Lys Thr

  Val Glu Phe Ala Glu Lys Pro Lys Gly Glu Gln Leu Lys Ala Met Met Val Asp Thr Thr 240
  Ile Glu Phe Val Glu Lys Pro Ala Asn Pro Pro Ser Met Pro Asn Asp Pro Ser Lys Ser

  Ile Leu Gly Leu Asp Asp Val Arg Ala Lys Glu Met Pro Tyr Ile Ala Ser Met Gly Ile 260
  Leu Ala Ser Met Gly Ile Tyr Val Phe Asp Ala Asp Tyr Leu Tyr Glu Leu Leu Glu Glu

  Thr Val Ile Ser Lys Asn Val Met Leu Gln Leu Leu Arg Glu Glu Phe Pro Gly Ala Asn 280
  Asp Asp Arg Asp Glu Asn Ser Ser His Asp Phe Gly Leu Asp Leu Ile Pro Lys Ile Thr

  Asp Phe Gly Ser Glu Val Ile Pro Gly Ala Thr Asn Ile Gly Met Arg Val Gln Ala Tyr 300
  Glu Ala Gly Leu Ala Tyr Ala His Pro Phe Pro --- Leu Ser Cys --- Val Gln Ser Asp

  Leu Tyr Asp Gly --- --- Tyr Trp Glu Asp Ile Gly Thr Ile Glu Ala Phe Tyr Asn Ala 318
  Pro --- Asp Ala Glu Pro Tyr Trp Pro Asp Val Gly Thr Leu Glu Ala Tyr Trp Lys Ala

  Asn Leu Gly Ile Thr Lys Lys Pro Val Pro Asp Phe Ser Phe Tyr Asp Arg Ser Ala Pro 338
  Asn Leu Asp Leu --- Ala Ser Val Val Pro Lys Leu Asp Met Tyr Asp Arg Asn Trp Pro

  Ile Tyr Thr Gln Pro Arg His Leu Pro Pro Ser Lys --- Val Leu Asp --- --- --- --- 353
  Ile Arg Thr Tyr Asn Glu Ser Leu Pro Pro Ala Lys Phe Val Gln Asp Arg Ser Gly Ser

  --- Ala Asp Val Thr Asp Ser Val Ile Gly Glu Gly Cys Val Ile Lys Asn Cys Lys Ile 372
  His Gly Met Thr Leu Asn Ser Leu Val Ser Gly Gly Cys Val Ile Ser Gly Ser Val Val

  His His Ser Val Val Gly Leu Arg Ser Cys Ile Ser Glu Gly Ala Ile Ile Glu Asp Ser 392
  Val Gln Ser Val Leu Phe Ser Arg Val Arg Val Asn Ser Phe Cys Asn Ile --- Asp Ser

  --- --- Leu Leu Met Gly Ala Asp Tyr Tyr Glu Thr Glu Ala Asp Lys Lys Leu Leu Gly 410
  Ala Val Leu Leu Pro Glu Val Trp Val Gly Arg Ser Cys Arg Leu Arg Arg Ser Val Ile

  Glu Lys Gly Gly --- Ile Pro Ile Gly Ile Gly Lys Asn Cys His Ile Arg Arg Ala Ile 429
  Asp Arg Ala Cys Val Ile Pro Glu Gly Met Val Ile Gly Glu Asn Ale Glu Glu Asp Ala

  Ile Asp Lys Asn Ala Arg Ile Gly Asp Asn Val Lys Ile Ile Asn Val Asp Asn Val Gln 449
  Arg Arg Phe Tyr --- Arg Ser Glu Glu Ile Val Leu Val Thr Arg Gly Met Leu Arg Lys

  Glu Ala Ala Arg Glu Thr Asp Gly Tyr Phe Ile Lys Ser Gly Ile Val Thr Val Ile Lys 469
  Leu Gly His Lys Gln Glu Arg STOP

  Asp Ala Leu Leu Leu Ala Glu Gln Glu Tyr Glu Val Ala Ala STOP                    483
```

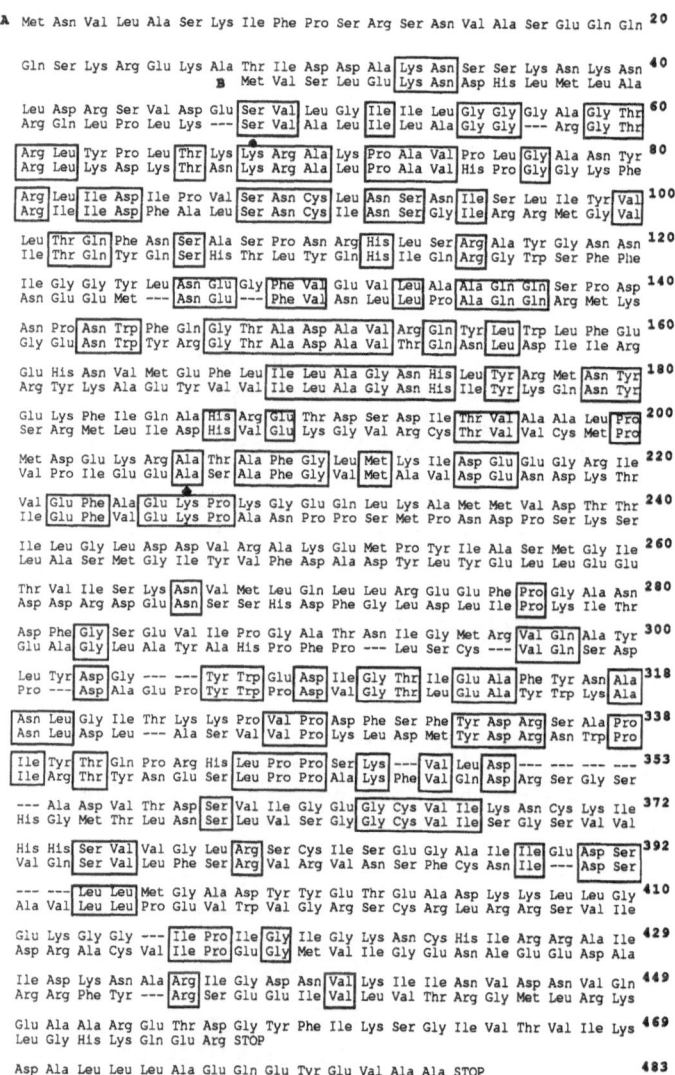

Fig. 9. Primary sequence conservation displayed by ADP-glucose pyrophosphorylases of rice endosperm (lanes A) and E. coli (lane B). Conserved residues are enclosed in boxes. The closed diamonds indicate the positions of lysine residues present in the allosteric (position 68) and ADP-glucose binding (position 226) sites of the bacterial enzyme. The putative allosteric site of the plant enzyme around the lysine residue at position 469.

question whether it represents the translational start. The size of the coded protein is consistent with the estimated size of the pyrophosphorylase subunit synthesized by translation *in vitro*. The first 20-30 amino acids, however, show little if any homology with known leader transit sequences functional in plastid transport of other nuclear encoded, chloroplast-localized proteins.[71] Thus, the N-terminus of the coded protein is either truncated (and does not represent the actual N-terminus of the unprocessed pyrophosphorylase protein), or represents an entirely different transit leader sequence. The latter explanation is quite possible since it may reflect the structural and biochemical differences displayed by chloroplasts and amyloplasts.

CONCLUSIONS

Genetic improvement of the protein quality in wheat and rice seeds is closer to reality. The genes of several different storage proteins have been isolated and preliminary studies are just beginning to unravel the molecular controls that subject them to developmental regulation. The transfer and analysis of these monocot genes in model dicot host systems will certainly provide some useful information on gene expression, but such knowledge must be viewed with some caution. Regulatory elements of some monocot genes may not be stringently recognized in a dicot background[72] as suggested by the low transcription of the wheat gliadin gene in transgenic tobacco. In addition, Ueng *et al.*[73] presented evidence that the gene for a maize storage protein was not developmentally regulated in transgenic petunia but appeared to be constitutively expressed at very low levels. Analysis of regulatory features of these genes must be evaluated in a homologous or closely related host system. DNA transfer and cell culture procedures for the transformation of rice have now been developed in several laboratories, and current efforts with wheat will undoubtedly be successful in the near future.

The synthesis of storage proteins also appears to be controlled at the translational level in many plant systems, particularly in rice. The more efficient recruitment of glutelin mRNAs over prolamine transcripts in membrane bound polysomes may be simply due to differences in the ability of these messages to undergo translational initiation. It should be pointed out, however, that the synthesis of these storage proteins is also dependent on the presence of rough ER membranes. The availability of membrane binding sites for polysomal complexes could also be a limiting factor in the synthesis cereal storage proteins. Furthermore, even though the prolamine protein bodies of rice are distended rough ER membranes, it is curious that glutelins are not present within these organelles as their surrounding membranes should provide a binding site for polysomes containing glutelin transcripts. It is clear that much has to be learned about the biochemical and cellular aspects of protein synthesis and targeting in plant cells.

The availability of molecular probes for ADPglucose pyrophosphorylase now permit an in-depth study on the regulation of this gene during normal endosperm development and in the variety of available mutants affecting starch metabolism. They will also allow detailed structure-function analysis of the enzyme with the potential for altering its allosteric behavior for genetic improvement in the conversion of photoassimilates into a renewable energy source.

ACKNOWLEDGMENTS

The senior author's investigations were supported by grant DCB-8702182 from the National Science Foundation, grant DE-FG06-87RL13699 from the Department of Energy, grant 86-CRCR-1-2132 from the United States Department of Agriculture, and grant RF-86058 #61 from the Rockefeller Foundation. We thank Dr. D.D. Kasarda for generously providing unpublished data. Special appreciation is extended to Dr. Jack Preiss for his advice and assistance and for the productive collaboration between our laboratories.

REFERENCES

1. FISHER, D.B., P.K. MACNICOL. 1986. Amino acid composition along the transport pathway during grain filling in wheat. Plant Physiol. 82: 1019-1023.

2. HARPER, L.A., P.P. SHARPE, G.W. LANGDALE, J.E. GIDDENS. 1987. Nitrogen cycling in a wheat crop: soil, plant, and aerial nitrogen transport. Agronomy J. 79: 965-973.
3. OSBORNE, T.B. 1907. The proteins of the wheat kernel. Carnegie Inst. Washington, Washington, D.C., 119 pp.
4. BRIGHT, S.W.J., P.R. SHEWRY. 1983. Improvement of protein quality in cereals. In Critical Reviews in Plant Sciences. (B.V. Conger, ed.), CRC Press, Inc., Boca Raton, Florida, Vol. 1, pp. 49-92.
5. NELSON, O.E. 1980. Genetic control of polysaccharide and storage protein synthesis in the endosperm of barley, maize, and sorghum. Adv. Cereal Sci. Technol. 2: 41-71.
6. KREIS, M., P.R. SHEWRY, B.G. FORDE, J. FORDE, B.J. MIFLIN. 1985. Structure and evolution of seed storage proteins and their genes with particular reference to those of wheat, barley and rye. In Oxford Surveys of Plant Molecular and Cell Biology. (B.J. Miflin, ed.), Oxford University Press, Oxford, Vol. 2, pp. 253-317.
7. SHOTWELL, M.A., B.A. LARKINS. 1989. The biochemistry and molecular biology of seed storage proteins. In The Biochemistry of Plants: A Comprehensive Treatise. (A. Marcus, ed.), Academic Press, New York, Vol. 15 (in press).
8. KASARDA, D.D., T.W. OKITA, J.E. BERNARDIN, P.A. BAECKER, C.C. NIMMO, E.T.-L. LEW, M.D. DIETLER, F.C. GREENE. 1984. Nucleic Acid (cDNA) and amino acid sequences of α-type gliadins from wheat (Triticum aestivum). Proc. Natl. Acad. Sci. USA 81: 4712-4716.
9. OKITA, T.W., V. CHESSBROUGH, C.D. REEVES. 1985. Evolution and heterogeneity of the α-/β-type and γ-type gliadin DNA sequences. J. Biol. Chem. 260: 8203-8213.
10. RAFALSKI, J.A., K. SCHEETS, M. METZLER, D.M. PETERSON, C. HEDGEOTH, D.G. SÖLL. 1984. Developmentally regulated plant genes: the nucleotide sequence of a wheat gliadin genomic clone. EMBO J. 3: 1409-1415.
11. SUMMER-SMITH, M., J.A. RAFALSKI, T. SUGIYAMA, M. STOLL, D. SÖLL. 1985. Conservation and variability

of wheat α/β-gliadin genes. Nucleic Acids Res. 13: 3905-3916.
12. ANDERSON, O.D., J.C. LITTS, M.-F. GAUTIER, F.C. GREENE. 1984. Nucleic acid sequence and chromosome assignment of a wheat storage protein gene. Nucleic Acids Res. 12: 8129-8144.
13. REEVES, C.D., T.W. OKITA. 1987. Analyses of α/β-type gliadin genes from diploid and hexaploid wheats. Gene 52: 257-266.
14. OKITA, T.W. 1984. Identification and DNA sequence analysis of a γ-type gliadin cDNA plasmid from winter wheat. Plant Mol. Biol. 3: 325-332.
15. SCHEETS, K., J.A. RAFALSKI. C. HEDGCOTH, D.G. SÖLL. 1985. Heptapeptide repeat structure of a wheat γ-gliadin. Plant Sci. Lett. 37: 221-225.
16. RAFALSKI, J.A. 1986. Structure of wheat gamma gliadin genes. Gene 43: 221-229.
17. FORDE, J., J.-M. MALPICA, N.G. HALFORD, P.R. SHEWRY, O.D. ANDERSON, F.C. GREENE, B.J. MIFLIN. 1985. The nucleotide sequence of a HMW glutelin subunit gene located on chromosome IA of wheat (<u>Triticum aestivum</u> L.). Nucleic Acids Res. 13: 6817-6832.
18. SUGIYAMA, T., A. RAFALSKI, D. PETERSON, D. SÖLL. 1985. A wheat HMW glutenin subunit gene reveals a highly repeated structure. Nucleic Acids Res. 13: 8729-8737.
19. THOMPSON, R.D., D. BARTELS, N.P. HARBERD. 1985. Nucleotide sequence of a gene from chromosome ID of wheat encoding a HMW-glutenin subunit. Nucleic Acids Res. 13: 6833-6846.
20. KASARDA, D.D., J.-C. AUTRAN, E.J.-L. LEW, C.C. NIMMO, P.R. SHEWRY. 1983. N-terminal amino acid sequences of ω-gliadin and ω-secalins: implications for the evolution of prolamin genes. Biochim. Biophys. Acta 747: 138-150.
21. COLOT, V., L.S. ROBERT, T.A. KAVANAGH, M.W. BEVAN, R.D. THOMPSON. 1987. Localization of sequences in wheat endosperm protein genes which confer tissue-specific expression in tobacco. EMBO J. 6: 3559-3564.
22. AN, G., P.R. EBERT, B.-Y. YI, C.-H. CHOI. 1986. Both TATA box and upstream regions are required for the nopaline synthase promoter activity in transformed tobacco cells. Mol. Gen. Genet. 203: 245-250.

23. DYNAN, W.S., R. TJIAN. 1985. Control of eukaryotic messenger RNA synthesis by sequence-specific DNA binding proteins. Nature 316: 774-778.
24. MAIER, U.-G., J.W.S. BROWN, L.M. SCHMITZ, M. SCHWALL, G. DIETRICH, G. FEIX. 1988. Mapping of tissue-dependent and independent protein binding sites to the 5' upstream region of a zein gene. Mol. Gen. Genet. 212: 241-245.
25. CHEN, Z.-L., N.-S. PAN, R.N. BEACHY. 1988. A DNA sequence element that confers seed-specific enhancement to a constitutive promoter. EMBO J. 7: 297-302.
26. YAMAGATA, H., T. SUGIMOTO, K. TANAKA, Z. KASAI. 1982. Biosynthesis of storage proteins in developing rice seeds. Plant Physiol. 70: 1094-1100.
27. KRISHNAN, H.B., T.W. OKITA. 1986. Structural relationships among rice glutelin polypeptides. Plant Physiol. 81: 748-753.
28. LUTHE, D.S. 1983. Storage protein accumulation in developing rice (Oryza sativa L.) seeds. Plant Sci. Lett. 32: 147-158.
29. ZHAO, W.M., J.A. GATEHOUSE, D. BOULTER. 1983. The purification and partial amino acid sequence of a polypeptide from the glutelin fraction of rice grains; homology to pea legumin. FEBS Lett. 162: 96-102.
30. WEN, T.-N., D.S. LUTHE. 1985. Biochemical characterization of rice glutelin. Plant Physiol. 78: 172-177.
31. OKITA, T.W., H.B. KRISHNAN, W.T. KIM. 1988. Immunological relationships among the major seed proteins of cereals. Plant Science (Shannon) 57: 103-111.
32. TAKAIWA, F., S. KIKUCHI, K. OONO. 1987. A rice glutelin gene family - A major type of glutelin mRNAs can be divided into two classes. Mol. Gen. Genet. 208: 15-22.
33. HIGUCHI, W., C. FUKAZAWA. 1987. A rice glutelin and a soybean glycinin have evolved from a common ancestral gene. Gene 55: 245-253.
34. STASWICK, P.E., M.A. HERMODSON, N.C. NIELSEN. 1981. Identification of the acidic and basic subunit complexes of glycinin. J. Biol. Chem. 256: 8752-8755.

35. ARGOS, P., S.V.L. NARAYANA, N.C. NIELSEN. 1985. Structural similarity between legumin and vicilin storage proteins from legumes. EMBO J. 4: 1111-1117.
36. KIM, W.T., T.W. OKITA. 1988. Nucleotide and primary sequence of a major rice prolamine. FEBS Lett. 231: 308-310.
37. KIM, W.T., T.W. OKITA. 1988. Structure, expression, and heterogeneity of the rice seed prolamines. Plant Physiol. 88: 649-655.
38. TAKAIWA, F., H. EBINUMA, S. KIKUCHI, K. OONO. 1987. Nucleotide sequence of a rice glutelin gene. FEBS Lett. 221: 43-47.
39. LYCETT, G.W., R.R.D. CROY, A.H. SHIRSAT, D. BOULTER. 1984. The complete nucleotide sequence of a legumin gene from pea (*Pisum sativum* L.). Nucleic Acids Res. 12: 4493-4506.
40. REEVES, C.D., H.B. KRISHNAN, T.W. OKITA. 1986. Gene expression in developing wheat endosperm. Plant Physiol. 82: 34-40.
41. FABIJANSKI, S., I. ALTOSAAR. 1985. Evidence for translational control of storage protein biosynthesis during embryogenesis of *Avena sativa* L. (oat endosperm). Plant Mol. Biol. 4: 211-218.
42. KRISHNAN, H.B., V.R. FRANCESCHI, T.W. OKITA. 1986. Immunochemical studies on the role of the Golgi complex in protein-body formation in rice seeds. Planta 169: 471-480.
43. LARKINS, B.A., W.T. HURKMAN. 1978. Synthesis and deposition of protein bodies of maize endosperm. Plant Physiol. 62: 256-263.
44. YAMAGATA, H., K. TANAKA. 1986. The site of synthesis and accumulation of rice storage proteins. Plant Cell Physiol. 27: 135-145.
45. CHRISPEELS, M.J. 1985. The role of the Golgi apparatus in the transport and post-translational modification of vacuolar (protein body) proteins. *Op. cit.* Reference 6, pp. 43-68.
46. TAYLOR, J.R.N., L. SCHUSSLER, N.v.d.W. 1985. Protein body formation in the starch endosperm of developing *Sorghum bicolor* (L.) moench seeds. S. Afr. J. Bot. 51: 35-40.
47. CAMPBELL, W.P., J.W. LEE, T.P. O'BRIEN, M.G. SMART. 1981. Endosperm morphology and protein body formation in developing wheat grain. Aust. J. Plant Physiol. 8: 5-19.

48. MIFLIN, B.J., S.R. BURGESS, P.R. SHEWRY. 1981. The development of protein bodies in the storage tissues of seeds: subcellular separation of homogenates of barley, maize and wheat endosperm and pea cotyledons. J. Exp. Bot. 32: 199-219.
49. PARKER, M.L., C.R. HAWES. 1982. The Golgi apparatus in developing endosperm of wheat (Triticum aestivum L.). Planta 154: 277-283.
50. BECHTEL, D.B., B.D. BARNETT. 1986. A freeze-fracture study of storage protein accumulation in unfixed wheat starch endosperm. Cereal Chem. 63: 232-240.
51. BECHTEL, D.B., R.L. GAINES, Y. POMERANZ. 1982. Protein secretion of wheat endosperm: formation of the matrix protein. Cereal Chem. 59: 336-343.
52. KIM, W.T., V.R. FRANCESCHI, H.B. KRISHNAN, T.W. OKITA. 1988. Formation of wheat protein bodies: involvement of the Golgi apparatus in gliadin transport. Planta 173: 173-182.
53. MUNRO, S., H.R.B. PELHAM. 1987. A C-terminal signal prevents secretion of luminal ER proteins. Cell 48: 811-907.
54. MARKS, M.D., J.S. LINDELL, B.A. LARKINS. 1985. Nucleotide sequence analysis of zein mRNAs from maize endosperm. J. Biol. Chem. 16: 451-459.
55. PFEFFER, S.R., J.E. ROTHMAN. 1987. Biosynthetic protein transport and sorting by the endosplasmic reticulum and golgi. Annu. Rev. Biochem. 56: 829-852.
56. HOFFMAN, L.M., D.D. DONALDSON, R. BOOKLAND, K. RASHKA, E.M. HERMAN. 1987. Synthesis and protein body deposition of maize 15-kd zein in transgenic tobacco seeds. EMBO J. 6: 3213-3221.
57. TAGUE, B.W., M.J. CHRISPEELS. 1987. The plant vacuolar protein, phytohemagglutinin, is transported to the vacuole of transgenic yeast. J. Cell Biol. 105: 1971-1979.
58. PERNOLLET, J.C., J. MOSSÉ. 1983. Secondary structure prediction of seed storage protein genes. Int. J. Pept. Protein Res. 22: 456-463.
59. PREISS, J. 1982. Regulation of the biosynthesis and degradation of starch. Annu. Rev. Plant Physiol. 33: 431-454.
60. HANNAH, L.C., D.M. TUSCHAU, R.J. MANS. 1980. Multiple forms of maize endosperm ADP-glucose

pyrophosphorylase and their control by Shrunken-2 and Brittle-2. Genetics 95: 961-970.

61. LIN, T.-P., T. CASPAR, C. SOMERVILLE, J. PREISS. 1988. Isolation and characterization of a starchless mutant of Arabidopsis thaliana (L.) Heynh lacking ADPglucose pyrophosphorylase activity. Plant Physiol. 86: 1131-1135.

62. HELDT, H.W., C.J. CHON, D. MARONDE, Z.S. STANKOVIC, D.A. WLAKER, A. KRAMINER, M.R. KIRK, U. HEBER. 1977. Role of orthophosphate and other factors in the regulation of starch formation in leaves and isolated chloroplasts. Plant Physiol. 59: 1146-1155.

63. DICKINSON, D.B., J. PREISS. 1969. ADPglucose pyrophosphorylase from maize endosperm. Arch. Biochem. Biophys. 130: 119-128.

64. PREISS, J., M. BLOOM, M. MORELL, V.L. KNOWLES, W.C. PLAXTON, T.W. OKITA, R. LARSEN, A.C. HARMON, C. PUTNAM-EVANS. 1987. Regulation of starch synthesis: enzymological and genetic studies. In Tailoring Genes for Crop Improvement. (G. Bruening, J. Harada, T. Kosuge, A. Hollaender, eds.), Plenum Publishing Corp., New York, pp. 133-152.

65. SOWOKINOS, J.R., J. PREISS. 1982. Pyrophosphorylases in Solanum tuberosum. II. Purification, physical, and catalytic properties of ADPglucose pyrophosphorylase in potatoes. Plant Physiol. 69: 1459-1466.

66. PREISS, J. 1984. Bacterial glycogen synthesis and its regulation. Annu. Rev. Microbiol. 38: 419-458.

67. MORELL, M.K., M. BLOOM, V. KNOWLES, J. PREISS. 1987. Subunit structure of spinach leaf ADP-glucose pyrophosphorylase. Plant Physiol. 85: 182-187.

68. MORELL, M., M. BLOOM, R. LARSEN, T.W. OKITA, J. PREISS. 1987. II. Biochemistry and molecular biology of starch synthesis. In Plant Gene Systems and Their Biology. (J. Key, L. McIntosh, eds.), Alan R. Liss, Inc., pp. 227-242.

69. KRISHNAN, H.B., C.D. REEVES, T.W. OKITA. 1986. ADPglucose pyrophosphorylase is encoded by different mRNA transcripts in leaf and endosperm of cereals. Plant Physiol. 81: 642-645.

70. BAECKER, P.A., C.E. FURLONG, J. PREISS. 1983. Biosynthesis of bacterial glycogen: primary structure of Escherichia coli ADPglucose synthetase as deduced from the nucleotide sequence of the glgC gene. J. Biol. Chem. 258: 5084-5088.

71. KARLIN-NEUMANN, G.A., E.M. TOBIN. 1986. Transit peptides of nuclear-encoded chloroplast proteins share a common amino acid framework. EMBO J. 5: 9-13.

72. KEITH, B., N.-H. CHUA. 1986. Monocot and dicot pre-mRNAs are processed with different efficiencies in transgenic tobacco. EMBO J. 5: 2419-2425.

73. UENG, P., G. GALILI, V. SAPANARA, P.B. GOLDSBROUGH, P. DUBE, R.N. BEACHY, B.A. LARKINS. 1988. Expression of a maize storage protein gene in petunia plants is not restricted to seeds. Plant Physiol. 86: 1281-1285.

74. FORDE, B.G., A. HEYWORTH, J. PYWELL, M. KREIS. 1985. Nucleotide sequence of a B1 hordein gene and the identification of possible upstream regulatory elements in endosperm storage protein genes from barley, wheat and maize. Nucleic Acids Res. 13: 7327-7339.

75. LANGRIDGE, P., G. FEIX. 1983. A zein gene of maize is transcribed from two widely separated promoter regions. Cell 34: 1015-1022.

Chapter Ten

PRIMARY AND SECONDARY METABOLISM OF POLYAMINES IN PLANTS

HECTOR E. FLORES, CALIXTO M. PROTACIO
AND MARK W. SIGNS

Biotechnology Institute
Pennsylvania State University
University Park, Pennsylvania 16802

INTRODUCTION

The diamine putrescine and the polyamines spermidine and spermine are amino acid-derived, aliphatic nitrogenous compounds of wide distribution among plant cells (Fig. 1). The earliest reference to polyamines in the scientific literature is van Leeuwenhoek's classic letter to the Royal Society of London in 1678. During his studies

$H_2N-(CH_2)_3-NH_2$
1,3-Diaminopropane

$H_2N-(CH_2)_4-NH_2$
Putrescine

$H_2N-(CH_2)_5-NH_2$
Cadaverine

$H_2N-(CH_2)_4-NH-(CH_2)_3-NH_2$
Spermidine

$H_2N-(CH_2)_3-NH-(CH_2)_3-NH_2$
Norspermidine (Caldine)

$H_2N-(CH_2)_4-NH-(CH_2)_4-NH_2$
Homospermidine

$H_2N-(CH_2)_3-NH-(CH_2)_4-NH-(CH_2)_3-NH_2$
Spermine

$H_2N-(CH_2)_3-NH-(CH_2)_3-NH-(CH_2)_3-NH_2$
Norspermine (Thermine)

$H_2N-(CH_2)_3-NH-(CH_2)_3-NH-(CH_2)_3-NH-(CH_2)_3-NH_2$
Caldopentamine

Fig. 1. Chemical structures of di- and polyamines found in bacterial and plant cells.

describing spermatozoa, he observed the gradual formation of colorless crystals upon drying the samples. The correct structure of these crystals, which corresponded to spermine, was not determined until over 250 years later.[1] Spermidine was later found in mammalian tissues. The related diamines putrescine and cadaverine were found in decomposing animal and vegetable matter as a result of microbial activity. Within this historical frame, it is not surprising that earlier studies on these compounds were done mostly on non-plant systems. Recent interest in the function of polyamines in higher plants is in good part derived from discoveries in microbial and animal cells.

In the 1950s and '60s a general correlation was established between high titers of di- and polyamines and rapid cell division in microorganisms.[2,3] This correlation was extended to mammalian cells, and more recently to plant cells. The development of protocols for the growth of mammalian cells in defined media facilitated numerous studies on the cell cycle.[2,4] It was found that depletion of polyamines led to arrest of the cells in the G1-S phase of the cell cycle. The resumption of cycling upon addition of polyamines to the arrested cells strongly indicated the involvement of these compounds in cell division. The availability of enzyme activated, irreversible inhibitors of polyamine biosynthesis in the early 1980s helped strengthen the case for this causal link.[4]

The observations in bacterial and animal systems have thus affected in a major way the direction for polyamine research in plants in the last two decades. As we will describe below, there is considerable evidence indicating that polyamines may in fact play a major role in plant growth and development. It will become clear, however, that polyamines are also involved in plant responses which have no precedent in bacterial or animal cells. For example, the original studies on putrescine in plants were related to nutrient deficiencies.[5] It is now apparent that polyamine metabolism in plants may be more complex than previously expected. Although polyamine metabolism in plants follows the major course established for bacterial and animal cells, several features are unique to plants. In addition to the arginine/ornithine dichotomy for putrescine production, polyamines share a common precursor with a major plant hormone (ethylene), and in turn serve as precursors for a wide variety of plant secondary metabolites.

In this review we will outline the major features of polyamine metabolism unique to plant cells, and their involvement in stress responses, cell division, and plant development. We will emphasize research done in the last decade. Recent reviews have covered various aspects of plant polyamine research,[6-8] and provide excellent introductions to this subject.

METABOLIC PATHWAYS

Occurrence and Distribution of Polyamines

In view of their biosynthetic and catabolic relationships and as a matter of convenience, the aliphatic diamines will be considered along with the polyamines proper. Appropriate distinctions will be made, however, when discussing their functional significance. Of the structures shown in Figure 1, putrescine (Put) and spermidine (Spd) are found, without exception, in all prokaryotic and eukaryotic cells investigated so far. Although the concentrations and ranges of Put and Spd may vary widely with cell type and physiological state, no plant is known in nature which lacks these compounds. The tetraamine spermine (Spm) is found only in eukaryotic cells with few exceptions, although it is utilizable by prokaryotes. Cohen[3] has pointed out that evolution of Spm biosynthesis is not tied in any obvious way to the evolution of the utilization of molecular oxygen, which closely follows the prokaryote:eukaryote transition. The presence of Spm, however, may be accompanied in animal and plant cells by evolution of specific oxidases.[9] The recent finding of spermine and higher order polyamines in thermophilic bacteria suggests that this may not be a unique eukaryotic attribute.[10]

Diaminopropane (Fig. 1) is an oxidation product of Spd and Spm and occurs in monocotyledonous[11] and dicotyledonous species.[12] Cadaverine (Cad, 1,5-diaminopentane) also shows a restricted distribution, having been reported in the flowers of _Arum_ lilies[13] and in legume species.[14] Cad has not been reported in algae, and is not common in other families of higher plants; even within the Leguminosae, its distribution among tribes has not been thoroughly studied.

An interesting problem is presented by the evolution of polyamine homologs. For example, sym-homospermidine, the C4-C4 homolog of spermidine (C4-C3), is present in such disparate organisms as bacteria,[15] _Lathyrus_,[16] and sandalwood.[17] In bacteria, the natural occurrence of this symmetrical polyamine is confined to species in which Spd is absent or present in trace amounts, while in _Lathyrus sativus_ this biosynthetic ability must coexist with two pathways for Spd synthesis.[18] Because these symmetrical polyamines occur in plants grown in extreme habitats, it has been speculated that _Lathyrus_, being a drought-resistant plant, may use the homo-Spd synthetic pathway under adverse conditions. This idea receives support from recent studies (see section on polyamines and plant stress).

Several other polyamine and guanidino analogs are of more limited occurrence than the above compounds. Arcain (1,4-diguanidinobutane), the guanidino analog of Put, is found in Basidiomycetes and in the seed of _Luffa_ *cylindrica*, and is formed by transamidination from arginine to agmatine.[19] Galegin (3-methylbut-2-enyl-guanidine) occurs in the legume *Galega officinalis* and is also derived from arginine, but the transamidinase acceptor is not known.[20] Tetramethyl-Put is found in shoots of Rubiaceae and Solanaceae.[21,22]

Polyamine Biosynthesis

The currently accepted scheme for di- and polyamine biosynthesis is shown in Figure 2. The ultimate precursors for polyamine synthesis in plant cells are the amino acids ornithine and arginine. Ornithine, a non-protein amino acid derived from glutamate, is the precursor for arginine by way of the Krebs-Henseleit[23] urea cycle. The arginine pool can be increased through protein degradation, but ornithine is the only precursor for "*de novo*" formation of arginine. Ornithine and urea formation from arginine allows recovery of nitrogen and carbon from the cycle (Fig. 2). Arginine residues may be a significant storage source of nitrogen during germination and senescence.[24] Because it has been suggested that polyamines are a nitrogen source for plant cells,[25] and in periods of rapid cell division the demand for arginine/ornithine may cause a significant drain on their metabolic pools, it is surprising that so little is known about the dependence of polyamine synthesis on precursor availability (see section on polyamines and plant stress).

Putrescine is the obligate precursor for the synthesis of Spd and Spm (Fig. 2). In animal and fungal cells,[2] this diamine is formed only by decarboxylation of ornithine by ornithine decarboxylase (ODC). In contrast, bacterial cells have an alternative enzyme which decarboxylates arginine (ADC) to form agmatine.[26] Both ODC and ADC are enzymes dependent on pyridoxal phosphate. Plants share with bacteria the widespread coexistence of arginine and ornithine decarboxylation.[6,8] The relative contributions of these enzymes to Put synthesis is still a controversial subject. Since arginine and ornithine can be interconverted through the urea cycle, it is clear that assessing the contribution of each pathway by feeding either labeled amino acid cannot by itself give an unambiguous answer. Correlation of enzyme activity with a specific developmental process or plant

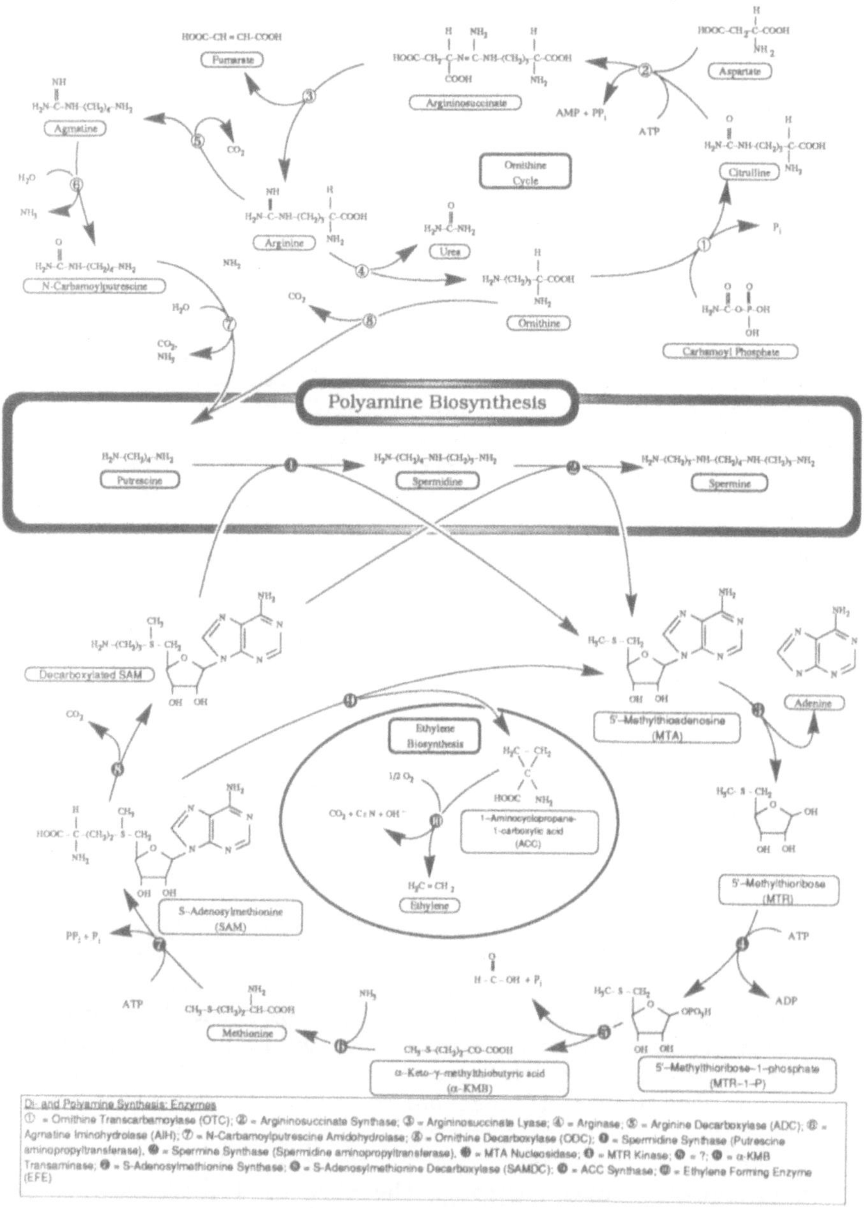

Fig. 2. Integration of the polyamine biosynthetic pathway with intermediary nitrogen metabolism.

response may also be used. For example ADC, but not ODC, decline is correlated with a decrease in Spd content in senescing oat leaves.[27] In contrast, ODC and not ADC activity rises during potato tuber sprouting, in young tomato fruits, and in exponentially growing tobacco cell suspensions (see Ref. 8). While these contrasting situations suggest that rate-limiting roles in Put biosynthesis may have evolved for different developmental processes, a cause and effect relationship cannot be established from these studies alone. The availability of enzyme-activated irreversible inhibitors for ODC and ADC has greatly improved our chances of arriving at a definitive answer. Examples for the use of such inhibitors in various situations will be given throughout this chapter.

ADC is widespread in higher plants and has been purified from oat,[9] rice[28] and other sources (see Ref. 8). The enzyme appears to be a hexamer of identical subunits about 39 KD in molecular weight. ODC activity has also been reported for many plant species, but has not yet been purified to homogeneity. In contrast to ADC which is a cytosolic enzyme, ODC can also be found associated with chromatin in germinating barley seedlings.[29]

Agmatinase, the enzyme converting agmatine to putrescine in bacteria, has never been reported in plant cells. The current evidence indicates that agmatine formed by ADC is metabolized to Put in a two-step conversion. Agmatine iminohydrolase catalyzes the formation of N-carbamoylputrescine (Fig. 2). This enzyme has been reported in maize, safflower, and groundnut,[8] and purified to homogeneity from maize shoot.[30] Although N-carbamoylputrescine (NCP) may also be formed by decarboxylation of citrulline[31] in sugarcane and <u>Sesamum</u>,[32] this enzyme is not well characterized. Labeling experiments so far support the notion that NCP is mainly derived from agmatine. The final step in Put formation involves NCP amidohydrolase,[33,34] which is also widespread among plants.[30,35]

A single multifunctional "putrescine synthase" has been reported in the legume <u>L. sativus</u>[36] and recently purified from cucumber seedlings.[37] Put synthase has agmatine iminohydrolase activity. The carbamyl moiety of the resulting product, NCP, is then eliminated by Put carbamoyl transferase activity. Carbamoyl phosphate can also be used to form citrulline in the presence of ornithine transcarbamoylase, or metabolized to CO_2 and NH_3 by carbamate kinase. All of these activities are present in the _Lathyrus_ enzyme. The enzyme from cucumber also has ADC activity.[38]

An additional pathway for Put synthesis has been suggested.[39] Citrulline decarboxylase catalyzes the formation of NCP from citrulline in _Helianthus tuberosus_. However, when _Helianthus_ tubers are fed with [carbamoyl ^{14}C]-citrulline, some label is also found in arginine in addition to NCP, suggesting that this compound can be formed at least in part via ADC.

Relatively little is known about the synthesis of Spd and Spm from Put in plant cells. In most species studied so far, the sequence appears to be the same as for animal and bacterial cells.[40] The aminopropyl moiety added to Put to form Spd and Spm is ultimately derived from methionine via S-adenosylmethionine (SAM), as shown with radiolabeling experiments.[41] SAM decarboxylase (SAMDC) has been purified from corn seedlings,[42] Chinese cabbage,[43] and tobacco cell suspensions.[44] Because the decarboxylation reaction is essentially irreversible, this is the key step committing the pool of SAM to polyamine synthesis.

Spd and Spm are synthesized by the addition of one or two aminopropyl groups to putrescine, respectively (Fig. 2). The aminopropyltransferase commonly known as spermidine synthase is clearly separate from SAMDC, and has been partially purified from _Lathyrus_[45] and Chinese cabbage.[43] It has been isolated from leaf protoplasts, and is present in 20-fold excess over SAMDC. Thus, in at least this case, SAMDC appears to be the rate-limiting factor for the production of Spd and Spm. Spm synthase activity has also been detected in Chinese cabbage,[46] but is not well characterized in any plant system. Based on information from bacterial and animal cells, it is currently assumed that this enzyme is distinct from Spd synthase.

An alternative pathway for spermidine biosynthesis has been reported in bacteria[15] and recently described in *L. sativus*.[47] In this case, the propylamino moiety is contributed by beta-aspartic semialdehyde. The sequence involves Schiff-base formation between aspartic semialdehyde and Put, reduction to carboxyspermidine, and decarboxylation. The last enzyme in this sequence has been partially purified. In *Lathyrus*, this pathway coexists with the route involving SAMDC and the polyamine synthases.

It is unique to plant cell metabolism that SAM, in addition to being a methyl donor and a polyamine precursor, is also a precursor for the plant hormone ethylene. The key step committing SAM to ethylene biosynthesis is catalyzed by 1-aminocyclopropane-1-carboxylic acid (ACC) synthase. Ethylene is generally considered a senescence-promotor, and polyamines have been shown to delay the senescence syndrome in detached leaves.[48] Whether polyamines are in fact antisenescence factors in the whole plant is still an open question, but this intriguing possibility has led to the proposal that the antisenescence properties of polyamines are based on their inhibition of ethylene biosynthesis and simultaneous promotion of their own synthesis.[6,49]

Several experiments are consistent with this view. For example, polyamine addition to orange peel discs can block ethylene formation; this is paralleled by an increased incorporation of labeled SAM into polyamines.[50] In senescing carnation flowers, inhibition of ACC synthesis inhibits ethylene production and increases polyamine titers; this is paralleled by a delay in the onset of senescence;[51] conversely, inhibition of polyamine synthesis results in enhanced rates of ethylene formation and triggering of senescence. Taken at face value, however, these results only show that when one of two alternative pathways is blocked, more of the common precursor is available to follow the other route. There is yet no experimental evidence for feedback regulation by metabolic intermediates of the key enzymes diverting the flow of SAM, SAMDC, and ACC synthase.

The 5'-methylthioadenosine generated by the action of polyamine synthases or ACC synthase is degraded to methylthioribose and recycled back to methionine and SAM (Fig. 2). This is an important sequence ensuring the availability of methione and SAM for protein synthesis

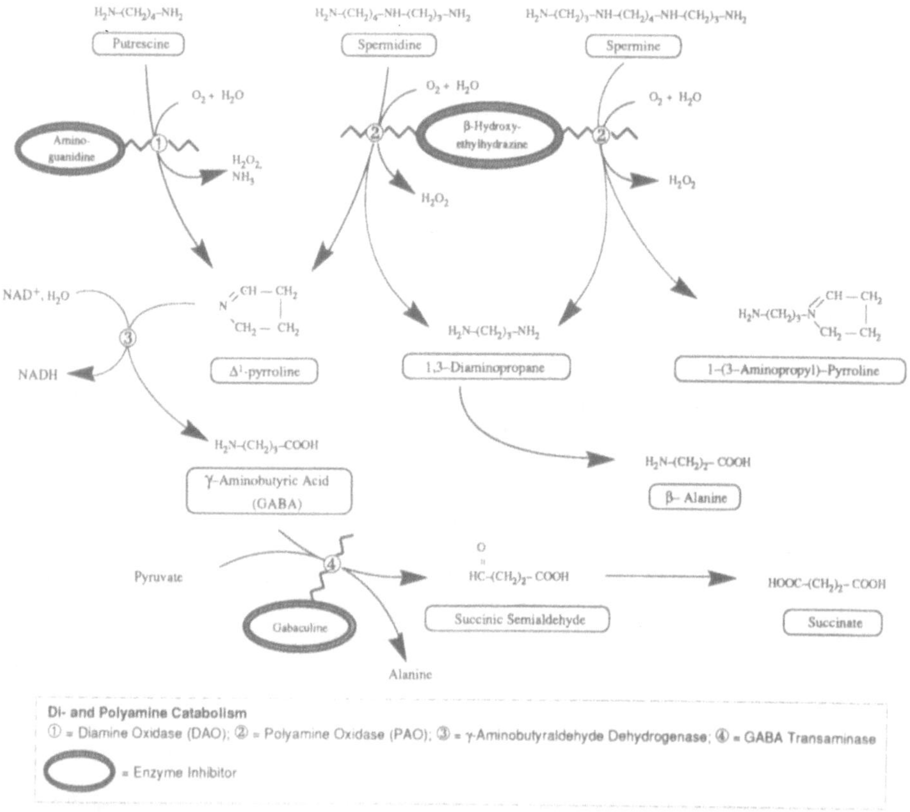

Fig. 3. Catabolism of plant di- and polyamines.

and methylation reactions, respectively. It is surprising that so little is known about the regulation of metabolic flow through SAM, polyamines and ethylene, considering their importance in plant growth and development.

Polyamine Catabolism

Figure 3 summarizes what is known about the degradation of di- and polyamines in plant cells. Two major classes of oxidases are known. Diamine oxidases (DAO) are widespread in the Leguminosae. The pea enzyme has been known for many years[52] and has been purified to homogeneity.[53,54] This copper-containing enzyme is most active on Put and Cad, but can also oxidize Spd and Spm.

The polyamine oxidases (PAO), so far found only in Gramineae, are specific for polyamines. PAO has FAD as a cofactor and has been purified from oat and maize.[55,56]

The most intriguing biological question posed by DAO and PAO concerns their localization. It was shown early on that PAO activity from barley leaves was associated with the vascular strands.[57] In oat leaves, PAO activity appears loosely associated with the apoplast; a significant amount of activity leaks into the medium when detached peeled leaves are placed on buffered solution.[58] Freshly digested oat protoplasts show no PAO, whereas at least 80% of the total activity is found in the cell wall digest.[59,60] As shown in barley, significant activity is also associated with the stomatal guard cells and vascular bundles. The presence of DAO in the cell wall has been confirmed in lentil seedlings by immunohistochemistry.[61]

What is the physiological significance of a cell wall locale for DAO and PAO? Assuming the substrates are also present in the cell wall, they could be bound to negatively charged sites.[62] Oxidation of these organic polycations may be a way to regulate their interaction with inorganic cations such as Ca^{++}, and may have implications for the proposed role of polyamines on membrane stabilization (see Ref. 49). Other products of di- and polyamine oxidation may also be significant. Hydrogen peroxide is generated by DAO and PAO activity. The fact that these enzymes appear to be tightly bound to lignified cell walls suggests that H_2O_2 may provide a necessary substrate for the peroxidase-mediated synthesis of this cell wall polymer.[58] The fate of the NH_3 generated by DAO/PAO activity is so far unknown.

The 4-aminobutyraldehyde formed by the oxidation of Put or Spd is further oxidized by a NAD-dependent pyrroline dehydrogenase.[60] This enzyme has been found in every case where DAO/PAO activity is present. The product of this reaction, γ-aminobutyric acid (GABA), can be transaminated and the resulting succinic acid incorporated into the Krebs cycle. This metabolic sequence can thus account for the complete recycling of the carbon and nitrogen from di/polyamines. It may also explain the fact that, in at least some cases, di- and polyamines can be used as organic nitrogen sources by plant cells. A tobacco cell line capable of utilizing putrescine as a nitrogen source appears to metabolize this diamine to GABA.[60]

Although PAO activity has also been found in Amaranthus seedlings[58] and DAO has been purified from Euphorbia,[63] these enzymes have not been reported in families such as Compositae, Cruciferae and Solanaceae.[58] Alternative pathways of polyamine catabolism, such as oxidation of conjugated substrates, should be considered. In animal cells, acetylated polyamines appear to be the major substrates for oxidative deamination. In tobacco cells, Put oxidation to GABA may occur via cinnamoyl intermediates.[64]

Inhibitors of Polyamine Metabolism

In the last decade, a number of metabolic inhibitors have been developed which affect various steps in the biosynthesis and catabolism of polyamines. Their application to plant systems has been reviewed recently.[65] The inhibitors with the best potential as physiological tools are probably the enzyme-activated (suicide type) irreversible inhibitors of ODC and ADC.[4]

When ODC was established as the rate-limiting step in polyamine biosynthesis in animal and bacterial cells, it became the major target in efforts to design a specific inhibitor. The expectations were to unravel the physiological role of polyamines and to control rapid cell proliferation; ODC is extremely active in neoplastic cells. A major effort at Merrel-Dow resulted in several effective inhibitors, the most promising of which was α-difluoromethyl ornithine (DFMO, Fig. 4). When acted upon by ODC, which is a pyridoxal-dependent enzyme, DFMO is decarboxylated in the same way as the methyl analog.[66] The latter is a competitive inhibitor, and is decarboxylated to methyl Put. In the case of DFMO, however, the reactivity of the halogen atoms is unmasked during catalysis, resulting in the formation of a highly reactive electrophile which can bind covalently to a neighboring nucleophilic residue at the active site.[66]

In plant and bacterial cells, Put can also be made via ADC. A similar difluoromethyl analog of arginine

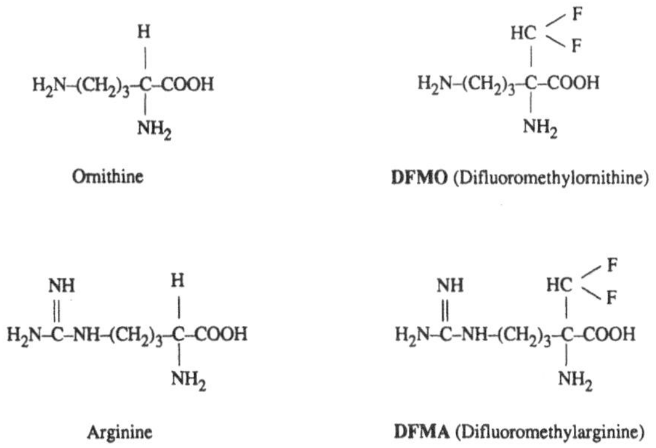

Fig. 4. Enzyme-activated irreversible inhibitors of putrescine biosynthesis.

(DFMA) was therefore developed.[4] The mechanism of action is the same as that of DFMO. Both inhibitors are effective in plant cells "in vivo" and "in vitro".[65] DFMA has been used to probe stress responses,[67] somatic embryogenesis[68] and alkaloid metabolism.[69] Both inhibitors are highly specific, and in most cases do not appear to be metabolized by plant cells, the only exception being tissues with high arginase activity. In these tissues, DFMA is partially metabolized to urea and DFMO, thereby indirectly inhibiting ODC.[70]

Methylglyoxal bis-guanylhydrazone (MGBG, Fig. 5) is the most commonly used inhibitor of SAM decarboxylase, which catalyzes the step committing SAM to the synthesis of polyamines. MGBG is a competitive inhibitor of SAMDC purified from Chinese cabbage.[43] Tobacco cell lines resistant to this inhibitor have been selected, and have been shown to overproduce a 35-kd protein,[71] which has been identified as SAMDC. MGBG-treated plant cells also show dramatic increases in enzyme activity.[44] The effect appears to result from stabilization of the enzyme by the bound MGBG, leading to a doubling of the enzyme half-life. The use of MGBG, however, may be limited by its potential side effects. In rats, for example, high doses of MGBG lead to a decrease in Spd and a rise in Put, consistent with inhibition of SAMDC. However, the effect is mostly accounted for by inhibition

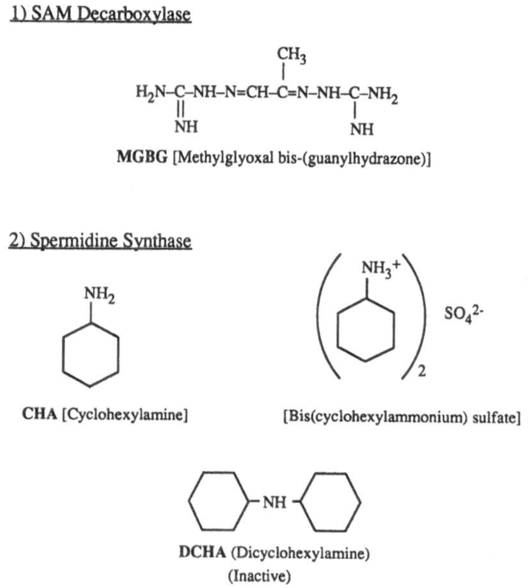

Fig. 5. Inhibitors of polyamine biosynthesis.

of diamine oxidase and not SAMDC.[72] In view of this, it would be important to determine if plant PAO/DAO are also MGBG-sensitive.

There have been several reports indicating that a compound named dicyclohexylammonium sulfate (DCHA, Fig. 5) is a very effective inhibitor of Spd synthase.[73] In all cases showing positive results, the compound used had been purchased from Sigma Chemical Co. and labeled as DCHA. More recently, DCHA synthesized in the laboratory has been shown to be inactive.[73] The compound sold as DCHA by Sigma was actually shown to be the sulfate salt of cyclohexylamine [bis(cyclohexylammonium) sulfate]. The earlier literature must therefore be revised accordingly. Cyclohexylamine (CHA) is a very effective and specific inhibitor of Spd synthase in Chinese cabbage.[74] It also inhibits Spd synthesis in tobacco flower explants.[75]

Aminoguanidine (Fig. 6) is a carbonyl reagent effective against plant DAO.[60] Hydroxyethyl-hydrazine

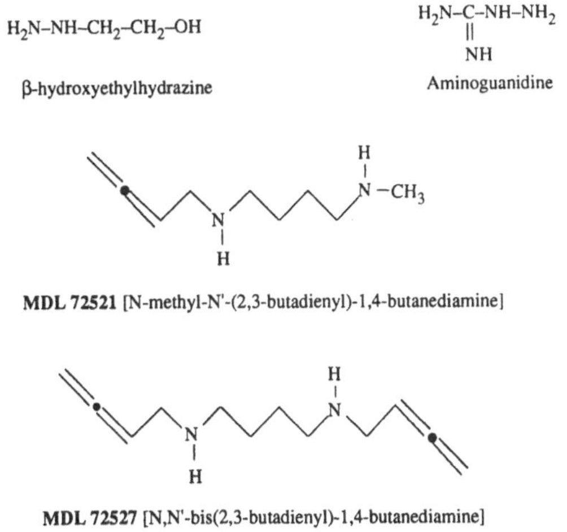

Fig. 6. Inhibitors of di- and polyamine oxidases.

has been used against PAO.[59,60] Both inhibitors block the oxidation of di- or polyamines to GABA. Gabaculine (Fig. 3) is an irreversible inhibitor of GABA transaminase derived from Streptomyces toyocaensis.[76] Irreversible inhibitors of polyamine oxidase have been developed (Fig. 6), and may be promising probes for plant cells.

Artifactual "Polyamine Metabolism"

Well established protocols are now available for the assay of the major enzymes involved in polyamine metabolism in plant cells. As this is not a review of techniques, the reader is referred to a recent compendium for details.[77] Most of these assays were originally developed for animal or bacterial cells, and in adapting them to plant systems we must not overlook the unique aspects of plant metabolism.

The routine assays for ODC and ADC involve measurement of labeled CO_2 released upon incubation of ^{14}C-ornithine or arginine, respectively. Ornithine labeled in the carboxyl position is available, and has been used in most cases, the assumption being that release of $^{14}CO_2$

is a direct measure of OD activity leading to Put formation. However, it has been found that ornithine can undergo oxidative decarboxylation by wheat leaf extracts, with the production of 4-aminobutanamide and 4-aminobutyric acid.[78] This reaction is also PLP-dependent. The major active component has a molecular weight of ca 4 kd. It is not known how widespread this activity is among higher plants, but in view of these results it is suggested that assays involving carboxyl-labeled ornithine must be followed by stoichiometric confirmation of Put production.

Similar precautions should apply to ADC assays. First, cells containing high arginase activity may rapidly convert labeled arginine to labeled ornithine, and the release of $^{14}CO_2$ will actually result from ODC, not ADC activity.[70] Second, the most commonly used labeled substrate for ADC assays is U-^{14}C-arginine. Under such conditions, arginase activity or oxidative decarboxylation such as that reported for ornithine can also lead to an overestimation of ADC activity. In this case too, measurement of labeled CO_2 release must be confirmed by stoichiometric confirmation of agmatine production.

Artifacts may also arise during the assay of SAMDC. The usual assay substrate, [1,^{14}C]S-adenosylmethionine, can yield $^{14}CO_2$ in reactions other than direct decarboxylation of SAM when added to crude plant extracts. The presence of an artifactual Put and Mg^{++}-dependent "SAMDC" activity has been reported in *Lathyrus*.[45] It can be inhibited by catalase, and has been attributed to the presence of a relatively unspecific diamine oxidase which is also capable of decarboxylating various amino acids including SAM. The real SAMDC activity is Put-independent. In addition to this artifactual enzymatic activity, a nonenzymatic decarboxylation of SAM can be obtained in the presence of PLP and Cu^{++} or Mn^{++} ions.[79]

Because $^{14}CO_2$ liberation from [1-^{14}C]SAM may not be sufficient evidence for SAMDC activity in crude extracts, the actual conversion of SAM to dSAM needs to be shown. An HPLC system has been developed which allows good resolution and sensitive determination of SAM and dSAM.[74] However, as the enzyme has a very high K_m for SAM and is subject to feedback inhibition by dSAM, stoichiometry can only be shown by coupling this reaction to spermidine synthase and to compare the yield of triamine to that of CO_2.[80]

As interest in the function of polyamines in plant cells spreads, it is expected that the above assays will be used by an increasing number of researchers. Most of the artifacts outlined here can be avoided by verification of results with an alternative assay. If stoichiometry is confirmed, then a standard procedure can be adapted for the system under study, even if the extracts are crude or partially purified.

POLYAMINES AND PLANT STRESS

Since the classic studies of Schulze and Prianishnikov (reviewed in Ref. 81), it has been known that nitrogenous compounds can accumulate in plants as a result of drastic changes in the environment. For example, seedlings or detached leaves grown in darkness without an external nitrogen supply show striking increases of asparagine and glutamine; these amides may function to detoxify ammonia released by amino acid oxidation.

Various mineral deficiencies can also cause increases in soluble nitrogen. Potassium deficiency in the same plant leads to a rapid disappearance of protein, and a marked increase in amino and amide nitrogen.[82] While studying the changes in free amino acids in K-deficient barley, Richards et al.[83,84] observed the presence of an unknown ninhydrin positive spot. This compound, present only in trace amounts under normal conditions, accumulated in K-deficient plants to become the predominant soluble amino compound; it was identified as putrescine by Richards and Coleman.[83]

It was later shown[85] that leaves of K-deficient wheat and red clover (Trifolium pratense) also accumulate Put. The buildup of this diamine under K stress has been established in every plant species examined so far (see Ref. 7); thus, the response appears to be universal.

Other mineral nutrient deficiencies may also cause significant increases in nitrogenous compounds. Ornithine, arginine and citrulline levels increase in sulphur-deficient flax (Linum usitatissimum).[86] The severity of K-deficiency symptoms in barley is strongly influenced by the external supply of phosphorus.[87] Whenever K supply is reduced, Put content rises above normal, but it accumulates in large amounts only when phosphorus supply is in excess of K. In tobacco, Put increases under deficiency of K, P, Ca, Mg, Fe, Mn, S and B,[88] but to a much larger extent in the K+ and P-deficient plants. Magnesium deficiency causes putrescine accumulation in barley, pea, bean and radish leaves, but is without effect on spinach leaves.[89]

Early studies in barley and white clover[90] showed that Put became rapidly labeled if K-deficient leaves were supplied with [2,^{14}C]-ornithine, suggesting that a direct decarboxylation mechanism is responsible for the increase in the diamine. However, several workers later found that agmatine is present in the K-deficient shoots, and that feeding ^{14}C-arginine results in a more rapid production of labeled putrescine than that observed feeding labeled ornithine.[87,91] Increases in ADC and NCP amidohydrolase parallel the rise in putrescine in K-deficient barley.[33,92,93] These results strongly suggest that Put formation under K deficiency occurs via arginine decarboxylation. This was the first evidence for the existence of a putrescine biosynthetic pathway previously known only in bacteria.[94]

The correlation between the increase in putrescine and ADC "activation" has also been found under other types of stress, such as low pH, ammonium nutrition and osmotic shock.[95-97] It is clear from these studies that polyamine metabolism is extremely sensitive to changes in the external environment, especially ionic type stresses. In recent years, the availability of specific inhibitors has allowed us to probe the mechanisms involved, and may help us uncover the physiological significance of these responses.

In our attempts to culture oat mesophyll protoplasts, we found them to contain significantly higher putrescine levels than the mesophyll cells they were derived from.[67,97] This change occurred during the 2- to 4-hour period involved in protoplast isolation, and appeared to be the result of osmotic shock. If detached oat leaf segments were floated over the same osmoticum (0.4-0.6 M sorbitol) used to isolate protoplasts, but in the absence of cell-wall digesting enzymes, Put levels rose rapidly, with a concomitant rise in ADC, but not ODC, activity. Addition of DFMA prevented the rise in putrescine and ADC, while DFMO treatment had no such effect. We concluded from these results that the osmotic shock-induced synthesis of Put was mediated via the ADC pathway. Experiments performed on oat leaves treated with low pH buffers gave the same results.

Two features of the stress response mentioned above are of interest in the context of this review. First, the incorporation of label from ^{14}C-arginine into Put in the osmotically stressed cereal leaves is not accompanied by a comparable increase in the label of Spd or Spm.[97] This result is in agreement with the finding that under other types of ionic stress, there is little if any change in the polyamines proper (see Ref. 7). It further suggests that the stress response involves not only a rise in Put biosynthesis, but also a block in polyamine synthesis. Second, the Put increase is strongly dependent on precursor availability. Leaf segments stressed in the dark reach only a third of the Put levels of stressed, light-incubated tissue.[97] If these dark-stressed leaves are fed with arginine, the level of stress-induced putrescine then becomes comparable to that of light-incubated stressed leaves. Thus, the size of the arginine pools is limiting to the response.

The striking similarities in the kinetics and magnitude of Put accumulation under various types of ionic stresses suggests that there may be a common mechanism involved. In the case of K-deficiency, the response is so massive as to result in a final Put concentration of 1.2% dry wt.[7] This represents at least 20% of the total nitrogen. We must ask, if for only this reason, what the physiological significance of these dramatic changes may be.

Coleman and Richards[85] suggested that alkali metal deficiency shifts the internal balance between inorganic anions and cations in the direction of increased acidity. Amine accumulation would function as a homeostatic mechanism to keep intracellular pH at a constant value. It was estimated that at least 29% and 30% of the cation deficit due to potassium deficiency can be restored by

Put in barley and black currant, respectively.[98] Smith[7] has estimated that Put and agmatine together may compensate between 45-140% for suboptimal K levels in the nutrient medium. It is cautioned, however, that interpretations based on these estimates may be of limited use considering that the full balance of cations and anions in K-deficient plants is not completely understood.

The conditions of ion imbalance under K-deficiency can be mimicked by an external supply of excess hydrogen ions.[95] When barley seedlings are fed with 0.025 N HCl or H_2SO_4, arginine, agmatine, and Put contents increase, as well as the activity of ADC and NCP amidohydrolase (2-fold increase). When cotyledons of Cucumis sativus are exposed to 5-10 mM HCl,[99] the Put titer and ADC activity increase at least 2-fold. Sulfur dioxide fumigation of pea seedlings causes a significant increase in free and bound putrescine.[100] Since SO_2 absorption into the cells results in higher acidity as a result of the formation of HSO_3^- and $SO_3^=$ ions, and eventually $SO_4^=$, amine accumulation may also in this case compensate for the relative deficit of cations.

It has also been shown that Put, and to a lesser extent Cad and Spd, accumulate when soybean seedlings are grown with ammonium as the sole nitrogen source.[101] In dark- as well as light-grown seedlings, there is a direct relation between the external ammonium supply and the endogenous putrescine titer.[96] ADC and agmatine iminohydrolase activities also increase, but no change is observed in ODC activity.[35 96] Put accumulation under ammonium nutrition has also been found in pea, corn and wheat.[102] Uptake and assimilation of ammonium also leads to the production of H^+ ions, part of which diffuse into the nutrient medium (physiologically acid reaction of ammonium salts), reducing cation uptake by competition effects.[103] Thus, amines synthesized under ammonium "stress" may bind excess H^+ ions.

The results with acid stress and ammonium nutrition are consistent with the hypothesis that Put accumulation compensates for lowered pH in the cytoplasm of the stressed cells. That such a change does occur and precedes and/or triggers Put accumulation, however, remains to be shown. The recent development of non-destructive (NMR) techniques for monitoring cytoplasmic changes in pH should make such studies possible.[104]

Other experiments, however, suggest that ion stress-induced Put accumulation is a deleterious response. Coleman and Richards[85] suggested that K-deficiency symptoms in barley are a consequence of Put buildup, since upon feeding Put through cut barley leaves, the white necrotic areas characteristic of potassium deficiency were produced. Agmatine and Put were most highly concentrated at the tip of the leaf, where degeneration typically began.[87] A similar time course has been reported in black currant leaves.[99]

Accumulation of diamines has also been observed during salinization (see Ref. 105). Feeding NaCl in excess of 50 mM leads to a sharp increase in the titer of putrescine and cadaverine in peas and horse bean plants.[106] ^{14}C-Arginine is rapidly incorporated into putrescine in the leaves of cotton plants grown under excess Na_2SO_4 (42-84 mM). In barley, horse bean, cotton and peas, giving putrescine and cadaverine in excess of 1 mM leads to loss of turgor and causes formation of necrosis, mimicking the symptoms of extreme salinization.[106,107]

Recent evidence also argues for Put accumulation being a toxic response. Oat protoplasts incubated in the presence of DFMA show marked reductions in Put compared to the untreated controls.[108] DFMA-treated protoplasts contain higher levels of Spd and Spm, and show improved rates of macromolecular synthesis. DiTomaso (personal communication) found that putrescine fed to corn seedlings causes membrane depolarization and increased potassium leakage. Overall, the available evidence tends to indicate that increases in Put induced by ionic stress may be injurious to the plant cell. However, the hypothesis of pH adjustment has not yet been disproved; several findings are consistent with it, and the question is still open.

Another intriguing stress response involving polyamines has been recently reported. It is well known that thermophilic bacteria synthesize Spd under normal conditions, but when grown at temperatures over 50°C they produce a new series of polyamines. These compounds have been implicated in the adaptation of thermophilic organisms to extreme environments.[109,110] The extreme expression of this adaptation to high temperature is the appearance of caldopentamine in cells grown at 80° C.[111] Rodriguez-Garay et al. have reported the presence of norspermidine (caldine, Fig. 1) and norspermine (thermine) in shoot meristems of alfalfa subject to drought stress.[112] Thus, it is an intriguing possibility that adaptation of higher plants to drought or high temperature stress may be mediated by these polyamines previously reported only in microorganisms.

INVOLVEMENT OF POLYAMINES IN PLANT GROWTH AND DEVELOPMENT

Correlations With Growth and Cell Division

In animal systems, polyamines have been shown to be necessary for growth of microorganisms such as *Haemophilus parainfluenzae*,[113] *Aspergillus nidulans*,[114] and polyamine-auxotrophic mutants of *Escherichia coli* and *Saccharomyces cerevisiae*.[115] Polyamines have also been implicated as growth factors in microorganisms and mammalian cells.[2,3]

Interest in the relation of polyamines to plant cell division began with Bagni's work in *Helianthus tuberosus*.[116] It was shown that aliphatic amines stimulated cellular proliferation in tuber explants grown *in vitro*. Subsequently, much evidence has accumulated indicating that polyamines are important in plant growth and development. The best established correlation in most reports is that polyamine titer and activity of their biosynthetic enzymes are usually high in cells where active growth and cell division occurs (see Refs. 6, 8, 177). Studies involving analysis of whole seedlings or plant parts in which meristematic cells comprise a minor portion may fail to reveal clear correlations.[118] There is an obvious problem in the interpretation of results obtained with such systems. It may be difficult to distinguish between "growth" by cell enlargement and that dependent on cell division. As a result, any correlations obtained cannot be clearly attributed to an effect of polyamines on a specific process (cell division vs. cell elongation).

In attempting to establish and understand these correlations, we should also be aware of the existence of organ-specific patterns of free and bound polyamines, and the presence of reverse gradients of di- and polyamines. For example, Spd and Spm are the predominant polyamines present in pea buds; Put is present in small amounts, and Cad is absent.[58] In contrast, Put and Cad are the major amines in pea internodes which contain only trace amounts of Spm.

The presence of reverse gradients of di- and polyamines is now well established in several systems and may in fact be of general occurrence. It was first reported in etiolated pea seedlings.[58] If the third internode is cut into sections and analyzed for polyamine content, Put titer follows a basipetal gradient, increasing from apical bud to the base of the internode. In contrast, Spd content is higher in the apical bud than in other regions of the epicotyl when expressed on a fresh weight basis (Fig. 7). However, on a protein basis, the distribution of Spd is relatively constant. The stable ratio of Spd:protein is consistent with the finding that this polyamine may facilitate macromolecular synthesis and stabilize nucleic acids in cell-free extracts.[2,14] Spm shows an even more dramatic acropetal pattern, and is not detectable below the sub-apical region. Similar reverse gradients have been shown in corn roots.[119] In corn coleoptiles, which lack meristematic region, these gradients are much less pronounced, if present at all.

How do reverse polyamine gradients arise? In the elongating zones of the root and shoot axis, where cell division has ceased, Put level may rise as a result of its limited conversion to Spd and Spm, assuming the rate of biosynthesis remains constant. Low Spd/Spm synthase activities in older cells may account for limited turnover of Put.

Whatever the mechanism underlying these reverse gradients, it is clear that Spd and Spm are the compounds most closely associated with cell division. Diamine gradients seem correlated with cell elongation rather than with meristematic activity.

The relation between polyamines and cell division has been recently studied in Jerusalem artichoke. Parenchyma explants from dormant tubers undergo a sequence of cellular changes over a period of 3-4 days in the presence of auxin and cytokinin.[120] The first 21-24 hrs comprise

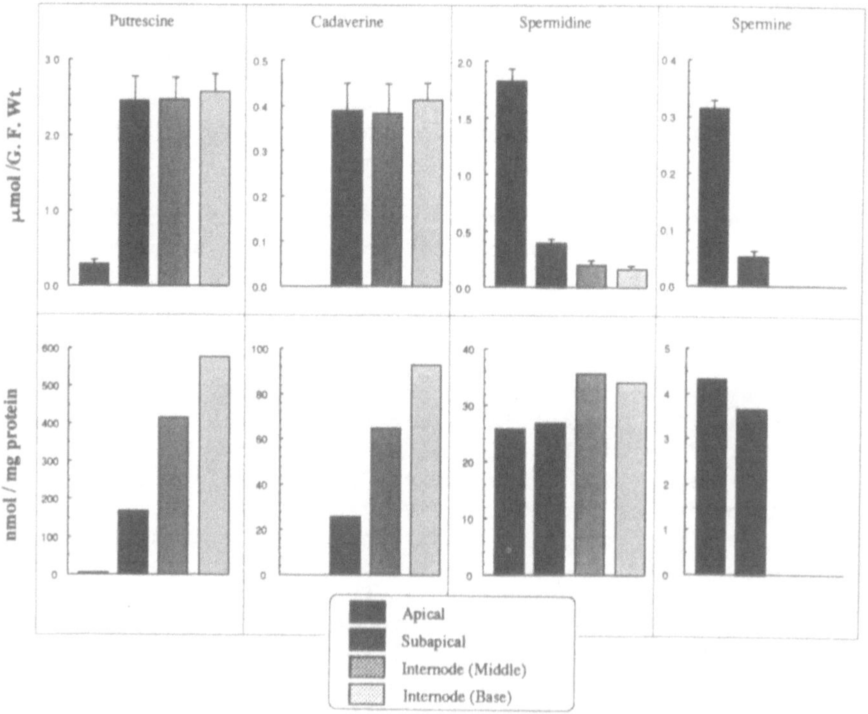

Fig. 7. Reverse polyamine gradients in pea epicotyls. Adapted from Flores, 1983.

an "activation" phase in which cell division is initiated, followed by a period of rapid and synchronous mitosis lasting another 24 hrs. The final phase is characterized by xylem differentiation. The titers of Spd, Dap and Cad increase about four-fold during the initial 24 hrs and decline thereafter. Thus, it appears that polyamine biosynthesis precedes the onset of cell division, results which are consistent with what is known in animal cells.[2,4] Put and ODC increase in the G_1 and S phase of the cell cycle, at least several hours before cell division (M phase) becomes apparent.

Do Polyamines Mediate Plant Hormone Response?

It is well known that plant hormones can stimulate cell division and/or elongation in many instances.

Several groups have followed changes in polyamine titers during hormone-mediated responses. The stimulation of internode growth in dwarf peas by gibberelic acid (GA) is correlated with enhanced ADC activity and accumulation of Put and Spd.[121] The rise in ADC is detected 6 hrs after GA treatment and before the start of internode growth. Experiments with inhibitors of polyamine biosynthesis support these findings. Treatment with DFMA or DFMO inhibits polyamine accumulation and internode elongation in normal and GA-stimulated dwarf pea mutants (nana).[122] Exogenous Put and agmatine restored growth and endogenous polyamine content. In lettuce hypocotyls, elongation induced by GA_3 was enhanced 179% when Put was added simultaneously with gibberelin.[123]

Similar results have been obtained in systems responsive to auxin treatment. Spd, Spm and Put increase in mung bean stem cuttings treated with indole butyric acid (IBA), prior to the development of any root primordia.[124] Addition of MGBG prevents the IBA-induced increase in Spd and Spm while increasing the level of Put. Spm supplied alone stimulates adventitious root formation but is far less effective than exogenously supplied auxin. Incorporation of label from ^{14}C-arginine or ^{14}C-ornithine into polyamines in auxin-stimulated mung bean cuttings is consistent with previous findings.[125] An early peak in polyamine synthesis is related to initiation of root primordia, and a later peak corresponds to root elongation.

Based on these and other findings, it has been postulated that polyamines are secondary messengers mediating the effects of plant hormones.[126] The correlations found so far are certainly intriguing and consistent with this hypothesis, but several problems remain. It has not been shown in any system that these putative second messengers, when added exogenously, are capable of inducing a response of the same magnitude as that which the growth hormone is capable of promoting. In at least some cases polyamines can be readily transported into plant cells;[127,128] therefore, lack of uptake cannot fully explain the limited inability of exogenous polyamines to induce a given response. The possibility remains that changes in polyamines are a result of plant hormone action unrelated to any second messenger which may be involved. It is hardly surprising, for example, that auxin- or cytokinin-stimulated cell division may result in Spd or Spm accumulation, as these compounds are required for continuous cell proliferation.

Transport and Localization

One of the problems in the interpretation of experiments involving exogenous polyamine treatment is our limited understanding of their uptake and localization in plant cells. Some progress in this area has been made recently.

The fate of labeled polyamines in seedlings appears to be strikingly different than that of their amino acid precursors. If labeled Put or Spd are injected into pea seedling cotyledons, little if any label is found in the embryo axis as Put or Spd after a few hours.[129] However, a significant amount of label from arginine or lysine shows up in Put and Cad, respectively. It was suggested that polyamines are not transported in pea seedlings, but instead made *in situ* from precursors transported from the storage organs of the germinating seedling. These results, however, may have been affected by the presence of a very active DAO present in the cotyledons. Considering the evidence that plant DAO is present in the cell wall, it is possible that the exogenously added Put and Spd may have been degraded before they could reach the transport site.

Bagni *et al*. have shown that polyamines can be taken up and translocated by isolated apple corymbs.[130] Put uptake can occur in petals of African violet (*Saintpaulia ionantha*) against a concentration gradient at low levels of Put in the external medium (0.5-100 μM), and following a concentration gradient at higher levels of external Put (0.1-100 μM).[127] Uptake of Put, Spd and Spm is dependent on the pH of the external medium.[131] Inorganic cations at low concentrations (17 μM) did not affect Put and Spd uptake; however, a 100-fold increase in concentration of Ca^{++} inhibited while K^{+} enhanced Spd uptake. Put and Spd are transported into carrot cells within 1 min following a biphasic system at low and high external polyamine concentrations.[128] Polyamine uptake is partially inhibited by 2,4-DNP and CCCP. Put, Spd and Spm have been detected in xylem and phloem exudates.[132] In summary, current evidence indicates that polyamines may indeed be translocated within the plant, but the mechanism involved is still not well understood.

Where are polyamines localized within the plant cell? Knowledge in this area is still scant. Put has been reported in the vacuole[131] and in the cytoplasm.[133] In carrot cell cultures, most of the Put can be recovered in the cytoplasmic soluble fraction; about 25% of the Put and most of the Spd pool is present in the cell walls.[128] Polyamines have been reported in isolated chloroplasts of Euglena gracilis and spinach leaves.[8]

Little is also known about the localization of polyamine biosynthetic enzymes. ODC activity has been found in the cell nuclei in germinating barley seeds.[29,134] ODC activity is also found in the nuclei-enriched fraction of tobacco cell cultures, but the largest activity is present in the cytosol.[135] ADC has only been found in the latter.[8] The enzyme for ornithine biosynthesis are found in the cytoplasm and plastid fraction of soybean protoplasts.[136] ODC, ADC and SAMDC activities have been detected in chloroplasts of Pinus radiata and in mitochondria isolated from tuber slices of H. tuberosus.[137] Spd synthase activity is found associated with chloroplasts of Chinese cabbage leaves.[46]

Polyamines and Somatic Embryogenesis

The "classic" system for the study of somatic embryogenesis is carrot cell culture.[138,139] Cells derived from carrot root phloem can be induced to undergo embryogenesis upon transfer from a medium containing 2,4-D to one from which the growth regulator has been omitted. It is assumed that carrot cells are competent to follow an embryogenic program and that 2,4-D prevents its expression. Because in this system previously undifferentiated cells undergo organization into embryonic axes upon a specific stimulus, we can thus follow the biochemical events associated with a developmental sequence.

Changes in polyamine titers were followed in carrot cells undergoing embryogenesis.[140] Within 24 hrs after transfer to an embryogenic medium, the level of Put increased to approximately twice that of the control. Incorporation of label from ^{14}C-arginine into Put by embryonic cells also occurred at twice the rate of control cells and was paralleled by increased ADC activity.[141] Feirer et al.[68] reported that DFMA (1 mM) inhibited embryo formation in carrot cells by 50%. This inhibition was reversible by the addition of Put, Spd or Spm.

Similar results have been obtained with other inhibitors of polyamine biosynthesis. Treatment with MGBG, an inhibitor of SAM-decarboxylase, resulted in lowered levels of Spd and Put, and was accompanied by a reduction in embryogenesis and growth.[142] Cyclohexylamine (CHA), an inhibitor of Spd synthase, also reduced Spd levels in the cells, with a parallel decrease in both processes. Fienberg et al.[143] confirmed that polyamine levels increased several fold in carrot cells upon transfer to embryogenic medium, but remained unchanged in control cultures grown in the presence of 2,4-D. The changes in ADC activity were also similar to those previously observed. Interestingly, these changes were not observed in a mutant cell line (W001) which is impaired in its ability to undergo embryogenesis. DFMA or CHA inhibited polyamine biosynthesis and decreased embryo formation by 95%. Addition of 5 mM Spd completely reversed the inhibition.

Androgenetic embryos can also be obtained *in vitro* from immature pollen grains.[144,145] Put and Spd levels rise during embryogenesis of pollen from *Nicotiana tabacum* and *Datura innoxia*. In the latter species, androgenetic induction is preceded by a peak in arginine. As with the carrot cell cultures, it appears that the ADC-mediated pathway for Put biosynthesis is directly correlated with embryogenesis. The activity of ornithine carbamoyl transferase, an enzyme involved in arginine biosynthesis, is also increased during embryogenesis of wild carrot cell cultures.[146]

Excess of endogenous Put derived from arginine may have a significant effect in cells competent to undergo embryogenesis. When carrot cells are transferred to an embryogenic medium containing 40 mg/l arginine, they reach the globular embryo stage but fail to develop further.[147] Exogenous Put also appears to arrest the embryos at an earlier stage. When arginine is deleted from the embryogenic medium, however, synchronous development of the arrested embryos resumes. This finding indicates that precise regulation of endogenous polyamine levels is required during critical stages in plant development.

The above results strongly suggest that ADC and polyamines are required for plant cell embryogenesis. Still, care must be taken in the interpretation of results involving inhibitors. In most cases there is a significant reduction in growth in addition to the decrease in embryo development. It is important to find conditions under which the growth of disorganized cell clumps remains unaffected, but which lead to arrest at specific developmental stages. The use of developmental mutants in these studies would be a powerful complement to experiments with specific inhibitors of polyamine biosynthesis.[148]

The Free vs. Bound Polyamine Dichotomy

Until recently, research on plant polyamines has focused mostly on the pools of "free" compounds. It is now apparent, however, that in many cases covalently bound forms of di- and polyamines can account for a significant portion of the metabolic pools. For example, Put in seeds of Amaranthus is present 100% as a bound form. In the same species, 89% of Spd and 60% of Spm are bound in the seed.[58] A full understanding of polyamine biochemistry and physiology must therefore account for the presence and function of these "bound" pools.

The routine procedures for polyamine analysis involve extraction in the cold with 5% perchloric acid (PCA) or trichloroacetic acid (TCA).[8] Polyamines which are freely soluble in PCA or TCA remain in the supernatant fraction after centrifugation and can be identified as such after analysis by HPLC or TLC. These are referred to as free polyamines, and presumably are present inside the cell in ionized form ("physiological" pH). Polyamines can also be released from the supernatant upon hydrolysis in 6N HCl, and are presumed to be linked with low molecular weight compounds. The amides formed between di- or polyamines and hydroxycinnamic acids (see below) can be found in this fraction. The pellet which remains after centrifugation of the PCA or TCA extract can also release polyamines upon hydrolysis; these may represent covalent linkages with high molecular weight compounds (e.g., proteins).[149]

The best characterized conjugated polyamines are the hydroxycinnamic acid amides (HCAs, Fig. 8). These compounds are widely distributed and have been reported from at least 13 families of higher plants, where they occur as the main phenolic constituents of the reproductive parts.[150] There are two major classes of HCAs: water-soluble basic amides of the aliphatic di- and polyamines, and water-insoluble neutral amides which include both aliphatic and aromatic amines (tyramine, dopamine, octopamine, and tryptamine).[150] The neutral amides are predominant in the male parts of tobacco flowers while the basic amides are usually found in the ovaries.[151] Caffeoyl-Put has also been found in the seeds of *Pentaclethra macrophylla*[152] and in tobacco flowers.[153] Feruloyl-Put was first detected in *Salsola subaphylla*[154] and subsequently in leaves and fruits of orange and grapefruit.[155] Feruloyl, p-coumaroyl, and caffeoyl putrescine have also been reported in tobacco callus.[156]

Besides their association with flowers, HCAs have also been linked to disease resistance. Large amounts of feruloyl-Put, diferuloyl-Put and feruloyltyramine accumulate during the hypersensitive response of tobacco after TMV infection.[150] These HCAs accumulate at very high levels in the living cells which surround the hypersensitive zone. The application of coumaroyl-, dicoumaroyl- and caffeoyl-Put to tobacco leaf discs causes a reduction of 90% in the number of local lesions caused by TMV.[157] Dimers of 4-coumaroylagmatine, known as hordatines, are found in barley shoots.[158,159] Hordatines inhibit the germination of a wide range of fungal spores and are related to resistance of barley seedlings to infection by *Helminthosporium sativum*.[160,161] One of the earliest reactions of potato tubers inoculated with avirulent isolates of *Phytophtora infestans* is the accumulation of amides of p-coumaric and ferulic acids with tyramine or octopamine.[162] Cinnamoylamine-conjugates may form complex structures such as marocyclic alkaloids. These are also proposed to function as chemical defenses against pathogens.[163]

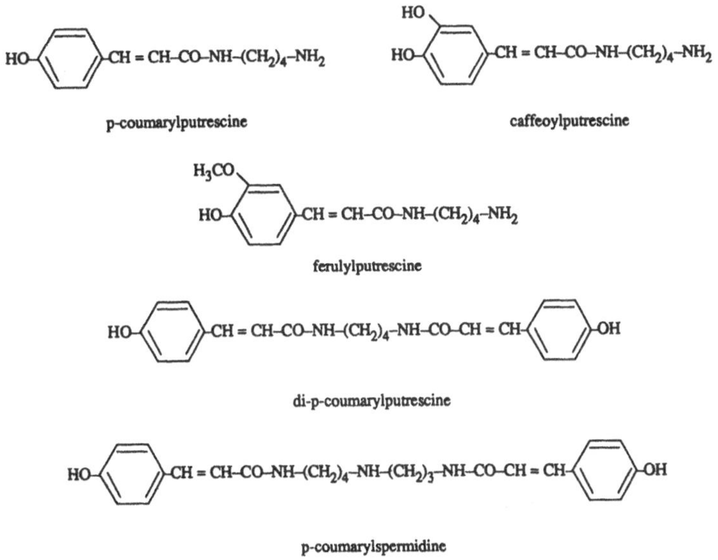

Fig. 8. Structures of some hydroxycinnamic acid amides.

Hydroxycinnamic Acid Amides and Flowering

One of the most striking features of HCAs is that they first appear in apical leaves of plants approaching the flowering stage, before the first observable sign of a floral apex (Fig. 9).[164] The highest levels, however, are reached in fully developed flowers. Several reports indicate a tight correlation between HCA accumulation and ripeness to flower.[151,164-166] In tobacco, this state is reached at about 50 days after planting. Basic HCAs such as caffeoyl-Put and caffeoyl-Spd accumulate even at temperatures inhibitory to flowering.[164] Phenylalanine ammonia lyase (PAL) activity, which may be limiting in the synthesis of cinnamic acids,[167] also increases in temporal coincidence with the onset of the floral stage. In contrast, HCAs have never been detected at any stage of development in a mutant of tobacco (RMB_7) which lacks the ability to flower and remains vegetative even under inductive conditions.[168] Interestingly, this mutant is also extremely susceptible to TMV infection. The suggestion from the above results is that HCAs may be involved in the development of competence to flower.

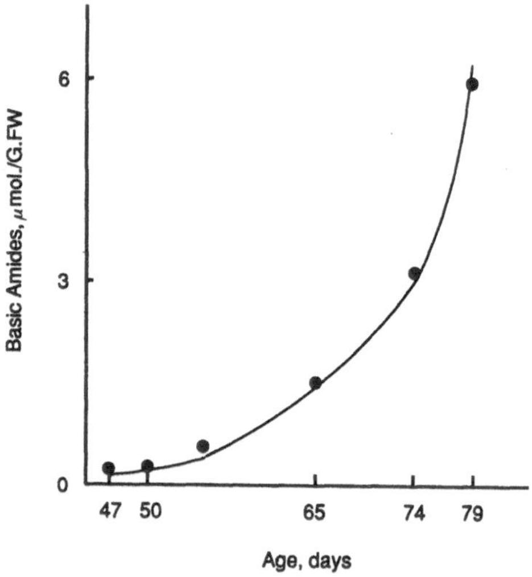

Fig. 9. Accumulation of basic amides in tobacco shoot apex. Adapted from Martin-Tanguy, 1985.

Polyamine levels have been studied during flower induction. Dai and Wang[169] found that Put declined after photoperiodic induction of Pharbitis nil, a short-day species. The content of Spd and Spm showed dramatic fluctuations in the induced cotyledons compared with the non-induced controls. Agmatine increased at the time of floral induction in leaves of Sechium edule.[170] Put and Spd declined before the appearance of the first floral buds. Only free polyamines were measured in this study, and it is not known if the above changes are related to conjugation with phenolics or other compounds.

The distribution of HCAs in higher plant families supports their association with flower development.[151,171,172] HCAs are found in large amounts in the male and female flowers of Araceae, but are absent from sterile flowers.[172] In maize, HCAs also occur at considerable levels in anthers of male fertile lines but are absent, or present in trace amounts, in anthers of cytoplasmic male sterile lines (Table 1). When fertility is regained after crossing with a restorer line, normal HCA levels reappear.[173] HCAs have been proposed as biochemical markers of male fertility.[150]

Table 1. Hydroxycinnamic acid amides (HCA) content in the anthers of male fertile and male sterile lines of maize.

HCA Compounds	Male Fertile Lines F7N	FC3	Male Sterile Lines F7T	F7C
	(nmol/g fresh weight)			
BASIC				
Ferulylputrescine	6000	6500	50	90
Tyramine	2000	1800	--	--
NEUTRAL				
Diferulylputrescine	2600	3000	90	50
Diferulylspermidine	130	100	--	--
Diferulylspermine	40	70	--	--
Ferulytyrosine	1000	1200	20	50

Adapted from Martin-Tanguy et al. (1982), Phytochemistry 21: 1939-1945.

Considering the strong correlations between the presence of HCAs and normal flower development, it is surprising how little is known about their metabolism in plant cells. The biosynthesis of coumaroyl agmatine has been shown to proceed via a coumaroyl-CoA precursor. The enzyme catalyzing the transfer of the coumaroyl moiety to agmatine has been characterized in barley seedlings.[160,174] At present, it is not known if a similar mechanism applies to HCAs. HydroxycinnamoylCoA: putrescine transferase was not detected in tobacco cell lines which overproduce cinnamoyl-Put.[175]

Polyamine Metabolism in Cell and Organ Cultures

Cell cultures with a simplified pattern of HCAs may be used to understand the physiology of these compounds.[176] Cell lines with high levels of cinnamoyl-Put levels have been established.[177] Selection of tobacco cells resistant to the amino acid analog p-fluorophenylalanine (PFP) resulted in a cell line (TX4) which shows dramatic increases in cinnamoyl-Put, up to 10% dry wt. This is correlated with a 10- to 20-fold rise in the activity of PAL, ADC and ODC.

Changes in nitrogen source can also lead to the accumulation of cinnamoyl-Put in tobacco XD cells. This cell line, derived from _N. tabacum_ cv. Xanthi, is able to grow in a variety of organic nitrogen sources in addition to nitrate or ammonium.[178] These cells can accumulate cinnamoyl-Put to levels similar to those in flower parts by simply switching the nitrogen source.[178] Cells grown in nitrate have very low levels of free and conjugated putrescine. The high-urease (14U) cell line grown on urea contains caffeoyl-Put as the major organic nitrogen compound. A similar pattern is found in XD cells grown on γ-aminobutyric acid (GABA). A cell line which can grow on Put as the sole nitrogen source (PUT) has been selected from XD cells and shows intermediate levels of free and found Put.[178] These Put-utilizing (PUT) cells appear to behave biochemically like flower cells. Cinnamoyl-Put are not only made but also turned over in PUT cells. Both the 14U and PUT cells take up labelled putrescine at the same rate. Most of the label appears as cinnamoyl-Put in urea cells, while PUT cells convert over 60% to GABA within 2-4 hrs.[178] Hydroxycinnamoyl-Put and conjugated GABA are suggested as obligatory intermediates in the assimilation of Put by PUT cells. Caffeoyl-GABA, a presumed intermediate in the catabolism of Put, is present in both PUT cells and tobacco flowers.[64] It can be considered a flower-specific metabolite, as it has not been detected in vegetative tissue.

The relationship between polyamines and flowering has also been studied in MGBG-resistant tobacco cell lines and other variants resistant to polyamine biosynthetic inhibitors developed by Malmberg _et al_.[71,179] Plants resistant to MGBG or DFMO are developmentally abnormal and exhibit phenotypes ranging from the inability to regenerate, arrested development in shoot culture, dwarf plants and developmental switches in flowers. Plants regenerated from MGBG-resistant cell cultures showed a variety of floral phenotypes, including stamenoid ovules, stigmoid anthers, petaloid anthers and sepals, shrunken anthers, long styles and shifts in temporal development.[179]

Cell lines resistant to polyamine inhibitors show dramatic differences in metabolic pools compared to wild-type cells. The elevated titer of free Put in DFMO-resistant tobacco cells (Dfrl) is not accompanied by a corresponding increase in Put conjugates.[180] Mgr3 and Mgr12, two MGBG-resistant lines, have elevated levels of conugated Put but normal levels of polyamines and their biosynthetic enzymes.

Wild-type tobacco cells and some resistant lines utilize arginine-derived Put preferentially. Relative utilization of ornithine or arginine is measured by following precursor label into Put and its conjugates. Mgr12, in contrast, utilized ornithine-derived Put for conjugation to cellular constituents.[180] Arginine-derived Put is incorporated into conjugated forms at a much lower rate in these cells. These results suggest that two forms of Put are present in the cell and are independently regulated.

Tobacco cell cultures exposed to acidic shock synthesize Put mainly via arginine, but this is not reflected in an increased conversion of Put into conjugates (A.C. Hiatt, unpublished). It appears that formation of Put conjugates is not affected by acid stress. During the log phase, levels of conjugated Put rises proportionately to increased synthesis and accumulation of Put. In contrast, Put accumulation in acid stressed cells was accompanied by reduction of 40% in cell division and inhibition of conjugate synthesis. These results suggest that free and bound Put may serve entirely different purposes during growth and development of plants.

While cell cultures may give some insights into the biosynthesis and function of HCAs, we still need to relate this information to the whole plant. A system of intermediate complexity would be useful in this respect. The development of the thin cell layer (TCL) system of tobacco by Tran Thanh Van[181] has provided a tool for the study of flower morphogenesis and development. Epidermal peels are excised from stalks or pedicels carrying mature green fruits and grown in a medium composed of macro and micro

Table 2. Effect of exogenous spermidine on flower floral morphogenesis in thin cell layers of tobacco cv. Samsun.

Concentration (mM)	Number of Flower Buds/Explant
0	7.1[a]
0.5	10.6[ab]
1.0	16.2[b]
2.5	8.9[a]
5.0	1.2[c]

Means followed by the same letter are not significantly different at 5% level of significance (C. Protacio and H. Flores, unpublished).

nutrients, vitamins, sucrose, agar and varying amounts of auxin and cytokinin. After 2 weeks in culture under the right conditions, flower buds can be observed on the explant surface. The morphogenetic program can be changed at will by varying the hormonal balance.[182] Thus, depending on the amount and ratio of auxin:cytokinin, epidermal peels can be induced to form flower buds, vegetative buds, roots, or undifferentiated callus.

Tiburcio et al.[75] have shown that DFMA inhibits floral bud initiation while DFMO inhibits subsequent development in TCLs treated at the beginning of the experiment and after 10 days. These results suggest that Put may be required for floral bud initiation and development. It is not known, however, if this effect is mediated by Spd, Spm or conjugated polyamines.

We have found that addition of Spd to a flower-inducing medium (FIM) increases the number of flower buds per explant (Table 2). In addition, preliminary results show that Spd and Spm addition stimulate floral bud formation even in shoot-inducing medium (SIM) (Table 3). In particular, exogenous Spm induces floral bud formation to

Table 3. Effect of polyamine addition on flower and vegetative bud formation from thin cell layers of tobacco cv. Samsun.

Treatment	Number of Flower Buds Per Explant	Number of Vegetative Buds Per Explant
FIM	10.3[a]	7.3[a]
SIM	1.2[b]	7.9[a]
SIM + 0.5 mM Put	0.4[b]	14.4[b]
SIM + 1.0 mM Put	2.0[b]	19.1[b]
SIM + 5.0 mM Put	0.0[b]	0.4[c]
SIM + 0.5 mM Spd	2.1[b]	16.5[b]
SIM + 1.0 mM Spd	2.8[b]	14.0[b]
SIM + 5.0 mM Spd	3.3[b]	13.7[b]
SIM + 0.5 mM Spm	7.5[a]	14.1[b]
SIM + 1.0 mM Spm	8.3[a]	15.1[b]
SIM + 5.0 mM Spm	11.2[a]	13.3[a]

Means followed by the same letter are not significantly different at 1% level of significance (C. Protacio and H. Flores, unpublished).

the same levels obtained with FIM. Put is without effect. Vegetative bud formation is also stimulated by Put, Spd, and Spm. Stimulation of floral bud formation by Spd in tobacco TCLs grown in SIM has been reported by Kaur-Sawhney *et al.*[183] In this case, however, no comparison was shown with flower-inducing medium.

Are conjugated polyamines relevant in the TCL system? Tiburcio *et al.* have shown that the conjugated

Table 4. Effect of AOPP on bud formation in tobacco thin cell layers.

Treatment	Number of Flower Buds Per Explant	Number of Vegetative Buds Per Explant
CONTROL	6.0[a]	2.2[a]
AOPP	1.4[b]	2.7[a]

Means followed by the same letter are not significantly different at 5% level of significance (C. Protacio and H. Flores, unpublished).

levels of Spd in explants forming flower buds are more than 7-fold higher than in vegetative explants. In contrast, Torrigiani et al.[184] found that titers of conjugated polyamines were comparable in both developmental programs; conjugated Put was consistently higher than conjugated Spd. The identity of the polyamine conjugates reported by these groups was not established, and it is therefore difficult to explain their similar contrasting results.

In our TCL system, dramatic changes are apparent in the patterns of free and conjugated Put. Caffeoyl-Put is not detectable in freshly excised epidermal peels but accumulates with the onset of floral bud differentiation. At the same time, free Put decreases by about the same magnitude (Fig. 10). It thus appears that endogenous Put is conjugated to a large extent in developing flower buds. Aminooxyphenyl propionic acid (AOPP), an inhibitor of PAL, causes a significant reduction of flower bud formation, with little or no effect on the number of vegetative shoots (Table 4). These results suggest that polyamine conjugates are in fact necessary for differentiation of flower buds *in vitro*.

Fruit Set and Development

Cell division occurs at a high rate during early development of fruit. In view of the observed correlations between polyamines and cell division, one would

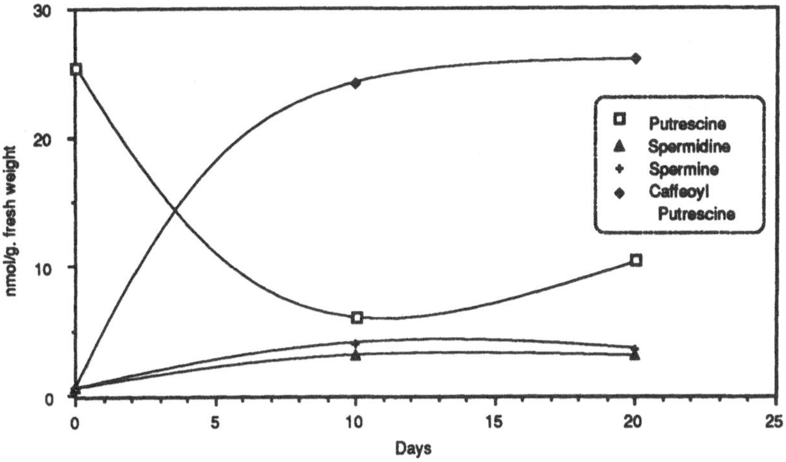

Fig. 10. Changes in free and conjugated putrescine in tobacco thin cell layers (C. Protacio and H. Flores, unpublished).

expect high polyamine titer during this developmental stage. If this correlation were biologically significant, one would further expect that exogenous polyamine application would improve fruit set. Costa and Bagni[185] sprayed apple trees with Put, Spd, and Spm at 10^{-4}M and 10^{-5}M. The number of fruits set per tree as well as final fruit weight increased.

Pollination of the tomato ovary is followed by increased levels of auxin in this organ.[186] Exogenously applied auxin can induce fruit setting in tomato[187] and mimic the effects of natural pollination. ODC is present in pollinated tomato ovaries[188] and increases dramatically during the early stages of fruit growth.[189] Auxin application to non-pollinated tomato ovaries results in a comparable increase in ODC activity. ADC is also present, but its activity does not appear to change during the logarithmic phase of fruit growth.[190] DFMO treatment reduces post-fertilization growth of tomato ovaries, and ODC activity decreases to only 3% of the control level.[190] The effect is completely reversed by exogenous Put. It appears that Put synthesized via ODC is essential during the first 10 days of tomato fruit development. Similar experiments in tobacco ovaries confirm the suggestion that ODC and Put are important during fruit development.[191]

While polyamine biosynthesis may occur at high rates during early fruit development, it appears to decline as maturation approaches. Such a situation has been found in avocado, pears and mandarin.[192-194]

Polyamines and Plant Senescence

Plant senescence involves a series of degradative events that lead to death of an organ or of the whole organism.[195] Leaf senescence involves the breakdown of nucleic acids and proteins, disintegration of chloroplast structure, and loss of chlorophyll and photochemical activities. These changes are accompanied by a rise in hydrolytic enzymes.[196] Several studies have shown that exogenous polyamine treatment can delay senescence. Protoplasts isolated from oat leaves show sharp rises in the activity of RNAse and proteases which mimic the senescence syndrome of detached leaves.[49] Addition of Cad, Put or Spd stabilizes the protoplasts, increases mitosis and delays senescence.[197] Polyamine addition results in net increases in synthesis of proteins, RNA, and DNA in the protoplasts. Polyamines also stabilize chloroplast thylakoids[198] and retard chlorophyll loss in leaves of barley, radish and *Hydrilla verticillata*.[199-201] Furthermore, polyamines retard or prevent the increase in RNAse and protease in detached oat and radish leaves.[48,202] Spm is more active than Spd which in turn is more active than Put or Cad in inhibiting the rise in RNAse and protease levels. In suspension cultures of Paul's Scarlet rose, Put and Cad at 1 mM and Spd and Spm at 0.1 μM delayed senescence.[203] Polyamine-treated cultures remained 50-100% viable after 60 days of culture.

Polyamines may retard senescence by inhibiting ethylene production[204] or by stabilizing nucleic acids and cell membranes against enzymatic degradation and solute leakage.[2,197,205,206] However, these effects are still far from being understood. For instance, while exogenous treatment with polyamines reduced the amount of ethylene produced by senescing petals of *Tradescantia*, anthocyanin leakage was not prevented from the same petals.[204] Roberts *et al.*[207] found that endogenous levels of Put rose while Spd and Spm levels remained the same during senescence without any inhibitory effect to ethylene production in cut carnation flowers. When inhibitors of polyamine synthesis were applied to the carnation flowers, ethylene production and the onset of senescence was promoted. Inhibition of ethylene production by aminooxyacetic acid, on the other hand, increased the level of Spm. Downs and Lovell[208] found no protective effect of Put or Spd (0.1-10 mM) added to cut carnation flowers; nor did they inhibit ethylene production. Polyamine treatment may retard chlorophyll loss in detached leaves incubated in darkness, but accelerates it during light incubation.[48] A similar effect has been observed in leaves of *Hydrilla*.[201]

The apparent inverse relation between polyamines has been tested in ripening fruits. Inhibition of ethylene biosynthesis by AVG increased the incorporation of ^{14}C-methionine into Spd and Spm in orange peel discs.[50] Addition of Put or Spd also resulted in increased incorporation of label from methionine. Addition of Put or Spm to avocado fruit inhibits ethylene biosynthesis by inhibiting both ACC synthase and conversion of ACC to ethylene.[209] In pear discs, Spm is a very effective inhibitor of ethylene synthesis.[210] Cultivar differences in polyamine content have also been related to fruit ripening. Summer pears cv. Barlett begin to synthesize ethylene immediately after harvest, in contrast to Comice pears, which usually require 40-45 days of low temperature storage.[193] In both cultivars, the decrease in polyamine content in the fruit corresponds to the time when ethylene synthesis begins. The titer of Spm, in particular, is much lower in the cultivar which ripens early.

Overall, these studies suggest that polyamines may be involved in the control of plant senescence. Although there is a close inverse correlation, it is not yet clear if polyamine and ethylene are in fact antagonistic in the intact plant.

POLYAMINE-DERIVED ALKALOIDS

As shown above, conjugated forms of di- and polyamines such as the hydroxycinnamic acid amides may be important

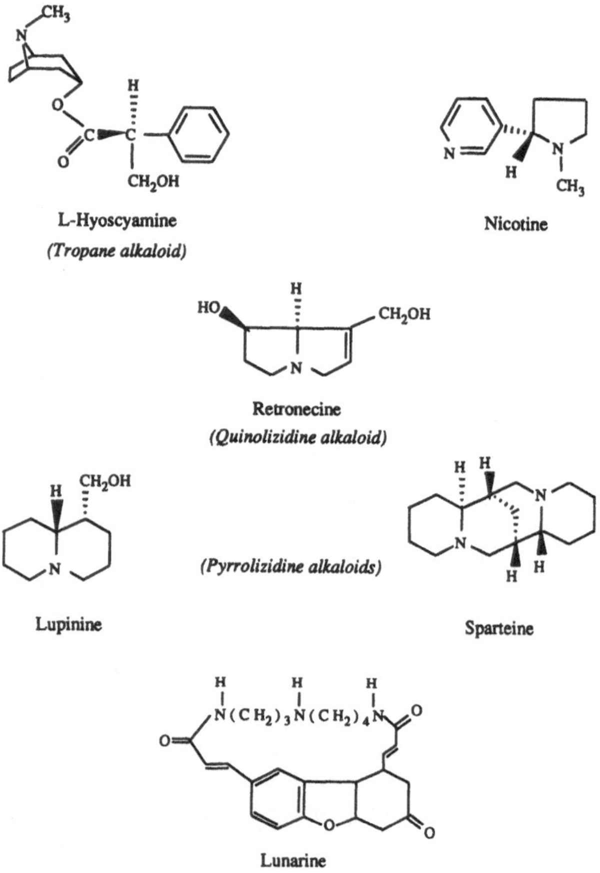

Fig. 11. Structures of di- and polyamine-derived alkaloids.

in the regulation of flower development or as storage forms of polyamines. In addition to these, a wide variety of secondary metabolites can be synthesized from di- or polyamines (Fig. 11) and are also of biological interest. Put derivatives include the nicotine, tropane, and pyrrolizidine alkaloids. Cadaverine is a precursor of the quinolizidine alkaloids, while lunarine alkaloids are derived from Spd.[211] A number of other distinct alkaloid classes are derived from polyamines, including further transformations of the cinnamic acid and fatty acid conjugates.[212] Representatives of these alkaloids are widespread in the plant kingdom.

The study of the nicotine and tropane alkaloids has provided the main body of knowledge concerning secondary metabolites derived from Put. Nicotine, the major biologically active component of tobacco leaves, is a feeding deterrent for many animals and was used as an insecticide for many years.[213] It is found in the genus Nicotiana and other Solanaceae, but has also been documented in many plant orders, including the primitive lycopods and horsetails.[214] The tropane alkaloids are related to nicotine, and include the medically important anticolinergic compounds hyoscyamine (Fig. 11) and scopolamine. In addition to the Solanaceae, tropane alkaloids are also found in the Convolvulaceae, Cruciferae, and Rhizophoraceae.[215]

A striking similarity has been shown in the plant organ where synthesis of these two classes of alkaloids occurs. In a series of classic grafting experiments between tobacco and tomato, Dawson showed that the pattern of nicotine alkaloids in the shoots was dependent on the source of the rootstock.[216] Thus, the shoots of tomato grafted over tobacco roots accumulated nicotine, but the reciprocal graft did not. Similar results were obtained for tropane alkaloids (see Ref. 214). Subsequent experiments with root cultures have conclusively shown that this organ is the site of synthesis of nicotine and tropane alkaloids.[217,218]

The biosynthetic pathway for nicotine and tropane alkaloids is shown in Figure 12. The base moiety of these compounds, N-methylpyrroline, is derived from N-methyl-Put via 4-methylaminobutyraldehyde. N-methyl-Put can be produced through direct methylation of Put, or by decarboxylation of δ-N-methylornithine. The contribution of these alternative routes has been studied by following the incorporation of label from ornithine and arginine.

In the case of nicotine, the label from [2-^{14}C] ornithine is incorporated symmetrically into the carbons adjacent to the N-methyl moiety of the pyrroline ring.[211] The biosynthesis of this alkaloid in tobacco callus can be inhibited by treatment with the ADC inhibitor DFMA.[219] These results indicate that Put derived via ADC is the

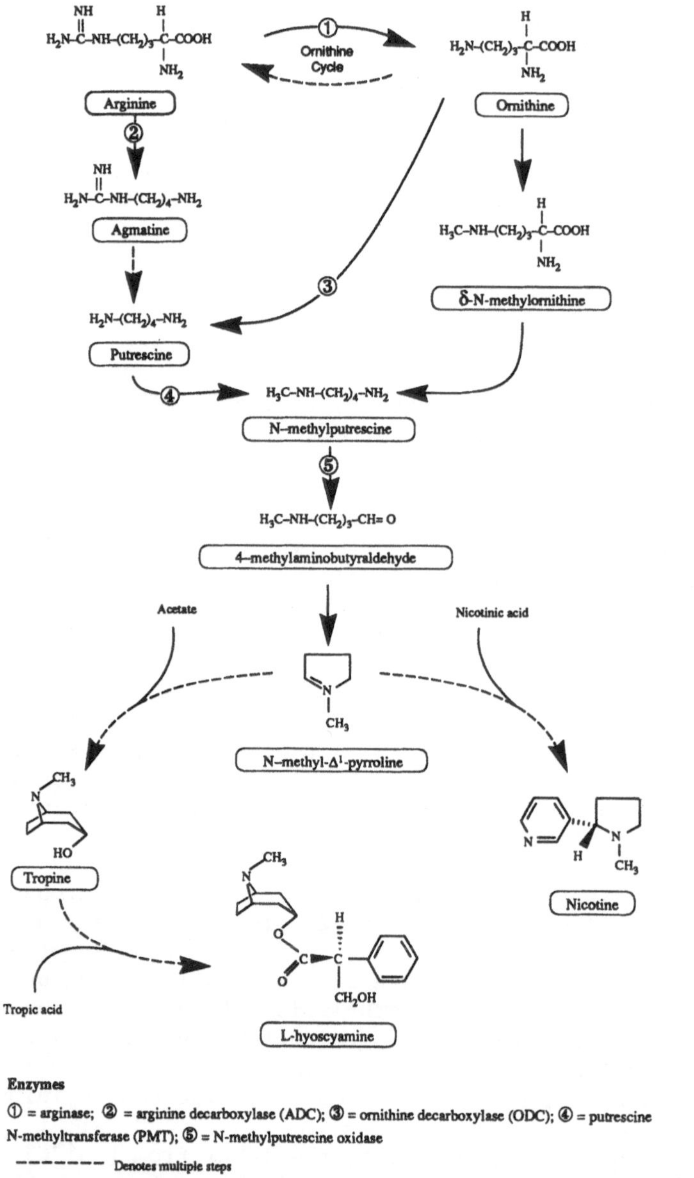

Fig. 12. Schematic pathway of nicotine and tropane alkaloid biosynthesis.

direct precursor, and therefore the symmetrical intermediate, in the synthesis of the N-methylpyrroline ring of nicotine. Spd appears to be an indirect precursor of this ring via oxidation to Put.[220]

The biosynthetic origin of the tropane alkaloids is less clear. Leete[221] found that when [2-^{14}C]ornithine was fed to plants of Datura stramonium, the label was found in the C-1 bridgehead of the tropane moiety in hyoscymaine, arguing for an asymmetric intermediate. He proposed that δ-N-methylornithine was the preferred intermediate (Fig. 12). However, Yamada and Hashimoto[222] could not detect the enzymatic activities postulated for these reactions in Hyoscymus albus root cultures, which also make tropane alkaloids. Instead, they found a strong correlation between the onset of hyoscyamine production and an earlier rise in Put N-methyltransferase (PMT). The activity of PMT reached a maximum at about the sixth day of culture, following earlier activity profiles for ADC, ODC, and arginase. Accordingly, they suggested that in Hyoscyamus at least, most if not all of the N-methyl-Put is synthesized via a symmetric intermediate, presumably Put. The biosynthetic origin of the tropane ring thus remains an open question.

The pyrrolizidine alkaloids are found in several genera of the Asteraceae, Boragianceae, and Leguminosae, with over 200 different structures known from natural sources. Alkaloids belonging to these classes are thought to serve as plant chemical defenses.[214,223] In addition, certain insects accumulate pyrrolizidine alkaloids in order to repel potential predators while others modify them into sex pheromones.[224]

The biosynthesis of the pyrrolizidine alkaloids has been investigated primarily in species of Crotalaria and Senecio. Radiolabelling experiments show that ornithine, arginine, Put, Spd, homospermidine and Spm are biosynthetic precursors of the necine base moiety via a symmetrical C_4 intermediate, presumably Put.[215] Hartmann et al.[223] demonstrated a close relationship between the biosynthesis of polyamines and senecionine N-oxide, the major pyrrolizidine alkaloid produced in root cultures of S. vulgaris. DFMA completely blocked the incorporation of labelled ornithine and arginine into polyamines and alkaloids while DFMO had no effect. Ornithine was rapidly converted into arginine, but the reverse did not occur. These results, combined with the fact that agmatine was as efficiently incorporated into senecionine N-oxide as Put, suggest that the pathway flows completely through arginine and agmatine while ODC makes no contribution. A reversible interconversion was observed between Put and Spd. It was proposed that a close coordination between the primary and secondary biosynthetic pathways allows the root tissue to maintain a constant alkaloid concentration independently of growth rate. Increases of polyamine levels in rapidly growing tissue may provide the substrate for maintaining the proper level of repellent compound, i.e., senecionine N-oxide.

The quinolizidine alkaloids are represented by about 70 related structures and are of common occurrence in the Leguminosae, particularly associated with the genus Lupinus, but are also found in the Chenopodiaceae, Ranunculaceae, Berberidaceae, Papaveraceae, and Solanaceae (Fig. 11).[224] Biosynthetic studies using an enzyme preparation from a suspension culture of L. polyphyllus showed the incorporation of three cadaverine molecules into a tetracyclic quinolizidine alkaloid, using pyruvate as a substrate for four required transaminations. In vivo, this enzyme system appears to be localized in the chloroplasts. Lysine, the precursor of cadaverine, is also made in this organelle.[223]

PERSPECTIVE

Polyamines are deceivingly simple compounds. Their relatively dull aliphatic structures belie the complexity of alternative biosynthetic pathways, their role as precursors of numerous secondary metabolites, and their possible functions in plant growth and development. It is now clear that polyamines occupy an important niche in the metabolic network of plant nitrogen metabolism. They can no longer be considered biochemical oddities, as their peripheral treatment in standard plant biochemistry texts indicates.

We still have a limited understanding of polyamine biology in plants. However, the progress made in recent years has been substantial, and we must be optimistic about the exciting prospects which lie ahead in this area of plant metabolism. We predict that specific inhibitors of polyamine metabolism will become standard tools for plant systems. We expect that the selection of developmental mutants which reflect specific lesions in the polyamine pathways will result in useful experimental systems. As the key enzymes in polyamine synthesis and catabolism become fully characterized and purified, we can look forward to molecular probes such as monoclonal antibodies and cDNA clones which will allow us to study the genetics and regulation of these pathways, and get us closer to understanding what polyamines do.

ACKNOWLEDGMENTS

We thank Marisa Protacio and Paula Sgrignoli for help with typing this manuscript, and Marleni Ramirez for editorial assistance.

REFERENCES

1. ROSENHEIM, O. 1924. The isolation of spermine phosphate from semen and testis. Biochem. J. 18: 1253-1263.
2. BACHRACH, U. 1973. Function of naturally occuring polyamines. Academic Press, New York, 211 pp.
3. COHEN, S.S. 1971. Introduction to the polyamines. Prentice-Hall, New Jersey, 179 pp.
4. McCANN, P.P., A.E. PEGG, A. SJOERDSMA. 1987. Inhibition of Polyamine Metabolism: Biological Significance and Basis for New Therapies. Academic Press, San Diego, 371 pp.
5. RICHARDS, F.J., R.G. COLEMAN. 1952. Occurrence of putrescine in potassium-deficient barley. Nature 170: 460.
6. SLOCUM, R.D., R. KAUR-SAWHNEY, A.W. GALSTON. 1984. The physiology and biochemistry of polyamines in plants. Arch. Biochem. Biophys. 235: 283-303.
7. SMITH, T.A. 1984. Putrescine and inorganic ions. Recent Adv. Phytochem. 18: 7-54.
8. SMITH, T.A. 1985. Polyamines. Annu. Rev. Plant Physiol. 36: 117-143.
9. SMITH, T.A. 1979. Arginine decarobylase of oat seedlings. Phytochemistry 18: 1447-1452.

10. OSHIMA, T. 1979. Molecular basis for unusual thermostabilities of cell constituents from an extreme thermophile, Thermus thermophilus. In Strategies of Microbial Life in Extreme Environments. (M. Shilo, ed.), Dahlem-Konferenzen, Berlin, pp. 455-469.
11. BAGNI, N. 1968. Spermine e spermidine nei semi. G. Bot. Ital. 102: 67-72.
12. FLORES, H.E., A.W. GALSTON. 1982. Analysis of polyamines in higher plants by high performance liquid chromatography. Plant Physiol. 69: 701-706.
13. SMITH, B.N., B.J.D. MEEUSE. 1966. Production of volatile amines in some arum lily species. Plant Physiol. 41: 343-347.
14. SMITH, T.A. 1975. Recent advances in the biochemistry of plant amines. Phytochemistry 14: 865-890.
15. TAIT, G.H. 1976. A new pathway for the biosynthesis of spermidine. Biochem. Soc. Trans. 4: 610-612.
16. SRIVENUGOPAL, K.S., P.R. ADIGA. 1980. Partial purification and properties of a transamidinase from Lathyrus sativa seedlings. Biochem. J. 189: 553-560.
17. KUTTAN, R., R. RADAKRISHNAN. 1972. Studies on the biosynthesis of sym-homospermidine in sandal (Santalum album L.). Biochem. J. 127: 61-67.
18. SRIVENUGOPAL, K.S., P.R. ADIGA. 1980. Enzymic synthesis of sym-homospermidine in Lathyrus sativa (grass pea) seedlings. Biochem. J. 190: 461-464.
19. SMITH, T.A. 1971. The occurrence, metabolism and functions of amines in plants. Biol. Rev. Camb. Philos. Soc. 46: 201-241.
20. REUTER, G. 1963. Arginine als vorstuffe von galegin in Galega officinalis L. Arch. Pharm. (Weinheim) 296: 516-522.
21. JOHNE, S., D. GROGER, R. RADEGLIA. 1975. Tetramethylputrescine from young plants of Ruellia rosea. Phytochemistry 14: 2635-2636.
22. GRIFFIN, W.J. 1967. The alkaloids of Duboisia leichhardtii: tetramethyl putrescine. Australas. J. Pharm. 48: S20-21 (Suppl. 51).
23. KREBS, H.A., K. HENSELEIT. 1932. Untersuchungen uber die harnstoffbildung im tierkorper. Hoppe-Seyler's Z. Physiol. Chem. 210: 33-66.

24. BEEVERS, L. 1976. Nitrogen Metabolism in Plants. Edward Arnold, London, 333 pp.
25. BAGNI, N., G.L. CALZONI, A. SPERANZA. 1978. Polyamines as sole nitrogen sources for Helianthus tuberosus explants in vitro. New Phytol. 80: 317-323.
26. MORRIS, D.R., A.B. PARDEE. 1966. Multiple pathways of putrescine synthesis in Escherichia coli. J. Biol. Chem. 241: 3129-3135.
27. KAUR-SAWHNEY, R., L.M. SHIH, H.E. FLORES, A.W. GALSTON. 1982. Relation of polyamine synthesis and titer to aging and senescence in oat leaves. Plant Physiol. 69: 405-410.
28. CHOUDHURI, M.M., B. GHOSH. 1982. Purification and partial characterization of arginine decarboxylase from rice embryos (Oryza sativa L.). Agric. Biol. Chem. 46: 739-743.
29. KYRIAKIDIS, D.A. 1983. Effect of plant growth hormones and polyamines on ornithine decarboxylase activity during the germination of barley seeds. Physiol. Plant. 57: 499-504.
30. YANAGISAWA, H., Y. SUZUKI. 1982. Purification and properties of N-carbamylputrescine amidohydrolase from maize shoots. Phytochemistry 21: 2201-2203.
31. MARETZKI, A., M. THOM, L.G. NICKELL. 1969. Products of arginine catabolism in growing cells of sugarcane. Phytochemistry 8: 811-818.
32. CROCOMO, O.J., L.C. BASSO, O.G. BRASIL. 1970. Formation of N-carbamyl putrescine from citrulline in Sesamum. Phytochemistry 9: 1487-1489.
33. SMITH, T.A. 1963. L-Arginine carboxylase of higher plants and its relation to potassium nutrition. Phytochemistry 2: 241-252.
34. SMITH, T.A. 1965. N-carbamylputrescine amidohydrolase of higher plants and its relation to potassium nutrition. Phytochemistry 2: 241-252.
35. LE RUDULIER, D., G. GOAS. 1980. Biogenese de N-carbamyl putrescine et de la putrescine dans les plantules de Glycine max (L.) Merr. Physiol. Veg. 18: 609-616.
36. SRIVENUGOPAL, K.S., P.R. ADIGA. 1981. Enzymic conversion of agmatine to putrescine in Lathyrus sativus. J. Biol. Chem. 256: 9532-9541.
37. PRASAD, G.L., P.R. ADIGA. 1986. Purification and characterization of putrescine synthase from cucumber seedlings. A multifunctional enzyme

involved in putrescine biosynthesis. J. Biosci. 10: 373-391.

38. PRASAD, G.L., P.R. ADIGA. 1987. Arginine decarboxylase is a component activity of the multifunctional enzyme putrescine synthase in cucumber seedlings. J. Biosci. 11: 571-579.

39. SPERANZA, A., N. BAGNI. 1978. Products of L-(^{14}C carbamoyl) citrulline metabolism in Helianthus tuberosus activated tissue. Z. Pflanzenphysiol. 88: 163-168.

40. TABOR, C.W., H. TABOR. 1985. Polyamines in microorganisms. Microbiol. Rev. 49: 81-99.

41. COHEN, S.S., R. BALINT, R.K. SINDHU. 1981. The synthesis of polyamines from methionine in intact and disrupted leaf protoplasts of virus-infected Chinese cabbage. Plant Physiol. 68: 1150-1155.

42. SUZUKI, Y., E. HIRASAWA. 1980. S-adenosylmethionine decarboxylase of corn seedlings. Plant Physiol. 66: 1091-1094.

43. YAMANOHA, B., S.S. COHEN. 1985. S-adenosylmethionine decarboxylase and spermidine synthase from Chinese cabbage. Plant Physiol. 78: 784-790.

44. HIATT, A.C., J. McINDOO, R.L. MALMBERG. 1986. Regulation of polyamine biosynthesis in tobacco. J. Biol. Chem. 261: 1293-1298.

45. SURESH, M.R., P.R. ADIGA. 1977. Putrescine-sensitive (artifactual) and insensitive (biosynthetic) S-adenosyl-L-methionine decarboxylase activities of Lathyrus sativus seedlings. Eur. J. Biochem. 79: 511-518.

46. SINDHU, R.K., S.S. COHEN. 1984. Subcellular localization of spermidine synthase in the protoplast of Chinese cabbage leaves. Plant Physiol. 76: 219-223.

47. SRIVENUGOPAL, K.S., P.R. ADIGA. 1980. Coexistence of two pathways of spermidine biosynthesis in Lathyrus sativus seedlings. FEBS Lett. 112: 260-264.

48. KAUR-SAWHNEY, R., A.W. GALSTON. 1979. Interaction of polyamines and light on biochemical processes involved in leaf senescence. Plant Cell Environ. 2: 189-196.

49. GALSTON, A.W., R.K. SAWHNEY. 1987. Polyamines and senescence in plants. In Plant Senescence: Its Biochemistry and Physiology. (W.W. Thomson, E.A.

Nothnagel, R.C. Huffaker, eds.), American Society of Plant Physiology, Rockville, Maryland, pp. 167-181.

50. EVEN-CHEN, Z., A.K. MATTOO, R. GOREN. 1982. Inhibition of ethylene biosynthesis by aminoethoxyvinylglycine and by polyamines shunts label from 3,4[^{14}C]methionine into spermidine in aged orange peel discs. Plant Physiol. 69: 385-388.

51. ROBERTS, D.R., M.A. WLAKER, J.E. THOMPSON, E.B. DUMBROFF. 1984. The effects of inhibitors of polyamine and ethylene biosynthesis on senescence, ethylene production and polyamine levels in cut carnation flowers. Plant Cell Physiol. 25: 315-322.

52. WERLE, E., E. PECHMANN. 1949. Uber die diaminoxydase der pflanzen und ihre adaptive bildung durch bakterien. Justus Liebigs Ann. Chem. 562: 44-60.

53. HILL, J.M. 1971. Diamine oxidase (pea seedlings). Methods Enzymol. 17B: 730-735.

54. YANAGISAWA, H., E. HIRASAWA, Y. SUZUKI. 1981. Purification and properties of diamine oxidase from pea epicotyl. Phytochemistry 20: 2105-2108.

55. SMITH, T.A. 1983. Polyamine oxidase (oat seedlings). Methods Enzymol. 94: 311-314.

56. SUZUKI, Y., H. YANAGISAWA. 1980. Purification and properties of maize polyamine oxidase: a flavoprotein. Plant Cell Physiol. 21: 1085-1094.

57. SMITH, T.A. 1970. Polyamine oxidase in higher plants. Biochem. Biophys. Res. Commun. 41: 1452-1456.

58. FLORES, H.E. 1983. Studies on the physiology and biochemistry of polyamines in higher plants. Ph.D. Dissertation, Yale University, 209 pp.

59. KAUR-SAWHNEY, R., H.E. FLORES, A.W. GALSTON. 1981. Polyamine oxidase in oat leaves: a cell wall localized enzyme. Plant Physiol. 68: 494-498.

60. FLORES, H.E., P. FILNER. 1985. Polyamine catabolism in higher plants: characterization of pyrroline dehydrogenase. Plant Growth Regul. 3: 277-291.

61. FEDERICO, R., R. ANGELINI, M.P. ARGENTO CERU, F. MANES. 1985. Immunohistochemical demonstration of lentil diamine oxidase. Cell. Mol. Biol. 31: 171-174.

62. DE MARTY, M., A. AYADI, A. MONIER, C. MORVAN, M. THELLIER. 1977. Electrochemical properties

of isolated cell walls of Lemna minor. In Transmembrane Ionic Exchanges in Plants. (M. Thellier, A. Monier, M. De Marty, J. Dainty, eds.), C.N.R.S., Rouen, pp. 61-73.

63. RINALDI, A., G. FLORIS, A. FINAZZI-AGRO. 1982. Purification and properties of diamine oxidase from Euphorbia latex. Eur. J. Biochem. 127: 417-422.

64. BALINT, R., G. COOPER, M. STAEBELL, P. FILNER. 1987. N- Caffeoyl-4-amino n-butyric acid, a new flower-specific metabolite in cultured tobacco cells and tobacco plants. J. Biol. Chem. 262: 11026-11031.

65. SLOCUM, R.D., A.W. GALSTON. 1987. Inhibition of polyamine biosynthesis in plants and plant pathogenic fungi. In Inhibition of Polyamine Metabolism. (P.P. McCann, A.E. Pegg, A. Sjoerdsma, eds.), Academic Press, San Diego, pp. 305-316.

66. BEY, P., C. DANZIN, M. JUNG. 1987. Inhibition of basic amino acid decarboxylases involved in polyamine biosynthesis. Op. cit. Reference 4, pp. 1-31.

67. FLORES, H.E., A.W. GALSTON. 1982. Polyamines and plant stress: activation of putrescine biosynthesis by osmotic shock. Science 217: 1259-1261.

68. FEIRER, R.P., G. MIGNON, J.D. LITVAY. 1984. Arginine decarboxylase and polyamines required for embryogenesis in wild carrot. Science 223: 1433-1435.

69. TIBURCIO, A.F., A.W. GALSTON. 1986. Arginine decarboxylase as the source of putrescine for tobacco alkaloids. Phytochemistry 25: 107-110.

70. SLOCUM, R.D., A.J. BITONTI, P.P. McCANN, R.P. FEIRER. 1988. [^{3}H]-DL-α-Difluoromethylarginine metabolism in tobacco and mammalian cells: inhibition of ornithine decarboxylase activity following arginase-mediated hydrolysis of DFMA to DFMO. Biochem. J. 255: 197-202.

71. MALMBERG, R.L., J. McINDOO. 1983. Abnormal flower development of a tobacco mutant with elevated polyamine levels. Nature 305: 623-625.

72. PEGG, A.E., H.G. WILLIAMS-ASHMAN. 1987. Pharmacologic interference with enzymes of polyamine biosynthesis and of 5'-methylthioadenosine metabolism. Op. cit. Reference 4, pp. 33-48.

73. BITONTI, A.J., P.P. McCANN. 1982. Inhibition of polyamine biosynthesis in microorganisms. Op. cit. Reference 4, pp. 259-275.
74. GREENBERG, M.L., S.S. COHEN. 1985. Dicyclohexylamine-induced shift of biosynthesis from spermidine to spermine in plant protoplasts. Plant Physiol. 78: 568-575.
75. TIBURCIO, A.F., R. KAUR-SAWHNEY, A.W. GALSTON. 1988. Polyamine biosynthesis during vegetative and floral bud differentiation in thin layer tobacco tissue cultures. Plant Cell Physiol. 29: 1241-1249.
76. KOBAYASHI, K., S. MIYASAWA, A. ENDO. 1977. Isolation and inhibitory activity of gabaculine, a new potent inhibitor of gamma-aminobutyrate aminotransferase produced by a Streptomyces. FEBS Lett. 76: 207-210.
77. TABOR, C.W., H. TABOR. 1983. Polyamines. Methods Enzymol. 94: 497.
78. SMITH, T.A., J.H.A. MARSHALL. 1987. The oxidative decarboxylation of ornithine by extracts of higher plants. Phytochemistry 27: 703-710.
79. ZAPPIA, V., C.R. CARTENI-FARINA, P. GALLETTI. 1977. Adenosylmethionine and polyamine biosynthesis in human prostrate. In The Biochemistry of Adenosylmethionine. (F. Salvatore, E. Borek, V. Zappia, H.G. Williams-Ashman, F. Schlenk, eds.), Columbia University Press, New York, pp. 473-492.
80. MANEN, C.M., D.H. RUSSEL. 1974. Comparative properties of rat liver and sea urchin eggs: S-adenosyl-L-methionine decarboxylase. Biochemistry 13: 4729-4735.
81. CHIBNALL, A.C. 1939. Protein metabolism in the plant. Yale University Press, New Haven, Connecticut, 306 pp.
82. RICHARDS, F.J., W.G. TEMPLEMAN. 1936. Physiological studies in plant nutrition. IV. Nitrogen metabolism in relation to nutrient deficiency and age in leaves of barley. Ann. Bot. 50: 367-402.
83. RICHARDS, F.J., R.G. COLEMAN. 1952. Occurrence of putrescine in potassium-deficient barley. Nature 170: 460.
84. RICHARDS, F.J., E. BERNER, JR. 1954. Physiological studies in plant nutrition. XVII. A general survey of the free amino acids of barley leaves

as affected by mineral nutrition with special reference to potassium supply. Ann. Bot. 18: 15-33.

85. COLEMAN, T.G., F.J. RICHARDS. 1956. Physiological studies in plant nutrition. XVIII. Some aspects of nitrogen metabolism in barley and other plants in relation to potassium deficiency. Ann. Bot. 20: 393-409.

86. COLEMAN, R.G. 1958. Occurrence of ornithine in sulphur-deficient flax and the possible place of ornithine and citrulline in the arginine metabolism of some higher plants. Nature 181: 776-777.

87. HACKETT, C., C. SINCLAIR, F.J. RICHARDS. 1965. Balance between potassium and phosphorus in the nutrition of barley. I. The influence on amine content. Ann. Bot. 29: 331-345.

88. TAKAHASHI, T., D. YOSHIDA. 1960. Relationship between the accumulation of putrescine and the nutrition of tobacco plant. J. Sci. Soil Manure (Japan) 31: 39-41.

89. BASSO, L.C., T.A. SMITH. 1974. Effect of mineral deficiency on amine formation in higher plants. Phytochemistry 13: 875-883.

90. COLEMAN, R.G., M.P. HEGARTY. 1957. Metabolism of DL-ornithine-2-^{14}C in normal and potassium deficient barley. Nature 172: 376-377.

91. SMITH, T.A., F.J. RICHARDS. 1962. The biosynthesis of putrescine in higher plants and its relation to potassium nutrition. Biochem. J. 84: 292-294.

92. SMITH, T.A. 1970. The biosynthesis and metabolism of putrescine in higher plants. Ann. N.Y. Acad. Sci. 171: 988-1001.

93. YOUNG, N.D., A.W. GALSTON. 1983. Putrescine and acid stress: induction of arginine decarboxylase activity and putrescine accumulation by low pH. Plant Physiol. 71: 767-771.

94. GALE, E.F. 1940. The production of amines by bacteria. III. The production of putrescine from (+)-arginine by Bacterium coli in symbiosis with Streptococcus faecalis. Biochem. J. 34: 853-857.

95. SMITH, T.A., C. SINCLAIR. 1967. The effect of acid feeding on amine formation in barley. Ann. Bot. 31: 103-111.

96. LE RUDULIER, D., G. GOAS. 1975. Influence des ions ammonium et potassium sur l'accumulation de la putrewcine chez les jeunes plantes de Soja hispida Moench privees de leurs cotyledons. Physiol. Veg. 13: 125-136.
97. FLORES, H.E., A.W. GALSTON. 1984. Osmotic stress-induced polyamine accumulation in cereal leaves. Plant Physiol. 75: 102-113.
98. MURTY, K.S., T.A. SMITH, C. BOULD. 1971. The relation between the putrescine content and potassium status of black currant leaves. Ann. Bot. 356: 687-695.
99. SURESH, M.R., S. RAMAKARISHNA, P.R. ADIGA. 1978. Regulation of arginine decarboxylase and putrescine levels in Cucumis sativus cotyledons. Phytochemistry 17: 57-63.
100. PRIEBE, A., H. KLEIN, H.J. JAGER. 1978. Effect of NaCl on the levels of putrescine and related polyamines in plants differing in salt tolerance. Plant Sci. Lett. 12: 365-369.
101. LE RUDULIER, D., G. GOAS. 1971. Mise en evidence et dosage de quelques amines dans les plantules de Soja hispida Moench privees de leurs cotyledons et cultivees en presence de nitrates, d'uree et de chlorure d'ammonium. Compt. Rend. Acad. Sci. (Paris) Ser. D. 279: 161-163.
102. LE RUDULIER, D., G. GOAS. 1979. Contribution a l'etude de l'accumulation de putrescine et de la putrescine dans les plantules de Glycine max (L.) Merr. Physiol. Veg. 18: 609-616.
103. BLAIR, G.J., M.H. MILLER, W.A. MITCHELL. 1970. Nitrate and ammonium as sources of nitrogen for corn and their influence on uptake of other ions. Agronomy J. 62: 530-532.
104. ROBERTS, J.K.M. 1984. Study of plant metabolism in vivo using NMR spectroscopy. Annu. Rev. Plant Physiol. 35: 375-386.
105. FLORES, H.E., N.D. YOUNG, A.W. GALSTON. 1985. Polyamine metabolism and plant stress. In Cellular and Molecular Biology of Plant Stress. (J.L. Key, T. Kosuge, eds.), Alan R. Liss, New York, pp. 93-114.
106. SHEVYAKOVA, N.I. 1981. Metabolism and the physiological role of diamines and polyamines in plants. Sov. Plant Physiol. 28: 1052-1061.

107. SHEVYAKOVA, N.I. 1966. On the stimulating and toxic effects of diamines on plants. Sov. Plant Physiol. 13: 472-475.

108. TIBURCIO, A.F., R. KAUR-SAWHNEY, A.W. GALSTON. 1986. Polyamine metabolism and osmotic stress. II. Improvement of oat protoplasts by an inhibitor of arginine decarboxylase. Plant Physiol. 82: 375-378.

109. DEROSA, M., S. DEROSA, A. GAMBACORTA, M. CARTENI-FARIA, V. ZAPPIA. 1976. Occurrence and characterization of new polyamines in the extreme thermophile Caldariella acidophila. Biochem. Biophys. Res. Commun. 69: 253-261.

110. OSHIMA, T. 1983. Novel polyamines in Thermus thermophilus: Isolation, identification, and chemical synthesis. Meth. Enzymol. 94: 401-411.

111. OSHIMA, T. 1982. A pentaamine is present in an extreme thermophile. J. Biol. Chem., 9913-9914.

112. RODRIGUEZ-GARAY, B., G.C. PHILLIPS, G.D. KUEHN. 1989. Detection of norspermidine and norspermine in Medicago sativa L. Plant Physiol. 89: 525-529.

113. HERBST, E.J. E.E. SNELL. 1948. Putrescine as a growth factor for Haemophilus parainfluenzae. J. Biol. Chem. 176: 989-990.

114. SNEATH, P.H.A. 1955. Putrescine as an essential growth factor for a mutant of Aspergillus nidulans. Nature 175: 818.

115. TABOR, C.W., H. TABOR. 1984. Polyamines. Annu. Rev. Biochem. 53: 749-490.

116. BAGNI, N. 1966. Aliphatic amines and a growth factor of coconut milk as stimulating cellular proliferation of Helianthus tuberosus (Jerusalem artichoke) in vitro. Experientia 22: 732-733.

117. GALSTON, A.W. 1983. Polyamines as modulators of plant development. BioScience 33: 382-388.

118. FELIX, H., J. HARR. 1987. Association of polyamines to different parts of various plant species. Physiol. Plant. 71: 245-250.

119. DUMORTIER, F.M., H.E. FLORES, N.S. SHEKHAWAT, A.W. GALSTON. 1983. Gradient of polyamines and their biosynthesis enzymes in coleoptiles and roots of corn. Plant Physiol. 72: 915-918.

120. PHILIPS, R., M.C. PRESS, A. EASON. 1987. Polyamines in relation to cell division and xylogenesis in cultured explants of Helianthus tuberosus: lack of

evidence for growth-regulatory action. J. Exp. Bot. 38: 164-172.
121. DAI, Y.R., R. KAUR-SAWHNEY, A.W. GALSTON. 1982. Promotion by gibberellic acid of polyamine biosynthesis in internodes of light-grown dwarf pea. Plant Physiol. 69: 103-105.
122. SMITH, M.A., P.J. DAVIES, J.B. REID. 1985. Role of polyamines in gibberellin-induced internode growth in peas. Plant Physiol. 78: 92-99.
123. CHO, S. 1983. Enhancement by putrescine of gibberellin-induced elongation in hypocotyls of lettuce seedlings. Plant Cell Physiol. 24: 305-308.
124. JARVIS, B.C., P.R.M. SHANNON, S. YASMIN. 1983. Involvement of polyamines with adventitous root development in stem cuttings of mung bean. Plant Cell Physiol. 24: 677-683.
125. FRIEDMAN, R., A. ALTMAN, U. BACHRACH. 1985. Polyamines and root formation in mung bean hypocotyl cuttings. Plant Physiol. 79: 80-83.
126. GALSTON, A.W., R.K. SAWHNEY. 1982. Polyamines: are they a new class of growth regulators? In Plant Growth Substances. (P.F. Wareing, eds), Academic Press, London, pp. 451-461.
127. BAGNI, N., R. PISTOCCHI. 1985. Putrescine uptake in Saintpaulia petals. Plant Physiol. 77: 398-402.
128. PISTOCCHI, R., N. BAGNI, J.A. CREUS. 1987. Polyamine uptake in carrot cell cultures. Plant Physiol. 84: 374-380.
129. YOUNG, N.D., A.W. GALSTON. 1983. Are polyamines transported in etiolated peas? Plant Physiol. 73: 912-914.
130. BAGNI, N., R. BARALDI, G. COSTA. 1984. Translocation and metabolism of aliphatic polyamines in leaves and fruitlets of Malus domestica (cv. Ruby Spur). Acta Hortic. (The Hague) 149: 173-178.
131. PISTOCCHI, R., N. BAGNI, J.A. CREUS. 1986. Polyamine uptake, kinetics, and competition among polyamines and between polyamines and inorganic cations. Plant Physiol. 80: 556-560.
132. FRIEDMAN, R., N. LEVIN, A. ALTMAN. 1986. Presence and identification of polyamines in xylem and phloem exudates of plants. Plant Physiol. 82: 1154-1157.

133. GOLDBERG, R., E. PEDRIZET. 1984. Ratio of free to bound polyamines during maturation in mung-bean hypocotyl cells. Planta 161: 531-535.

134. PANAGIOTIDIS, C.A., J.G. GEORGATSOS, D.A. KYRIAKIDIS. 1982. Superinduction of cytosolic and chromatin-bound ornithine decarboxylase activities of germinating barley seeds by actinomycin D. FEBS Lett. 146: 193-196.

135. WALKER, M.A., B.E. ELLIS, C.C.S. CHAPPLE, E.B. DUMBROFF. 1987. Subcellular localization of amines and activities of their biosynthetic enzymes in p-fluorophenylalanine resistant and wild type tobacco cell cultures. Plant Physiol. 85: 78-81.

136. JAIN, J.C., P.D. SHARGOOL, S. CHUNG. 1987. Compartmentation studies on enzymes of ornithine biosynthesis in plant cells. Plant Sci. (Shannon) 51: 17-20.

137. TORRIGIANI, P., D. SERAFINI-FRACASSINI, S. BIONDI, N. BAGNI. 1986. Evidence for the subcellular localization of polyamines and their biosynthetic enzymes in plant cells. J. Plant Physiol. 124: 23-29.

138. STEWARD, F.C. 1958. Growth and development of cultivated cells. III. Interpretations of the growth from free cell to carrot plant. Am. J. Bot. 45: 709-713.

139. SUNG, Z.R. 1985. Developmental states of embryogenic cultures. In Proceeding of the International Workshop on Somatic Embryogenesis in Carrots. (M. Terzi, L. Pitto, Z.R. Sung, eds.), Cons. Natl. Ricerche, Pisa, pp. 22-31.

140. MONTAGUE, M.J., J.W. KOPPENBRINK, E.G. JAWORSKI. 1978. Polyamine metabolism in embryogenic cells of Daucus carota. I. Changes in intracellular content and rates of synthesis. Plant Physiol. 62: 430-433.

141. MONTAGUE, M.J., T.A. ARMSTRONG, E.G. JAWORSKI. 1979. Polyamine metabolism in embryogenic cells in Daucus carota. II. Changes in arginine decarboxylase activity. Plant Physiol. 63: 341-345.

142. FEIRER, R.P., S.R. WANN, D.W. EINSPAHR. 1985. The effects of spermidine synthesis inhibitors on in vitro plant development. Plant Growth Regul. 3: 319-328.

143. FIENBERG, A.A., J.H. CHOI, W.P. LUBICH, Z.R. SUNG. 1984. Developmental regulation of polyamine metabolism in growth and differentiation of carrot culture. Planta 162: 532-539.

144. GUHA, S., S.C. MAHESWARI. 1964. In vitro production of embryos from anthers of Datura. Nature 204: 497.

145. BOURGIN, J.P., J.P. NITSCH. 1967. Obtention de Nicotiana haploides a partir de'etamines cultivees in vitro. Ann. Physiol. Veg. 9: 377-382.

146. BAKER, S.R., L.H. JONES, R.J. YON. 1983. Ornithine carbamoyltransferase activity and embryogenesis in a carrot cell suspension culture. Phytochemistry 22: 2167-2169.

147. BRADLEY, P.M., F. EL-FIKI, K.L. GILES. 1984. Polyamines and arginine affect embryogenesis of Daucus carota. Plant Sci. Lett. 34: 397-401.

148. THOMAS, H., D. GRIERSON, eds. 1987. Developmental Mutants in Higher Plants. Soc. Exp. Biol., Cambridge University Press, Cambridge, 287 pp.

149. SERAFINI-FRACASSINI, D., U. MOSSETTI. 1986. What is the function of conjugated polyamines in plants? In Biomedical Studies on Natural Polyamines. (C.M. Caldarera, C. Clo, C. Gaurniera, eds.), CLUEB, Bologna, pp. 197-202.

150. MARTIN-TANGUY, J. 1985. The occurrence and possible function of hydroxycinnamoyl acid amides in plants. Plant Growth Regul. 3: 381-400.

151. CABANNE, F., J. MARTIN-TANGUY, C. MARTIN. 1977. Phenolamines associes a l'induction florale et a l'etat reproducteur du Nicotiana tabacum var. Xanthi. Physiol. Veg. 15: 445-451.

152. MBADIWE, E.I. 1973. Caffeoylputrescine from Pentaclethra macrophylla. Phytochemistry 12: 2546.

153. BUTA, J.G., R.R. IZAC. 1972. Solanaceae: caffeoylputrescine in Nicotiana tabacum. Phytochemistry 11: 1188-1189.

154. RYABININ, A.A., E.M. IL'INA. 1949. The alkaloid of Salsola subaphylla. Drob. Dokl. Akad. Nauk. USSR 67: 513-616.

155. WHEATON, T.A., I. STEWART. 1965. Feruloylputrescine: isolation and identification from citrus leaves and fruits. Nature 206: 620-621.

156. MIZUSAKI, S., Y. TANABE, M. NUGUCHI. 1971. p-Coumaroylputrescine, caffeoylputrescine and feruloylputrescine from callus tissue culture of Nicotiana tabacum. Phytochemistry 10: 1347-1350.
157. MARTIN-TANGUY, J., C. MARTIN, M. GALLET, R. VERNOY. 1976. Sur de puissants inhibiteurs naturels de multiplication di virus de la mosaique du tubac. C.R. Seances Acad. Sci. (Paris) Ser. D. 282: 2231.
158. STOESSL, A. 1965. The antifungal factors in barley. III. Isolation of p-coumaroylagmatine. Phytochemistry 12: 973-977.
159. STOESSL, A., C.H. UNWIN. 1978. The antifungal factors in barley. V. Antifungal activity of the hordatines. Can. J. Bot. 48: 465-470.
160. BIRD, C.R., T.A. SMITH. 1981. The biosynthesis of coumaroylagmatine barley seedlings. Phytochemistry 10: 2345-2346.
161. SMITH, T.A., G.R. REST. 1978. Distribution of the hordatines in barley. Phytochemistry 17: 1093-1098.
162. CLARKE, D.D. 1982. The accumulation of cinnamic acid amides in the cell walls of potato tissue as an early response to fungal attack. In Active Defense Mechanism in Plants. (R.K.S. Wood, ed.), Plenum Press, New York.
163. SMITH, T.A., J. NEGREL, C.R. BIRD. 1983. The cinnamic acid amides of the di- and polyamines. In Advances in Polyamine Research. (U. Bachrach, A. Kaye, R. Chayen, eds.), Raven Press, New York, Vol. 4, pp. 347-370.
164. CABANNE, F., M.A. DALEBROUX, J. MARTIN-TANGUY, C. MARTIN. 1981. Hydroxycinnamic acid amides and ripening to flower of Nicotiana tabacum var. Xanthi nc. Physiol. Plant. 53: 399-404.
165. BELLIARD, J., J. PERNES, M. SANDMEIER. 1979. Les differentes phases du developpement chez le Mil (Pennisetum typhoides Stapf et Hubbard) et la recherche de marquers. Physiol. Veg. 17: 387-397.
166. MARTIN-TANGUY, J., J. MARGARA, C. MARTIN. 1984. Phenolamides et induction florale de Cichorium intybus dans differentes conditions de culture en serret et in vitro. Physiol. Plant. 61: 259-262.

167. CABANNE, F., M. PAYNOT, F. JAVELLE, J. MARTIN-TANGUY, C. MARTIN. 1977. Activite phenylalanine ammoniac lyase et etat floral du Nicotiana tabacum var. Xanthi nc. Physiol. Veg. 15: 445-451.
168. MARTIN, C., J. MARTIN-TANGUY. 1981. Polyamines conjugees et limitation de l'expansion virale chez les vegetaux. C.R. Seances Acad. Sci. (Paris) 292: 249-251.
169. DAI, Y., J. WANG. 1987. Relation of polyamine titer to photoperiodic induction of flowering in Pharbitis nil. Plant Sci. (Shannon) 51: 135-139.
170. VANSUYT, G., C. ZINSOU. 1986. Accumulation of agmatine in chayote (Sechium edule) leaves during development. Physiol. Plant. 67: 592-597.
171. MARTIN-TANGUY, J., F. CABANNE, E. PEDRIZET, C. MARTIN. 1978. The distribution of hydroxycinnamic acid amides in flowering plants. Phytochemistry 17: 1927-1928.
172. PONCHET, M., J. MARTIN-TANGUY, A. MARAIS, C. MARTIN. 1982. Hydroxycinnamoyl acid amides and aromatic amines in the inflorescence of some Araceae species. Phytochemistry 21: 2865-2869.
173. MARTIN-TANGUY, J., E. PERDRIZET, J. PREVOST, C. MARTIN. 1982. Hydroxycinnamic acid amides in fertile and cytoplasmic male sterile lines of maize. Phytochemistry 21: 1939-1945.
174. BIRD, C.R., T.A. SMITH. 1983. Agmatine coumaroyl-transferase from barley seedlings. Phytochemistry 22: 2401-2403.
175. BERLIN, J., K.H. KNOBLOCH, G. HOFLE, L. WITTE. 1982. Biochemical characterization of two tobacco cell lines with different levels of cinnamoyl putrescine. J. Nat. Prod. 45: 83-87.
176. FLORES, H.E. 1987. Use of plant cells and organ culture in the production of biological chemicals. In Biotechnology in Agricultural Chemistry. (H.M. LeBaron, R.O. Mumma, R.C. Honeycutt, J.H. Duesing, eds.), American Chemical Society, Washington, D.C., pp. 66-86.
177. BERLIN, J. 1980. p-Fluorophenylalanine resistant cell lines of tobacco. Z. Pflanzenphysiol. 97: 317-324.
178. FLORES, H.E., P. FILNER. 1985. Metabolic relationships of putrescine, GABA and alkaloids in cell and root cultures of Solanaceae. In Primary and Secondary Metabolism of Plant Cell Cultures.

(K.H. Neumann, W. Barz, E. Reinhard, eds.), Springer-Verlag, Berlin, pp. 174-185.

179. MALMBERG, R.L., J. McINDOO, A.C. HIATT, B.A. LOWE. 1985. Genetics of polyamine synthesis in tobacco: developmental switches in the flower. Cold Spring Harbor Symp. Quant. Biol. 50: 475-482.

180. HIATT, A.C., R.L. MALMBERG. 1988. Utilization of putrescine in tobacco cell lines resistant to inhibitors of polyamines synthesis. Plant Physiol. 86: 441-446.

181. TRAN THANH VAN, M. 1973. Direct flower neoformation from superficial tissue of small explants of Nicotiana tabacum. Planta 115: 87-92.

182. TRAN THANH VAN, M. 1977. Regulation of morphogenesis. In Plant Tissue Culture and Its Bio-technological Application. (W. Barz, E. Reinhard, N.H. Zenk, eds.), Springer-Verlag, Berlin, pp. 367-385.

183. KAUR-SAWHNEY, R., A.F. TIBURCIO, A.W. GALSTON. 1988. Spermidine and flower-bud differentiation in thin-layer explants of tobacco. Planta 173: 282-284.

184. TORRIGIANI, P., M.M. ALTAMURA, G. PASQUA, B. MONACELLI, D. SERAFINI-FRACASSINI, N. BAGNI. 1987. Free and conjugated polyamines during de novo floral and vegetative bud formation in thin cell layers of tobacco. Physiol. Plant. 70: 453-460.

185. COSTA, G., N. BAGNI. 1983. Effect of polyamines on fruit set of apples. Hort. Sci. 18: 59-61.

186. MAPELLI, S., C. FROVA, G. TORNI, G.P. SORESSI. 1978. Relationship between set, development and activities of growth regulators in tomato fruits. Plant Cell Physiol. 19: 1281-1288.

187. LEOPOLD, A.C., P.E. KRIEDEMANN. 1975. Plant Growth and Development. McGraw-Hill, Inc., New York, 545 pp.

188. HEIMER, Y.M., Y. MIZRAHI. 1982. Characterization of ornithine decarboxylase of tobacco cells and tomato ovaries. Biochem. J. 201: 373-376.

189. MIZRAHI, Y., Y.M. HEIMER. 1982. Increased activity of ornithine decarboxylase in tomato ovaries induced by auxin. Physiol. Plant. 54: 367-368.

190. COHEN, E., S. ARAD, Y.M. HEIMER, Y. MIZRAHI. 1982. Participation of ornithine decarboxylase in early stages of tomato fruit development. Plant Physiol. 70: 540-543.

191. SLOCUM, R.D., A.W. GALSTON. 1985. In vivo inhibition of polyamine biosynthesis and growth in tobacco overy tissues. Plant Cell Physiol. 26: 1519-1526.

192. APELBAUM, A. 1986. Polyamine involvement in the development and ripening of avocado fruit. Acta Hortic. 179: 779-785.

193. TOUMADJE, A., D.G. RICHARDSON. 1988. Endogenous polyamine concentrations during development, storage and ripening of pear fruits. Phytochemistry 27: 335-338.

194. NATHAN, R., A. ALTMAN, S.P. MONSELISE. 1984. Changes in activity of polyamine biosynthetic enzymes and in polyamine contents in developing fruit tissues of 'Murcott' mandarin. Sci. Hortic. 22: 359-364.

195. SALISBURY, F.B., C.W. ROSS. 1985. Plant Physiology. Wadsworth Publishing Co., Belmont, 348 pp.

196. THIMANN, K.V. 1980. The senescence of leaves. In Senescence in Plants. (K.V. Thimann, ed.), CRC Press, Boca Raton, pp. 85-115.

197. ALTMAN, A., R. KAUR-SAWHNEY, A.W. GALSTON. 1977. Stabilization of oat leaf protoplasts through polyamine-mediated inhibition of senescence. Plant Physiol. 60: 570-574.

198. POPOVIC, R.B., D.J. KYLE, A.S. COHEN, S. ZALIK. 1979. Stabilization of thylakoid membranes by spermine during stress-induced senescence of barley leaf discs. Plant Physiol. 64: 721-726.

199. SRIVASTAVA, S.K., D.J. VASHI, B.I. NAIK. 1983. Control of senescence by polyamines and guanidines in young and mature barley leaves. Phytochemistry 22: 2115-2154.

200. CHEN, C.T., C.H. KAO. 1986. Localized effect of 1,3-diaminopropane and benzyladenine on chlorophyll loss in soybean primary leaves. Bot. Bull. Acad. Sin. 27: 97-100.

201. KAR, R.K., M.A. CHOUDURI. 1986. Effects of light and spermine on senescence of Hydrilla and spinach leaves. Plant Physiol. 80: 1030-1033.

202. ALTMAN, A. 1982. Retardation of radish leaf senescence by polyamines. Physiol. Plant. 54: 189-193.

203. MUHITCH, M.J., LA.A EDWARDS, J.S. FLETCHER. 1983. Influence of diamines and polyamines on the senescence of plant suspension cultures. Plant Cell Rep. 2: 82-84.

204. SUTTLE, J. 1981. Effect of polyamines on ethylene production. Phytochemistry 20: 1477-1480.

205. KAUR-SAWHNEY, R., A. ALTMAN, A.W. GALSTON. 1978. Dual mechanisms in polyamine mediated control of ribonuclease activity in oat leaf protoplasts. Plant Physiol. 62: 158-160.

206. TABOR, C.W. 1962. Stabilization of protoplasts and sphreroplasts by spermine and other polyamines. J. Bacteriol. 83: 1101-1111.

207. ROBERTS, D.R., M.A. WALKER, J.A. THOMPSON, E.B. DUMBROFF. 1983. The effects of inhibitors of polyamine and ethylene biosynthesis on senescence, ethylene production and polyamine levels in cut carnations. Plant Cell Physiol. 25: 315-322.

208. DOWNS, C.G., P.H. LOWELL. 1986. The effect of spermidine and putrescine on the senescence of cut carnation. Physiol. Plant. 66: 679-684.

209. WINER, L., A. APELBAUM. 1986. Involvement of polyamines in the development and ripening of avocado fruits. J. Plant Physiol. 126: 223-233.

210. TOUMADJE, A., D.G. RICHARDSON. 1984. Inhibition of ethylene production by polyamines in pear discs and intact fruits. Plant Physiol. 75(S): 35.

211. MOTHES, K., H.R. SCHUTTE, M. LUCKNER, eds. 1985. Biochemistry of Alkaloids. Deutscher Verlag de Wissenschaften, Berlin, 406 pp.

212. GUGGISBERG, A., M. HESSE. 1983. Putrescine, spermidine, spermine and related polyamine alkaloids. In The Alkaloids: Chemistry and Pharmacology. (A. Brossi, ed.), Academic Press, New York, Vol. XXII, pp. 85-188.

213. JACOBSON, M., D.G. CROSBY. 1971. Naturally Occurring Insecticides. Marcel Dekker, New York, 573 pp.

214. WALLER, G.R., E.K. NOVACKI. 1977. Alkaloid Biology and Metabolism in Plants. Plenum Press, New York, 293 pp.

215. LIEBISCH, H.W., H.R. SCHUTTE. 1985. Alkaloids derived from ornithine. In Biochemistry of Alkaloids. (K. Mothes, H.R. Schutte, M. Luckner, eds.), Deutscher Verlag de Wissenschaften, Berlin,

pp. 106-127.

216. DAWSON, R.F. 1941. The localization of the nicotine biosynthetic mechanism in the tobacco plant. Science 94: 396-397.

217. DAWSON, R.F. 1942. Nicotine synthesis in excised tobacco roots. Am. J. Bot. 29: 813-815.

218. FLORES, H.E., M.W. HOY, J.J. PICKARD. 1987. Secondary metabolites from root cultures. Trends Biotechnol. 5: 64-69.

219. TIBURCIO, A.F., A.W. GALSTON. 1986. Arginine decarboxylase as the source of putrescine for tobacco alkaloids. Phytochemistry 25: 107-110.

220. LEETE, E. 1985. Spermidine: an indirect precursor of the pyrrolidine rings of nicotine and nornicotine in Nicotiana glutinosa. Phytochemistry 24: 957-960.

221. LEETE, E. 1962. The stereospecific incorporation of ornithine into the tropine moiety of hyoscyamine. J. Am. Chem. Soc. 84: 55.

222. YAMADA, Y., T. HASHIMOTO. 1988. Biosynthesis of tropane alkaloids. In Application of Plant Cell and Tissue Culture. (G. Bock, J. Marsh, eds.), CIBA Foundation Symposium, John Wiley and Sons, Sussex, pp. 199-212.

223. HARTMANN, T., H. SANDERS, R. ADOLPH, G. TOPPEL. 1988. Metabolic links between the biosynthesis of pyrrolizidine alkaloids and polyamines in root cultures of Senecio vulgaris. Planta 175: 82-90.

224. LUCKNER, M. 1984. Secondary Metabolism in Microorganisms, Plants and Animals. 2nd Edition, Springer-Verlag, 576 pp.

Chapter Eleven

CASTANOSPERMINE, SWAINSONINE AND RELATED POLYHYDROXY ALKALOIDS: STRUCTURE, DISTRIBUTION AND BIOLOGICAL ACTIVITY

LINDA E. FELLOWS, GEOFFREY C. KITE, ROBERT J. NASH, MONIQUE S.J. SIMMONDS, AND ANTHONY M. SCOFIELD*

Jodrell Laboratory
Royal Botanic Gardens
Kew, TW9 3DS
United Kingdom

*University of London Wye College
Wye, Ashford
Kent, TN25 5AH
United Kingdom

INTRODUCTION

Alkaloids which are inhibitors of glycosidases are now believed to be widespread in plants and microorganisms. Those isolated to date carry several hydroxyl substituents on a ring in a configuration that suggests a structural

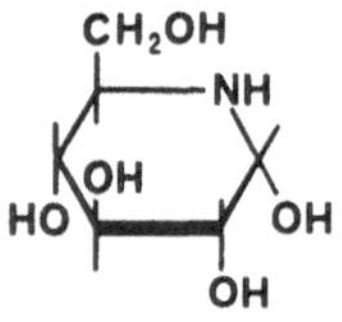

Nojirimycin

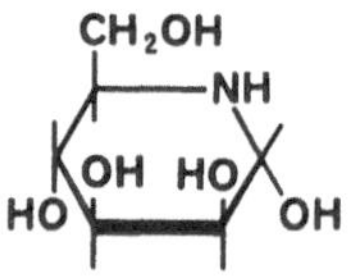

Nojirimycin B

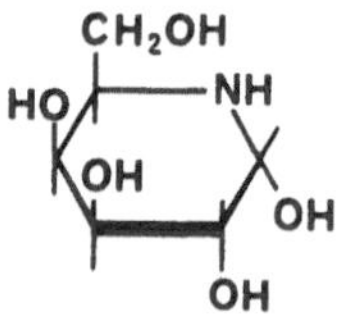

Galactostatin

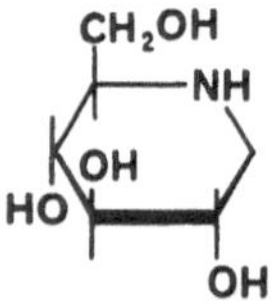

Deoxynojirimycin
DNJ

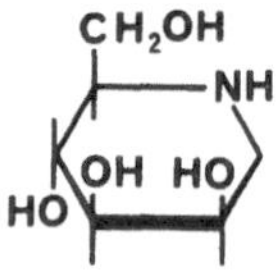

Deoxymannojirimycin
DMJ

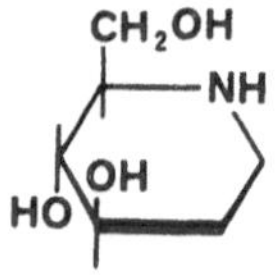

Fagomine

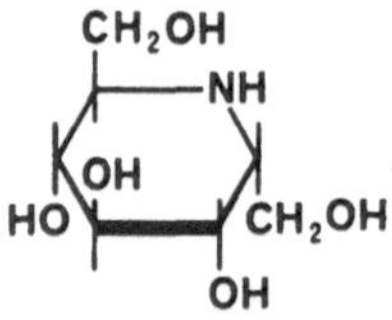

Homonojirimycin

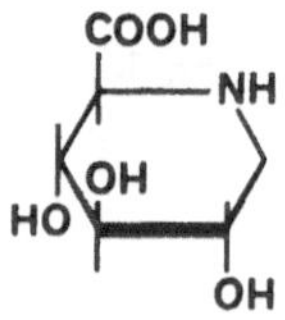

BR-1

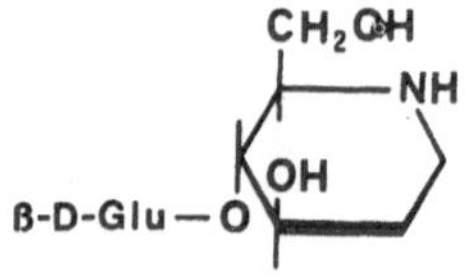

Fagomine glucoside
XZ-1

Fig. 1. Alkaloidal glycosidase inhibitors incorporating a 6-sided ring. Nojirimycin, 5-amino-5-deoxy-D-glucopyranose; nojirimycin B, 5-amino-5-deoxy-D-mannopyranose; galactostatin, 5-amino-5-deoxy-D-galactopyranose; deoxynojirimycin, DNJ, 1,5-dideoxy-1,5-imino-D-glucitol; deoxymannojirimycin, DMJ, 1,5-dideoxy-1,5-imino-D-mannitol; fagomine, 1,2,5-trideoxy-1,5-imino-D-arabinitol; homonojirimycin, 2,6-dideoxy-2,6-imino-D-glycero-L-gulo-heptitol; BR1, 2S-carboxy-3R,4R,5S-trihydroxypiperidine; fagomine glucoside, XZ-1, 1,2,5-trideoxy-4-O-(β-D-glucopyranosyl)-1,5-imino-D-arabinohexitol.

resemblance to monosaccharides. Despite their relatively recent detection (most have been discovered during the last ten years), they are now providing molecular biologists with tools with which to probe many hitherto intractable problems of molecular biology, including cancer metastasis, viral infectivity and the immune response. One, castanospermine, is under

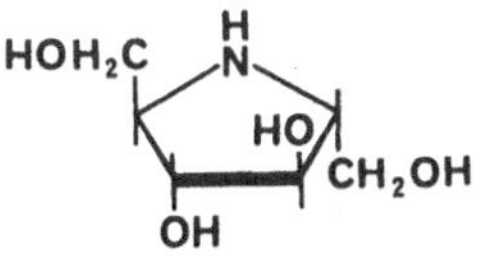

DMDP

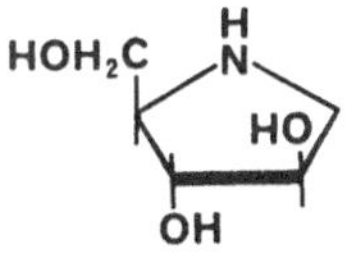

DAB-1

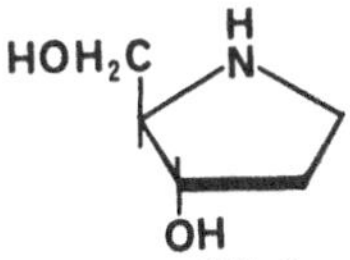

CYB-3

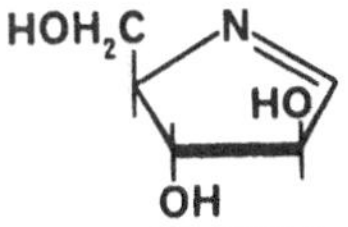

FR-900483

Fig. 2. Alkaloidal glycosidase inhibitors incorporating a 5-sided ring. DMDP, 2R,5R-dihydroxymethyl-3R,4R-dihydroxypyrrolidine; DAB-1, 1,4-dideoxy-1,4-imino-D-arabinitol; CYB-3, 2R-hydroxymethyl-3S-hydroxypyrrolidine; FR-900483, (3R,4R,5R)-3,4-dihydroxy-5-hydroxymethyl-1-pyrroline.

evaluation as a drug for the clinical management of HIV infection (AIDS). This paper outlines their structural range, natural distribution and biological activity.

STRUCTURAL RANGE

Natural alkaloidal glycosidase inhibitors (AGIs) discovered so far belong to one of five structural types, viz polyhydroxy derivatives of piperidine, pyrrolidine, pyrroline, octahydroindolizine and pyrrolizidine[1-5] (Fig. 1-3). The piperidines and pyrrolidines may be considered respectively as simple analogues of 1-deoxy pyranose and furanose sugars with the ring oxygen replaced by nitrogen. The bicyclic octahydroindolizines (swainsonine and castonospermine) and pyrrolizidines (alexine and 3,8-diepialexine) have a bicyclic ring structure and a less obvious structural relationship to monosaccharides but in each case the configuration of hydroxy substitutents on the ring can be compared to those of sugars. (For a list of systematic name(s), trivial names and common abbreviations see the legends to Figures 1-3).

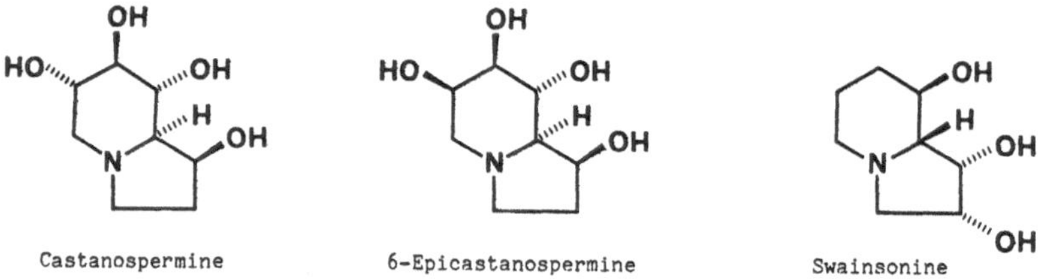

Castanospermine 6-Epicastanospermine Swainsonine

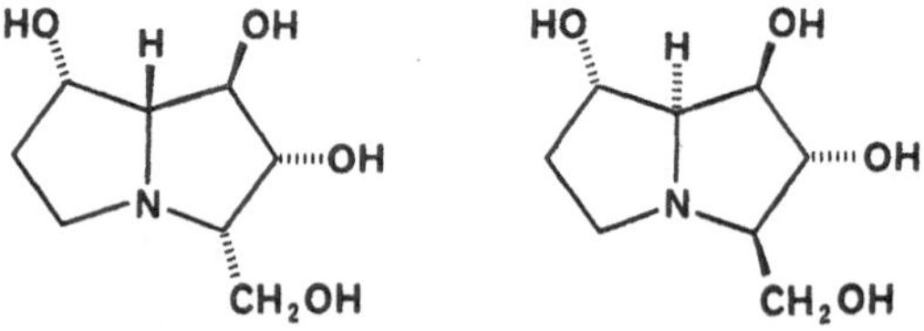

3,8-Diepialexine Alexine

Fig. 3. Alkaloidal glycosidase inhibitors with a bicyclic ring. Castanospermine, (1S,6S,7R,8R,8aR)-1,6,7,8-tetrahydroxyoctahydroindolizine; 6-epicastanospermine, (1S,6R,7R,8R,8aR)-1,6,7,8-tetrahydroxyoctahydroindolizine; swainsonine, (1S,2R,8R,8aR)-1,2,8-trihydroxyoctahydroindolizine; 3,8-diepialexine, (1R,2R,3S,7S,8R)-3-hydroxymethyl-1,2,7-trihydroxypyrrolizidine; alexine, (1R,2R,3R,7S,8S)-3-hydroxymethyl-1,2,7-trihydroxypyrrolizidine.

NATURAL DISTRIBUTION

In Microorganisms

The first AGI to be isolated from a natural source was nojirimycin, a simple analogue of D-glucose with the ring oxygen replaced by nitrogen. It was reported in 1966 to occur in a Streptomyces sp. and to be an inhibitor of α- and β-glucosidases from several plants and microbes.[6-9] The corresponding analogues of D-mannose and D-galactose, known by the trivial names of nojirimycin B and galactostatin respectively, were also recently isolated from Streptomyces strains[10,11] and shown to inhibit certain mannosidases and galactosidases. However, nojirimycin B was surprisingly as active as nojirimycin against almond β-glucosidase, despite the structural change at C2 (considered as a sugar). None of these three simple analogues of pyranose sugars with the anomeric hydroxyl at C1 retained has yet been found in higher plants. The 1-deoxy derivative of nojirimycin (deoxynojirimycin, DNJ), also a glucosidase inhibitor but with a different profile of activity from the parent compound, was first produced by chemical reduction but later shown to occur naturally in strains of both Bacillus and Streptomyces.[12,13] Recently a substituted pyrroline with α-glucosidase activity was isolated from the fungus Nectria lucida Hohnel.[4] Of the bicyclic alkaloids only swainsonine has yet been found in microorganisms. It was first isolated from a fungal forage contaminant, Rhizoctonia leguminicola Gough et E.S. Elliot, in 1973 but the structure was not reported correctly until some years later.[14,15] More recently it was found in a strain of the fungus Metarrhizium anisopliae Sorokin.[16]

In Higher Plants

Deoxynojirimycin (DNJ) was first found in higher plants in 1977 in the roots of certain Morus spp (Mulberries) by a Japanese group attempting to isolate a hypoglycemic agent. They isolated an alkaloid 'moranoline' which, when given with food to animals, prevented the post-prandial rise in blood glucose.[17,18] Moranoline and DNJ were found to be the same compound, a potent inhibitor of mammalian digestive α-glucosidase. The term deoxynojirimycin is used most widely today for convenience and because of historical precedent. (Researchers should be aware that a plethora of trivial names and abbreviations for this and other AGIs are commonly encountered. These difficulties are compounded by the fact that it is possible to describe many of the compounds systematically in different ways, e.g., DNJ may be correctly described as 1,5-dideoxy 1,5-imino-D-glucitol or 2S-hydroxymethyl-3R,4R,5S-trihydroxypiperidine). Recently we have isolated a derivative of DNJ, homonojirimycin (2,6-dideoxy-2,6-imino-D-glycero-L-gulo heptitol) from leaves of a species of Euphorbiaceae, Omphalea diandra L., a liana native to Central and South America. Preliminary studies indicate that it has similar properties as a glycosidase inhibitor to DNJ.[19]

In 1977 a novel pyrrolidine alkaloid, 2R,5R-dihydroxymethyl-3R,4R-dihydroxypyrrolidine (DMDP), an analogue of β-D-fructofuranose, was isolated from Derris elliptica Benth (Leguminosae).[20] It was shown to be an inhibitor of glucosidases[21] but differed from DNJ in its activity, e.g., the digestive α-glucosidase of bruchid beetles was more susceptible to DMDP than DNJ, but the reverse was true for the mouse.[22,23] DMDP was later found in O. diandra and in adults of the brightly colored day-flying moth Urania fulgens (Boisduval) Walker whose larvae feed on the plant.[24] There are several reports of the sequestration by Lepidoptera of chemicals from their host plant which subsequently protect the insect from predation by birds.[25] It is possible that the presence of DMDP does confer some as yet unknown advantage on U. fulgens, which clearly has adapted to levels of DMDP which are unacceptable to other insects (see below).

The 1-deoxy derivative of nojirimycin B, by analogy termed deoxymannojirimycin (DMJ), was isolated in 1979 from seed of Lonchocarpus sericeus (Poir) H.B. and K., a tropical legume.[26] Although it was shown to be a moderate but selective inhibitor of glycoprotein processing mannosidase I (see below), it was far more active against mammalian α-L-fucosidase.[27,28] This is one example of an empirically determined and unexpected property of an AGI which can nevertheless prove of considerable use as a research tool. (The equivalent galactose analogue has been synthesized but has not been reported to occur naturally.)[29]

Two of the most potent AGIs are swainsonine and castanospermine. The first recorded higher plant sources were both Australian legumes. Early European settlers in Australia noted that cattle grazing Swainsona Salisb. spp., herbaceous pea-like plants,[30] developed symptoms suggesting neurological damage, but the toxic agent was unknown. Almost a century later, biologists in Western Australia realized that the symptoms resembled those of a hereditary condition of mammals, known as mannosidosis, characterized by a loss of lysosomal α-mannosidase activity. They successfully demonstrated powerful inhibition of α-mannosidase by crude plant extracts and isolated the agent responsible, swainsonine, in 1979.[31] It was shown to be a potent and specific inhibitor of certain mannosidases, including glycoprotein processing mannosidase II, but not of those enzymes inhibited by DMJ.[32] No other type of glycosidase has so far been found to be susceptible to swainsonine. By 1981 it was also shown to occur in *Astragalus* L. and *Oxytropis* DC spp., plants poisonous to livestock in the western United States.[33] Castanospermine, a potent inhibitor of both α- and β-glucosidases of mammalian gut,[23] was purified in 1981 from seed of a tree, *Castanospermum australe* A. Cunn., long known to cause digestive disturbances if eaten by men or livestock.[34] It also inhibits glucosidase I of glycoprotein processing; this activity is believed to be responsible for some of its more spectacular effects, including *in vitro* inhibition of HIV infectivity.[35]

In the early 1980s a survey was made in this laboratory of the Leguminosae sub-family Papilionoideae for other alkaloids of this type. It was hoped that their distribution might shed light on evolutionary relationships within the group and lead to a more satisfactory delineation of some tribal and generic boundaries. Of particular interest was the relationship of the putatively primitive tribe Sophoreae to more advanced tribes, including the Galegeae (containing *Astragalus* and *Swainsona*) and the Tephrosieae (containing *Derris* and *Lonchocarpus*).[36] The Sophoreae as presently defined comprises 32 genera.[30] Some, such as *Ormosia* Jackson which contains quinolizidine alkaloids,[37] have been reported 'alkaloid positive' since the alkaloids which they contain may be detected in crude extracts using Dragendorff reagent and be extracted into chloroform from basic aqueous solution. *Castanospermum*, by this criterion, is an 'alkaloid negative' genus, since castanospermine and all other AGIs (with the possible exception of swainsonine)[31] do not react with Dragendorff reagent and must be isolated by ion-exchange chromatography.[2]

A re-examination of basic compounds from 'alkaloid-negative' Sophoreae genera led to the discovery of several novel compounds of the AGI type. In 1985 a pyrrolidine, 1,4-dideoxy-1,4-imino-D-arabinitol (DAB-1), structurally related to DMDP but lacking one hydroxymethyl group, was isolated from fruits of *Angylocalyx* Taub., a small genus of 10 species from tropical Africa.[38] DAB-1, which also occurs in a fern *Arachnoides standishii* (Moore) Ohwi,[39] is an inhibitor of glucosidases, although its specificity differs from that of DMDP; for instance it is more active against mammalian gut α- than β-glucosidases. (DMDP preferentially inhibits β-glucosidase activity).[23,40] In the same year two piperidine alkaloids related to DNJ were isolated from other Sophoreae genera from Africa. One, (2S-carboxy-3R,4R,5S-trihydroxypiperidine, BR1) carrying a carboxylic acid group in place of the hydroxymethyl group, and therefore an analogue of glucuronic acid rather than glucose, was found in seed of *Baphia* Afzel. spp. BR1 proved an inhibitor of human β-glucuronidase, but not of glucosidases.[41] Seeds of *Xanthocercis zambesiaca* Dumaz le-Grand were found to contain approximately 2% of XZ1, a 4-O-β-D-glucoside of fagomine (1,2,5-trideoxy-1,5-imino-D-arabinohexitol) together with traces of the aglycone.[42] Fagomine differs from DNJ or DMJ by the loss of the anomeric hydroxyl group at C2. It has been shown to have some activity against mammalian α-glucosidases, particularly isomaltase, but no other significant glycosidase inhibition has been reported. The glucoside XZ1 was a very weak inhibitor.[23] A simple pyrrolidine, 2R-hydroxymethyl-3S-hydroxypyrrolidine (CYB3) and also the 6-epimer of castanospermine, (1S,6R,7R,8R,8aR)-1,6,7,8-tetrahydroxyoctahydroindolizine, were both found to occur with castanospermine in seed of *C. australe*.[43,44] CYB-3 inhibits mammalian sucrase but at relatively high concentration;[23] 6-epicastanospermine, despite a configuration of hydroxy groups resembling those of mannose, was not found to inhibit α- or β-mannosidases, but, like castanospermine, it inhibits fungal amyloglucosidase (an exo 1,4-α-glucosidase). It has no effect on glycoprotein processing I.[44]

The taxonomic significance of these findings cannot be fully assessed until the biosynthetic routes to these compounds have been elucidated, but the alkaloid distribution within the Sophoreae as presently defined suggests that it is polyphyletic and should be subdivided.

Castanospermum and Alexa

Castanospermum is a monotypic genus restricted to northeastern Australia with no near relatives in that subcontinent. A survey was made of other Sophoreae genera in the hope that other sources of this potential anti-AIDS agent might be found. Records held at Kew showed that flowers and fruit of the South American sp. *Alexa leiopetala* Sandwith resembled those of *Castanospermum*, and that the pollen types were also similar.[30] A dried pod collected 40 years ago was taken from a herbarium specimen of *A. leiopetala* and found to contain castanospermine.[45] It is believed that the common ancestor of both *Alexa* and *Castanospermum* inhabited the ancient southern continent of Pangaea, a land mass which existed c. 50 million years ago and which encompassed modern South America, Antarctica and Australasia. The flora of areas destined to become South America and Australasia were eventually separated by the advancing ice and the breakup of the land mass.[46] It is likely that the antecedents of *Astragalus* and *Oxytropis* (not known in Australasia) and *Swainsona* (found only in Australia and New Zealand) similarly spanned this ancient continent. They too share many morphological characteristics, as well as their content of swainsonine.[30,31,33] Recent discoveries of novel polyhydroxy derivatives of pyrrolizidine alkaloids in both *Alexa* and *Castanospermum* also suggest that the two genera are related. Alexine has been found in this laboratory in pods of *Alexa leiopetala* and its 3,8 diepimer in *C. australe*.[3,5] [So far these compounds have only been found to be weak glycosidase inhibitors (see Table 1)].

BIOLOGICAL ACTIVITY

Inhibition of Glycoprotein Processing

Many secretory and membrane proteins carry oligosaccharide chains linked through an asparagine residue to the polypeptide chain. These glycoproteins are believed to play an important role in cell-cell recognition, tumorigenesis and metastasis, lymphocyte transformation, intracellular transport, etc., but the contribution of the glycan side chain to the biological

Table 1. Concentration (molar) of alkaloidal glycosidase inhibitors resulting in 50% inhibition of substrate hydrolysis by homogenates of the intestines of mouse and fifth and sixth instar larvae of *Spodoptera littoralis*.

	DMDP	DAB1	DNJ	Homonoj.	Cast	Alexine
Trehalase Mouse	3.2×10^{-4}	2.2×10^{-5}	8.7×10^{-5}	1.8×10^{-5}	9.8×10^{-6}	5.9×10^{-5}
Trehalase *Spodoptera*	1.3×10^{-4}	4.1×10^{-5}	5.4×10^{-5}	3.7×10^{-5}	2.3×10^{-5}	1.5×10^{-4}
Maltase Mouse	2.1×10^{-4}	4.7×10^{-5}	5.4×10^{-7}	7.2×10^{-7}	8.2×10^{-7}	NI
Maltase *Spodoptera*	7.9×10^{-6}	2.9×10^{-5}	8.6×10^{-5}	1.3×10^{-4}	NI	NI
Sucrase Mouse	4.2×10^{-5}	2.3×10^{-5}	6.0×10^{-8}	8.1×10^{-8}	4.2×10^{-8}	NI
Sucrase *Spodoptera*	2.2×10^{-6}	1×10^{-4}	NI	2.2×10^{-4}	NI	NI
Lactase Mouse	2.1×10^{-6}	NI	5.7×10^{-5}	5.5×10^{-5}	5.8×10^{-7}	1.3×10^{-4}
Lactase *Spodoptera*	2.7×10^{-6}	1×10^{-4}	2.3×10^{-4}	1.9×10^{-4}	1.8×10^{-5}	NI

NI = less than 50% inhibition of hydrolysis at 3.3×10^{-4}M.
Substrate concentration was 28 mM in 50 mM maleate buffer, pH 6. (Assay method, see Reference 23).

activity of the complete molecule is poorly understood. AGIs have been found to specifically inhibit different glycolytic enzymes, so-called processing glycosidases, involved in the elaboration of the side chains. Processing glucosidase I (inhibited by castanospermine and DNJ) hydrolyses α-1,2 glucoside linkages, and glucosidase II (inhibited by DNJ) α-1,3 linkages. Processing mannosidase I (inhibited by DMJ) cleaves α-1,2 mannoside linkages and mannosidase II (inhibited by swainsonine) cleaves α-1,3 and α-1,6 linkages. Judicious use of processing inhibitors can bring about selective changes in the final structure of the glycan chains; AGIs are therefore proving to be unique tools with which to probe the structure/activity relationship of glycoproteins and the complex biochemical processes in which they participate. (This topic has been extensively reviewed elsewhere).[47-49] Many of the most interesting effects of AGIs, outlined below, are believed to be the direct or indirect result of their inhibition of processing enzymes.

Antiviral Activity

Few properties of AGIs are attracting more topical interest than their effects on retroviruses, in particular on the human immunodeficiency virus (HIV type 1), the cause of AIDS. In 1987 the results of three studies were published indicating that at millimolar concentrations (189 μg/ml) castanospermine, and to a lesser extent DNJ and DMDP, reduce the infectivity of HIV at concentrations which are, surprisingly, not toxic to T-lymphocytes. Fagomine and AB1 had no effect.[50-52] HIVI displays a tropism for cells displaying the so-called CD4 surface protein, a specific cellular receptor for HIV. The infectious cycle is initiated by the interaction between an external glycoprotein, gp 120, on the surface of the HIV envelope and the CD4 receptors on T4 lymphocytes. The nature of the interaction between gp 120 (known to be heavily glycosylated and carrying possibly 31-36 N-linked glycans per molecule) and CD4 is not understood, but *in vitro* this can be seen to lead to syncytium formation by cell to cell fusion, and cell death. In the presence of castanospermine and other inhibitors of processing glucosidases, the virus produced is non-infectious and syncytium formation is inhibited. Evidence suggests that this is the result of impaired processing of gp 120, leading to aberrant glycan side chains and the loss of the ability to bind to CD4. DMJ had no effect.[51]

Castanospermine and DNJ were reported to have a similar effect at 1.2-2.5 μg/ml on Moloney murine leukaemia virus, also a retrovirus, in cultured cells. (Swainsonine and DMJ had no effect).[53] Recently they have been found to inhibit replication of a herpesvirus cytomegalovirus, a common pathogen of those having a compromised immune system, including AIDS patients.[54] These studies suggest that inhibition of the cleavage of high molecular weight precursors and of intracellular transport of viral glycoproteins prior to assembly of the mature virion are important consequences of the alkaloid inhibition.

Despite these results, which clearly suggest that processing glucosidase inhibitors may have a future as antiviral agents, earlier studies were less encouraging. Reports in 1982-1983 showed that while swainsonine and castanospermine modified the haemagglutinins of influenza virus at 1 and 5 μg/ml respectively the viral particles produced were full infective.[32,55] Subsequently it was shown that although DNJ and DMJ at 2 mM and 1 mM respectively caused changes to the glycoproteins of vesicular stomatitis virus and influenza, infectivity remained almost at control levels.[56] N-methyl DNJ, also a processing glucosidase inhibitor, caused abnormal glycoprotein production in Rous sarcoma and fowl plague virus without affecting infectivity.[57,58]

Cancer and the Immune Response

AGIs are now being used to investigate the role of glycoproteins in cancer, in particular in the transformation of normal to cancer cells and the processes of metastasis, and in the immune response.

Asparagine-linked oligosaccharides of glycoproteins show qualitative differences after neoplastic transformation with chemicals, oncogenic viruses or transfection with oncogenes.[59-61] Castanospermine (at 20 μg/ml) (and N-methyldeoxynojirimycin) have been shown to prevent the expression of a glycoprotein oncogene product encoded by a strain of feline sarcoma virus (SM-FeSV) in feline embryo cells.[62] Although there was no significant change in cell morphology of the transformed cells in the presence of castanospermine, growth of the cells became strictly serum dependent and growth in soft agar was drastically retarded despite the presence of 10% calf serum, indicating that the properties of the transformed cells were altered. Swainsonine had no effect. Also, nude mice injected subcutaneously with feline sarcoma virus-transformed rat cells (SM-FRE) had slower growing tumors (2.6 times smaller than controls after 24 days) when castanospermine was added to the diet at 2.4 mg/g, alkaloid consumption being approximately 0.84 mg/g mouse/day.[63] In contrast, swainsonine prevented phenotypic expression (growth in soft agar and glycoproteins characteristic of transformed cells) in mouse fibroblasts NIH 3T3 transfected with DNA from a human bladder cancer line. When swainsonine was withdrawn from the medium, the transformed phenotype was again expressed. It was suggested that glycosylation plays a more fundamental role in transformation than merely glycosylation of the oncogene product since in this case the product of the oncogene is not a glycoprotein, but clearly the maintenance of oligosaccharide processing is necessary for its expression.[64]

There is evidence that interference with maturation of the oligosaccharide subunits of tumor cell glycoproteins can reduce their metastatic capacity. The pretreatment of murine melanoma B16-F10 cells with castanospermine or swainsonine (at 1.0 and 3.0 μg/ml respectively) prior to i.v. injection into mice reduced pulmonary colonization by over 63%. This was shown not to be due to any cytotoxic effect, and cells showed normal tumorigenicity after sub-cutaneous implantation. Cells treated with the two inhibitors appeared to have nearly normal adhesive capacity and the reduced lung invasion was suggested to result from the blockage of a process distal to initial tumor arrest.[65,66] Recently, it has been reported that 80% reduction in pulmonary colonization was achieved by supplying drinking water containing swainsonine at 3 μg/ml <u>ad libitum</u> for 24 hr prior to injection of tumorigenic cells. Swainsonine was shown to elicit a 32% increase in spleen cell number 2 days after administration and to induce a 2-3 fold increase in splenic NK (natural killer) cell activity. Further, the effects of the alkaloid were not seen in mice depleted of their NK cell activity either experimentally or by mutation, and the authors conclude that its antimetastatic activity is mediated primarily by its ability to augment NK cell activity.[67] Swainsonine (0.3 μg/ml) has also been shown to reduce the metastatic potential of the lymphoid tumor line MDAY-D2. Growth of the tumor line *in vitro* was not affected but treated cells were shown to be more sensitive to the anti-proliferative effects of interferon both *in vitro* and *in vivo*.[68]

The first indication of the immunoregulatory potential of swainsonine was a report in 1985 that swainsonine could reverse the immunosuppressive action of a factor from the serum of sarcoma 180 tumor bearing mice on concanavalin A stimulated lymphocyte proliferation *in vitro*. In its presence the expression of concanavalin A receptors of spleen cells was enhanced.[16] The ability of immunodeficient mice (treated with immunosuppressive factor, antitumor drugs or inoculated with sarcoma 180 tumor) to produce antibody to intravenously administered red blood cells of sheep was restored by swainsonine (e.g., given intraperitoneally at 3.7 mg/kg daily for 5 days concomitantly with immunosuppressive factor prior to immunization). Further, swainsonine injected daily for 5 days at 10 or 30 mg/kg after inoculation of mice with S-180 tumor suppressed tumor growth by 50 and 100% respectively. These results strongly suggest a role for swainsonine as a research tool if not as a treatment of the immunocompromised host and in tumor management.[69]

In a recent study of the effects of processing inhibitors on the immune response, it was shown that treatment of cloned murine helper T cells with swainsonine or castanospermine prior to stimulation by antigen differentially affects the proliferative response of the cell. Swainsonine enhanced, and castanospermine suppressed, the response. Both inhibitors, but particularly swainsonine, enhanced the proliferative response to stimulation by the mitogen concanavalin A, and both suppressed the response to stimulation by interleukin 2. In each case a higher concentration of castanospermine than swainsonine was needed to exhibit maximal effect which may reflect difficulty in cellular uptake of castanospermine.[70] The results of such a study are particularly relevant to the evaluation of these alkaloids as therapeutic agents (see above).

It is possible that glycosidases other than processing glycosidases may be a suitable target for anti-cancer chemotherapy. Levels of several types of glycosidases have been shown to be elevated in the interstitial fluid of tumors, sera of animals and patients with tumors and in some tumor tissues when compared with controls. The cellular release of these enzymes probably results from tumor lysosomal exocytosis. It is likely that glycosidases assist tumor shedding from primary sites and local invasion with penetration of blood and/or lymph vessels, and further the invasion of the microvasculature with degradation of the basement membrane at secondary sites. Although there are no reports of AGIs being used in this way, a glucuronidase inhibitor, sodium glucaro-(1-4)-lactonate, has been successfully used to shrink bladder tumors.[71]

Effects on Plant Glycosidases

AGIs have been shown to inhibit processing glycosidases of plants; for example, castanospermine inhibited glucosidase I in suspension cultured soy bean cells and deoxymannojirimycin mannosidase I in mung beans.[72,73] Further, the action of DNJ and swainsonine on the processing of phytohemagglutinin of *Phaseolus vulgaris* L. was compatible with their inhibiting glucosidase I and mannosidase II respectively.[74] Three forms of soluble α-glucosidase extracted from suspension-cultured rice cells, each able to hydrolyze maltase or soluble starch, were found to be competitively inhibited by castanospermine.[75] Swainsonine has been shown to be a potent inhibitor of α-mannosidase activity in developing cotyledons of various legumes.[76] Recently castanospermine was found to inhibit the hydrolysis of a cyanogenic glucoside.[77] Castanospermine at 10^{-4}M was shown in this laboratory to inhibit established plant growth (lettuce, wheat). At 10^{-5}M it inhibited lettuce seedling growth by 50%; swainsonine, DAB1, DMDP, CYB-3 and DNJ were also inhibitory but to a lesser extent (RJN, unpublished).

Certain AGIs have been shown to inhibit plant cell wall extension and pollen germination. Early reports indicated that nojirimycin could prevent the auxin-induced extension growth of barley coleoptile segments, and that this resulted from the inhibition of a wall β-glucosidase involved in partial degradation of wall β-glucan.[78] It also prevented the spontaneous fragmentation of aggregations of rose cells in culture (at 2×10^{-4}M) and was shown to decrease the activity of both β-glucosidase and β-galactosidase associated with it.[79] Intact corn roots grown in castanospermine (20 μg/ml) were half the length of controls, the tip was curled and no secondary roots were formed; the authors suggested that this was the result of inhibition of β-glucosidase necessary for wall turnover and elongation. Swainsonine had no visible effect on growth although the walls were shown to contain a swainsonine-susceptible α-mannosidase.[90] Nojirimycin (at 10^{-5}M) inhibited a β-glucosidase in germinating pear pollen which was shown to increase after germination in proportion to tube wall development.[81] Castanospermine, DAB1, DMDP, DNJ and fagomine were found in this laboratory to inhibit the growth of pollen of *Sophora microphylla* Ait. at 10^{-4}M (RJN, unpublished).

Intestinal Glycosidase Inhibition

Intestinal α-glucosidase inhibitors have the potential to regulate carbohydrate digestion and absorption and thereby play a role in the clinical management of diabetes and some forms of hyperlipoproteinemia and adipositas.[82] Both castanospermine, DNJ and several synthetic derivatives of DNJ have been shown to delay the hyperglycaemic response to oral carbohydrate loading *in vivo* and are being assessed as potential drugs.[18,83,84]

In a comparative study of the action of AGIs on mouse digestive disaccharidases, castanospermine was shown to be a potent inhibitor of both α- and β-glucosidases, including trehalase. In contrast DNJ preferentially inhibited α- and DMDP β-glucosidases.[23] Similar results have recently been obtained using human amniotic fluid; chicken gut enzymes show some differences, e.g., the hydrolysis of maltose is markedly more resistant to both DNJ and castanospermine than in the human or mouse (AMS, inpublished). In marked contrast DMDP was found to be a potent competitive inhibitor of digestive α-glucosidase in the larvae of the bruchid beetle *Callosobruchus maculatus* F., a major pest of stored grain legumes, but DNJ had little effect.[22]

Castanospermine inhibits both α- and β-glucosidase activity, but in contrast to the mouse, the β-form is more susceptible.[85] A study of the effects of castanospermine on glycosidases from different insects showed that there is considerable variation in the susceptibility of enzymes from different species.[86] This selectivity might prove of value in the formulation of selective pesticides.

We have compared the susceptibility to some AGIs of insect (Spodoptera littoralis Boisduval.) and mammal (mouse) digestive disaccharidases using natural substrates (Table 1). Castanospermine had little effect on Spodoptera maltase or sucrase, despite being a strong inhibitor of the mammalian forms. DMDP, in contrast, was more active against the insect form of these enzymes. However, there were only marginal differences in the susceptibility of trehalase from both organisms to all the compounds tested. The differences in susceptibility of lactase to all AGIs, except castanospermine, were also small. Alexine, apart from some moderate inhibition of mouse trehalase, was a very weak inhibitor, as was 3,8-diepialexine.[5] (Note that these studies were carried out on homogenates of intestinal mucosa and insect larval gut, not purified enzymes, and that this should be taken into account when interpreting the data).

Insect Antifeedant Effects

It has been observed that DMDP deters feeding in Vth instar nymphs of the locust Locusta migratoria L. at 0.001% of an artifical diet,[87] and that castanospermine at 0.002% strongly deters the pea aphid Acyrthosiphon pisum Harris.[88] Natural chemicals which deter insects from feeding have potential as crop protection agents.

In this laboratory we have examined the effects of a range of AGIs on feeding behavior in 4 species of Lepidoptera. The behavioral response (antifeedancy) was compared with the electrophysiological response when taste receptors on sensilla on the insect's mouthparts were stimulated directly with solutions of each alkaloid. The use of these two bioassays, described below, enables us to correlate sensory input - the message which each alkaloid sends to the brain - with behavioral output.

In the behavioral bioassay, glass fiber discs (Whatman GF/A, 2.1 cm diameter), were presented in pairs (control and treatment) to individual final stadium larvae for up to 8 hrs. Control discs were made palatable by the application of 100 μl 0.05 M sucrose dissolved in an electrolyte, 0.05 M NaCl. For the treatment discs, alkaloids were added to the sucrose/NaCl solution in the concentration range 10^{-5}M to 10^{-2}M. (This yielded an approximate range of 0.001% to 1% of each alkaloid in 15% sucrose, as disc dry weight. NaCl was included so that the results could be more readily correlated with the electrophysiological assay). The amount of each disc eaten was determined by weighing, and an Antifeedant Index calculated according to the formula (C-T)/C+T) x 100%, where C is the weight of control and T of treatment disc eaten.

For the electrophysiological bioassay,[89] test compounds dissolved in 0.05 M NaCl over the same concentration range (10^{-5}M to 10^{-2}M) were used to stimulate individual maxillary styloconic sensilla. Each larva has four of these sensilla, two lateral and two medial, and each sensillum contains four chemosensory neurones. We can identify the activity of each neurone by the shape and height of its action potential, and for convenience each is designated by a letter A, B, C, D.

Figure 4 shows the result of correlating electrophysiological data (total firing of lateral styloconic sensillum) with Antifeedant Index for the polyphagous leafworm S. littoralis at 10^{-2}M. (Points shown as black circles). It can be seen that swainsonine, castanospermine, DMDP, and DAB1 were antifeedant but BR1, DNJ, DMJ, fagomine, XZ-1 and CYB-3 were not, despite the fact that some, in particular DNJ, DMJ and XZ-1, elicited a high rate of neuronal firing. An analysis of the spike train (not shown) revealed that the observed total firing rate is the sum of the firing rates of two neurones, neurone A and neurone C. The overall effect on behavior depends on which neurone fires at the higher rate when stimulated by a particular alkaloid. DMDP, swainsonine, castanospermine and AB-1 in fact stimulate both neurones, but C is stimulated more than A, resulting in antifeedant behavior. In contrast, CYB-3, fagomine and XZ-1 stimulate neurone A more than neurone C. The dotted line shows

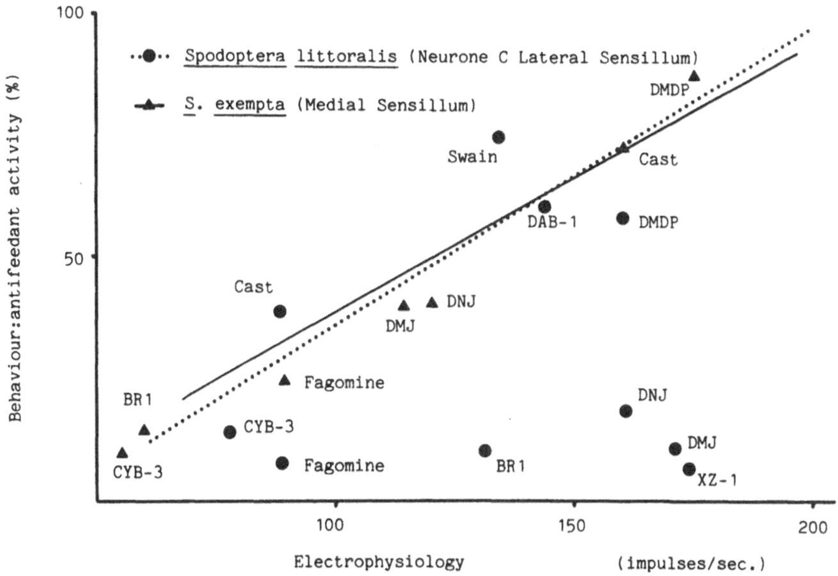

Fig. 4. Correlation of electrophysiological and behavioral responses of sixth stadium larvae of Spodoptera littoralis and Spodoptera exempta to test compounds at 10^{-3}M. Electrophysiological response: total firing, in first second of stimulation, of the lateral maxillary styloconic sensilla (S. littoralis) or medial styloconic sensilla (S. exempta), n = 5-10. Behavioral response: Antifeedant Index, (C-T)/(C+T) x 100%, n = 20.

the correlation of the firing of neurone C (only) with Antifeedant Index ($r = 0.8066$, $p < 0.001$).

Also indicated in Figure 4 (points shown as black triangels) are the results of a similar study using larvae of the related oligophagous armyworm (Spodoptera exempta Walker). It has been observed in previous studies[90] that S. exempta is more susceptible to the deterent effects of many types of plant secondary compounds than S. littoralis. In this study the electrophysiological response of the medial sensillum was correlated with the behavioral response since it has previously been observed that in this insect the medial sensillum contains a neurone (the 'deterrent' neurone)

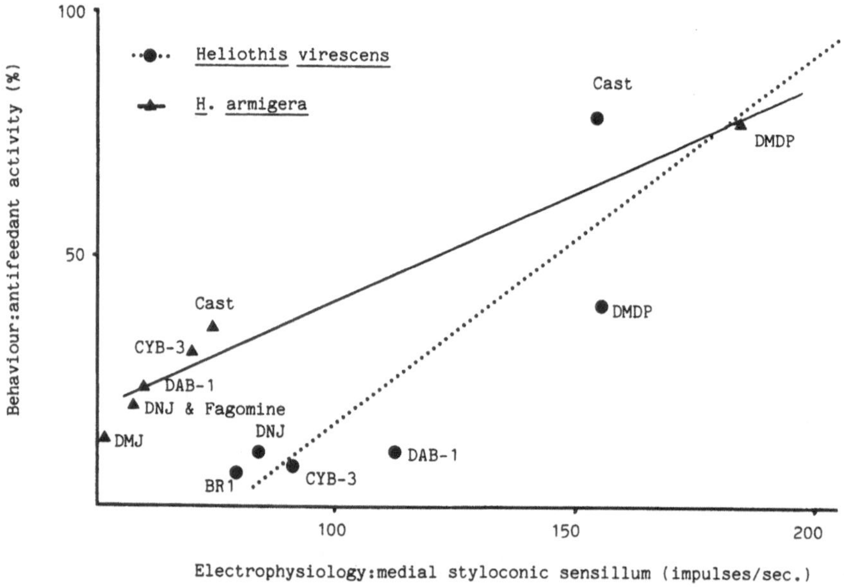

Fig. 5. Correlation of electrophysiological and behavioral response of sixth stadium larvae of *Heliothis virescens* and *Heliothis armigera* to test compounds at 10^{-3}M. Electrophysiological response: total firing in first second of stimulation of the medial maxillary styloconic sensilla, n = 5-10. Behavioral response: Antifeedant Index (C-T)/(C+T) x 100%, n = 20.

sensitive to many antifeedant compounds.[91] In contrast to *S. littoralis*, DNJ and DMJ also stimulated the deterrent neurone of *S. exempta*. The solid black line shows the correlation between the firing of the deterrent neurone of *S. exempta* and the Antifeedant Index for all AGIs tested; ($r = 0.9874$, $p < 0.0001$). (A shortage of XZ-1, DAB-1 and swainsonine prevented their being tested against *S. exempta*).

The results of a similar study on the larvae of two species of *Heliothis* are shown in Figure 5. In both species of *Heliothis* the AGIs stimulated neurones in the medial styloconic sensillum. This response, though low, did correlate significantly with changes in feeding behavior (*Heliothis virescens* F., $r = 0.8694$, $p < 0.001$;

Heliothis armigera Hubner, $r = 0.9852$, $p < 0.0001$). However, only DMDP stimulated the neurons of *H. armigera* at a rate that significantly decreased feeding, and only castanospermine those of *H. virescens*.

Previous studies have shown that neurone A in the lateral styloconic sensillum of *S. littoralis*, which responds to simple sugars which act as phagostimulants, has at least two types of receptor site, one responsive to glucose and one to fructose. Since DMDP is an analogue of fructose we compared the response of the sensillum to glucose, fructose and DMDP. Each elicited a high total rate of firing, but it was observed that after exposure to DMDP the ability of the sensillum to respond to glucose was reduced for up to one hour and to fructose for several hours. It seems likely that DMDP, because of its structural similarity to fructose, may interact with or block the neuronal receptor site for fructose on neurone A. This is shown diagramatically in Figure 6. Preliminary studies indicate that castanospermine may interact with the glucose receptor in a similar way. This suggests that alkaloids resembling sugars may protect the plants in which they accumulate by rendering insects 'blind' to the presence of phagostimulatory sugars.

SYNTHESIS AND BIOSYNTHESIS

Most of the known natural alkaloidal glycosidase inhibitors have been synthesized. All syntheses derive from carbohydrate starting material with the exception of a synthesis of swainsonine from *trans*-1,4-dichlorobutane and of swainsonine and DAB1 from (R)- and (S)-glutamic acid. This topic has recently been reviewed.[2] It is clear that many of the possible unnatural structural variants will also prove to have useful activity. Reference was made above to derivatives of DNJ with potential as anti-diabetic agents. Other examples are an iminopentitol more active against Jack Bean α-mannosidase than swainsonine, and an analogue of L-fucose with the ring O replaced by N which is a powerful fucosidase inhibitor.[40,92] The enantiomer of DAB1 proved a stronger inhibitor of mouse gut α-glucosidases than the natural form.[23]

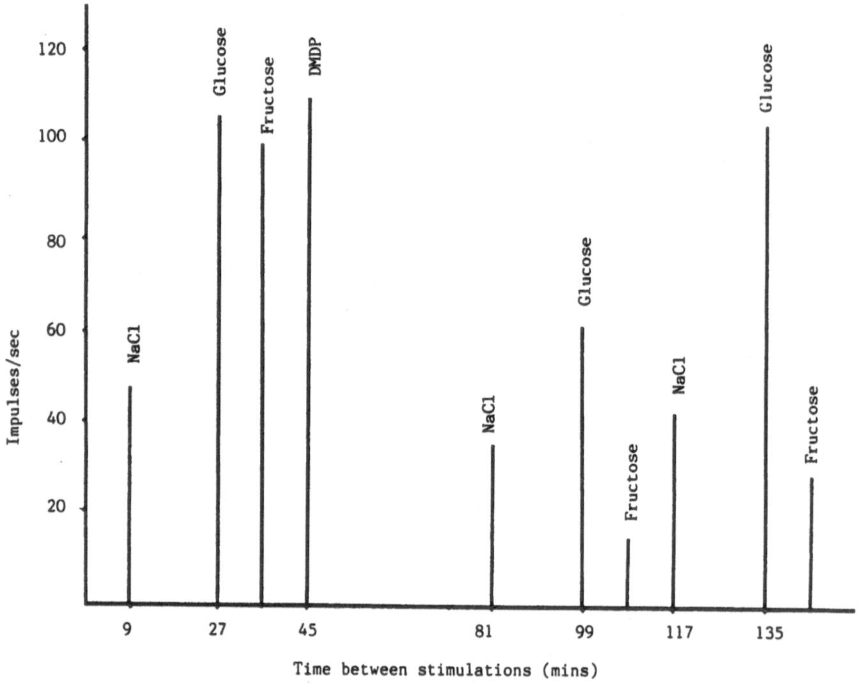

Fig. 6. Effect of stimulating the lateral styloconic sensilla of Spodoptera littoralis with glucose (0.1 M), fructose (0.1 M) and DMDP (0.001 M) in 0.05 M NaCl. Control: 0.05 M NaCl. Each point represents the mean firing rate of the first second's response to 10 larvae.

Biosynthetic studies have been restricted to microorganisms; to our knowledge there have been no studies of the pathways operating in higher plants. An early study of the biosynthesis of nojirimycin indicated that it is derived from glucose with head (C-1) and tail (C-6) inversion, and it was suggested that an amino group is introduced via 5-ketoglucose.[93] The biosynthesis of swainsonine has been studied in the mold Rhizoctonia leguminicola. The piperidine ring is derived from lysine via pipecolic acid. Two additional carbons are introduced to the side chain which is believed to cyclise and give rise to 1-hydroxyindolizidine and 1,2-indolizidine diol intermediates before being converted to swainsonine.[94]

SUMMARY

Alkaloidal glycosidase inhibitors have now been isolated from 5 families of higher plant (Leguminosae, Polygonaceae, Euphorbiaceae, Moraceae, Aspidiaceae)[1,24] and many microorganisms. Their ability to modulate the metabolism of mammals, insects and viruses has been clearly demonstrated and it seems likely that they are a widespread defensive strategy in nature. Their metabolic role in, and contribution to, the ecology of those species in which they accumulate needs further investigation. It is clear that mixtures of these alkaloids, each with different inhibitory properties, can occur in the same species, e.g., DMDP with homonojirimycin and DMJ in *Omphalea*; DMJ with DMDP in *Lonchocarpus*.[19,28] Varying the relative concentrations of the components of such mixtures could be one way in which such species respond to changing predator pressure.

The use of these compounds as tools in glycoprotein studies has opened up new frontiers in cancer, virus and immunological research. Much experimental work, including necessary comparative studies, has been hindered by the scarcity and high cost of these compounds. Demand for them is certain to remain high and natural sources a realistic alternative to chemical synthesis.

ACKNOWLEDGMENTS

We gratefully acknowledge the assistance of Dr. G. Fleet (Oxford University), Dr. W. Blaney (University of London) and members of the Photography Department and Herbarium, RBG Kew. Insect studies were carried out under MAFF licence no. PH F1020/10 issued under the Import and Export (Plant Health Great Britain) Order 1980 and Plant Pests (Great Britain) Order 1980.

REFERENCES

1. FELLOWS, L.E. 1986. The biological activity of polyhydroxyalkaloids from plants. Pestic. Sci. 17: 602-606.
2. FELLOWS, L.E., FLEET, G.W.J. 1988. Alkaloidal glycosidase inhibitors from plants. In Natural Products Isolation. (G.H. Wagman, R. Cooper, eds.), Elsevier, Amsterdam, pp. 540-560.
3. NASH, R.J., L.E. FELLOWS, J.V. DRING, G.W.J. FLEET, A.E. DEROME, T.A. HAMOR, A.M. SCOFIELD, D.J. WATKIN. 1988. Isolation from Alexa leiopetala and X-ray crystal structure of alexine, (1R,2R,3R,7S,8S)-3-hydroxymethyl-1,2,7-trihydroxypyrrolizidine [(2R,3R,4R,5S,6S)-2-hydroxymethyl-1-azabicyclo[3.3.0]octan-3,4,6-triol], a unique pyrrolizidine alkaloid. Tetrahedron Lett. 29: 2487-2490.
4. SHIBATA, T., O. NAKAYAMA, Y. TSURUMI, M. OKUHARA, H. TERANO, M. KOHSAKA. 1988. A new immunomodulator, FR-900483. J. Antibiot. 61: 296-301.
5. NASH, R.J., L.E. FELLOWS, A.C. PLANT, G.W.J. FLEET, A.E. DEROME, P.D. BAIRD, M.P. HEGARTY, A.M. SCOFIELD. 1988. Isolation from Castanospermum australe and X-ray crystal structure of 3,8-diepialexine, (1R,2R,3S,7S,8R)-3-hydroxymethyl-1,2,7-trihydroxypyrrolizidine [(2S,3R,4R,5S,6R)-2-hydroxymethyl-1-azabicyclo [3.30]octan-3,4,6-triol]. Tetrahedron 44: 5959-5964.
6. INOUYE, S., T. TSURUOKA, T. NIIDA. 1966. The structure of nojirimycin, a piperidinose sugar antibiotic. J. Antiobiot. Ser. A 19: 288-292.
7. ISHIDA, N., K. KUMAGAI, T. NIIDA, T. TSURUOKA, H. YUMOTO. 1967. Nojirimycin, a new antibiotic. II. Isolation, characterization and biological activity. J. Antibiot. 20: 66-71.
8. NIWA, T., S. INOUYE, T. TSURUOKA, Y. KOAZE, T. NIIDA. 1970. Nojirimycin as a potent inhibitor of glucosidase. Agric. Biol. Chem. 34: 966-968.
9. REESE, E.T., F.W. PARRISH. 1971. Nojirimycin and D-glucono-1,5-lactone as inhibitors of carbohydrases. Carbohydr. Res. 18: 381-388.
10. NIWA, T., T. TSURUOKA, H. GOI, Y. KODAMA, J. ITOH, S. INOUYE, Y. YAMADA, T. NIIDA, M. NOBE, Y.

OGAWA. 1984. Novel glycosidase inhibitors, nojirimycin B and D-mannonic-δ-lactam. J. Antibiot. 37: 1579-1586.
11. MIYAKE, Y., M. EBATA. 1987. Galactostatin, a new β-galactosidase inhibitor from Streptomyces lydicus. J. Antibiot. 40: 122-123.
12. SCHMIDT, D.D., W. FROMMER, L. MULLER, E. TRUSCHEIT. 1979. Glucosidase Inhibitoren aus Bazillen. Naturwissenschaften 66: 584-585.
13. MURAO, S., S. MIYATA. 1980. Isolation and characterization of a new trehalase inhibitor, S-GI. Agric. Biol. Chem. 44: 219-221.
14. GUENGERICH, F.P., S.J. DIMARI, H.P. BROQUIST. Isolation and characterization of a 1-pyridine fungal alkaloid. J. Am. Chem. Soc. 95: 2055-2056.
15. SCHNEIDER, M.J., F.S. UNGEMACH, H.P. BROQUIST, T.M. HARRIS. 1983. (1S,2R,8R,8aR)-1,2,8-trihydroyoctahydroindolizine (swainsonine), an α-mannosidase inhibitor from Rhizoctonia leguminicola. Tetrahedron 39: 29-32.
16. HINO, M., O. NAKAYAMA, Y. TSURUMI, K. ADACHI, T. SHIBATA, H. TERANO, M. KOHSAKA, H. AOKI, H. IMANAKA. 1985. Studies of an immunomodulator, swainsonine. 1. Enhancement of immune response by swainsonine in vitro. J. Antibiot. 38: 926-935.
17. YAGI, M., T. KOUNO, Y. AOYAGI, H. MURAI. 1976. The structure of moranoline, a piperidine alkaloid from Morus species. Nippon Nogeikagaku Kaishi 50: 571-572.
18. MURAI, H., K. OHATA, H. ENOMOTO, Y. YOSHIKUNI, T. KONO, M. YAGI. 1977. 2-Hydroxymethyl-3,4,5-trihydroxypiperidine and extraction process for its manufacture. Ger. Offen. 2,656,602, 30th June.
19. KITE, G.C., L.E. FELLOWS, G.W.J. FLEET, P.S. LIU, A.M. SCOFIELD, N.G. SMITH. 1988. Alpha-homonojirimycin (2,6-dideoxy-2,6-imino-D-glycero-L-gulo-heptitol) from Omphalea diandra L.: Isolation and glucosidase inhibition. Tetrahedron. Lett. 29: 6483-6486.
20. WELTER, A., J. JADOT, G. DARDENNE, M. MARLIER, J. CASIMIR. 1976. 2,5-Hydroxymethyl 3,4-dihydroxypyrrolidine dans les feuilles de Derris elliptica. Phytochemistry 15: 747-749.

21. FLEET, G.W.J., P.W. SMITH. 1985. Enantiospecific syntheses of deoxymannojirimycin, fagomine and 2R,5R-dihydroxymethyl-3R,4R-dihydroxypyrrolidine from D-glucose. Tetrahedron Lett. 26: 1469-1472.
22. EVANS, S.V., A.M.R. GATEHOUSE, L.E. FELLOWS. 1985. Detrimental effects of 2,5-dihydroxymethyl-3,4-dihydroxypyrrolidine in some tropical legumes on larvae of the bruchid _Callosobruchus_ _maculatus_. Entomol. Exp. Appl. 37: 257-261.
23. SCOFIELD, A.M. L.E. FELLOWS, R.J. NASH, G.W.J. FLEET. 1986. Inhibition of mammalian disaccharidases by polyhydroxy alkaloids. Life Sci. 39: 645-650.
24. HORN, J.M., D.C. LEES, N.G. SMITH, R.J. NASH, L.E. FELLOWS, E.A. BELL. 1986. The _Urania_/_Omphalea_ interaction: host plant secondary chemistry. _In_ Proceedings of the 6th International Symposium of Plant-Insect Relationships. (V. Labeyrie, G. fabres, D. Lachaise, eds.), Dr. W. Junk, Dordrecht, p. 394.
25. BROWER, L.P., J. VAN, Z. BROWER, J.M. CORVINA. 1967. Plant poisons in a terrestrial food chain. Proc. Nat. Acad. Sci. USA 57: 893-898.
26. FELLOWS, L.E., E.A. BELL, D.G. LYNN, F. PILKIEWICZ, I. MIURA, K. NAKANISHI. 1979. Isolation and structure of an unusual cyclic amino alditol from a legume. J. Chem. Soc. Chem. Commun. 977-978.
27. FUHRMANN, U., E. BAUSE, G. LEGLER, H. PLOEGH. 1984. Novel mannosidase inhibitor blocking conversion of high mannose to complex oligosaccharides. Nature 307: 755-758.
28. EVANS, S.V., L.E. FELLOWS, T.K.M. SHING, G.W.J. FLEET. 1985. Glycosidase inhibition by plant alkaloids which are structural analogues of monosaccharides. Phytochemistry 24: 1953-1955.
29. LEGLER, G., S. POHL. 1986. Synthesis of 5-amino-5-deoxy-D-galactopyranose and 1,5-dideoxy-1,5-imino-D-galactitol, and their inhibition of α- and β-D-galactosidases. Carbohydr. Res. 155: 119-129.
30. POLHILL, R.M. 1981. Galegeae. _In_ Advances in Legume Systematics. (R.M. Polhill, P.H. Raven, eds.), Royal Botanic Garden, Kew, Richmond, United Kingdom, pp. 357-363.

31. COLEGATE, S.M., P.R. DORLING, C.R. HUXTABLE. 1979. A spectroscopic investigation of swainsonine: an α-mannosidase inhibitor isolated from Swainsona canescens. Aust. J. Chem. 32: 2257-2264.
32. ELBEIN, A.D., P.R. DORLING, K. VOSBECK. M. HORISBERGER. 1982. Swainsonine prevents the processing of the oligosaccharide chains of influenza virus hemagglutinin. J. Biol. Chem. 257: 1573-1576.
33. MOLYNEUX, R.J., L.F. JAMES. 1981. Loco intoxication: indolizidine alkaloids of spotted locoweed Astragalus lentiginosus. Science 216: 190-191.
34. HOHENSCHUTZ, L.D., E.A. BELL, P.J. JEWESS, D.P. LEWORTHY, R.J. PRYCE, E. ARNOLD, J. CLARDY. 1981. Castanospermine, a 1,6,7,8-tetrahydroindolizine alkaloid from seeds of Castanospermum australe. Phytochemistry 20: 811-814.
35. SASAK, V.W., J.M. ORDOVAS, A.D. ELBEIN, R.W. BERNINGER. 1985. Castanospermine inhibits glucosidase I and glycoprotein secretion in human hepatoma cells. Biochem. J. 232: 759-766.
36. GEESINK, R. 1981. Tephrosieae. Op. cit. Reference 30, pp. 245-260.
37. KINGHORN, A.D., R.A. HUSSAIN, E.F. ROBBINE, M.F. BALADRIN, C.H. STIRTON. 1988. Alkaloid distribution in seeds of Ormosia, Pericopsis and Haplormosia. Phytochemistry 27: 439-444.
38. NASH, R.J., E.A. BELL, J.M. WILLIAMS. 1985. 2-Hydroxymethyl-3,4-dihydroxypyrrolidine in fruits of Angylocalyx boutiqueanus. Phytochemistry 24: 1620-1622.
39. JONES, D.W.C., R.J. NASH, E.A. BELL, J.M. WILLIAMS. 1985. Identification of the 2-hydroxymethyl-3,4-dihydroxypyrrolidine (or 1,4-dideoxy-1,4-iminopentitol) from Angylocalyx boutiqueanus and from Arachnoides standishii as the (2R,3R,4S) isomer by the synthesis of its enantiomer. Tetrahedron Lett. 26: 3125-3126.
40. FLEET, G.W.J., S.J. NICHOLAS, P.W. SMITH, S.V. EVANS, L.E. FELLOWS, R.J. NASH. 1985. Potent competitive inhibition of α-galactosidase and α-glucosidase activity by 1,4-dideoxy-1,4-iminopentitols: syntheses of 1,4-didieoxy-1,4-imino-D-lyxitol and of both enantiomers of

1,4-dideoxy-1,4-iminoarabinitol. Tetrahedron Lett. 26: 3127-3130.
41. MANNING, K.S., D.G. LYNN, J. SHABANOWITZ, L.E. FELLOWS, M. SINGH, B.D. SCHRIRE. 1985. A glucuronidase inhibitor from the seeds of Baphia racemosa: application of fast atom bombardment coupled with collision activated dissociation in natural product structure assignment. J. Chem. Soc. Chem. Commun. 127-129.
42. EVANS, S.V., A.R. HAYMAN, L.E. FELLOWS, T.K.M. SHING, A.E. DEROME, G.W.J. FLEET. 1985. Lack of glycosidase inhibition by, and isolation from Xanthocercis zambesiaca (Leguminoasae) of 4-O-(beta-D-glucopyranosyl)-fagomine, [1,2,5-trideoxy-4-O-(beta-D-glucopyranosyl)-1,5-imino-D-arabino-hexitol], a novel glucoside of a polyhydroxylated piperidine alkaloid. Tetrahedron Lett. 26: 1465-1468.
43. NASH, R.J., E.A. BELL, G.W.J. FLEET, R.H. JONES, J.M. WILLIAMS. 1985. The identification of a hydroxylated pyrrolidine derivative from Castanospermum australe. J. Chem. Soc. Chem. Commun. 738-740.
44. MOLYNEUX, R.J., J.N. ROITMAN, G. DUNNHEIM, T. SZUMILO, A.D. ELBEIN. 1986. 6-Epicastanospermine, a novel indolizidine alkaloid that inhibits α-glucosidase. Arch. Biochem. Biophys. 251: 450-457.
45. NASH, R.J., L.E. FELLOWS, J.V. DRING, C.H. STIRTON, D. CARTER, M.P. HEGARTY, E.A. BELL. 1988. Castanospermine in Alexa species. Phytochemistry 27: 1403-1404.
46. SMITH, A.G., J.C. BRIDEN, G.E. DREWRY. 1973. Phanerozoic world maps. Spec. Pap. Palaeontol. 12: 1-42.
47. SCHWARTZ, R.T., R. DATEMA. 1984. Inhibitors of trimming: new tools in glycoprotein research. Trends Biochem. Sci. 9: 32-34.
48. FUHRMANN, U., E. BAUSE, H. PLOEGH. 1985. Inhibitors of oligosaccharide processing. Biochim. Biophys. Acta 825: 95-110.
49. ELBEIN, A.D., R.J. MOLYNEUX. 1987. The chemistry and biochemistry of simple indolizidine and related polyhydroxy alkaloids. In Alkaloids:

Chemical and Biological Perspectives. (S.W. Pelletier, ed.), Wiley Interscience, New York. Vol. 5, pp. 1-54.

50. TYMS, A.S., E.M. BERRIE, T.A. RYDER, R.J. NASH, M.P. HEGARTY, D.L. TAYLOR, M.A. MOBBERLEY, J.M. DAVIS, E.A. BELL, D.J. JEFFRIES, D. TAYLOR-ROBINSON, L.E. FELLOWS. 1987. Castanospermine and other plant alkaloid inhibitors of glucosidase activity block the growth of HIV. Lancet, 1025-1026.

51. GRUTERS, R.A., J.J. NEEFJES, M. TERSMETTE, R.E.Y. DE GOEDE, A. TULP, H.G. HUISMAN, F. MIEDEMA, H.L. PLOEGH. 1987. Interference with HIV-induced syncytium formation and viral infectivity by inhibitors of trimming glucosidase. Nature 330: 74-77.

52. WALKER, B.D., M. KOWALSKI, W.C. GOH, K. KOZARSKY, M. KRIEGER, C. ROSEN, L. ROHRSCHNEIDER, W.A. HASELTINE, J. SODROSKI. 1987. Inhibition of human immunodeficiency virus syncytium formation and virus replication by castanospermine. Proc. Natl. Acad. Sci. USA 84: 8120-8124.

53. SUNKARA, P.S., T.L. BOLIN, P.S. LIU, A.A. SJOERDSMA. 1987. Antiretroviral activity of castanospermine and deoxynojirimycin, specific inhibitors of glycoprotein processing. Biochem. Biophys. Commun. 148: 206-210.

54. TAYLOR, D.L., L.E. FELLOWS, G.H. FARRAR, R.J. NASH, D. TAYLOR-ROBONSON, M.A. MOBBERLEY, T.A. RYDER, D.J. JEFFRIES, A.S. TYMS. 1988. Loss of cytomegalovirus infectivity after reaction with castanospermine or related plant alkaloids relates to aberrant glycoprotein synthesis. Antiviral Res. 10: 11-26.

55. PAN, Y.T., H. HORI, R. SAUL, B.A. SANFORD, R.J. MOLYNEUX, A.D. ALBEIN. 1983. Castanospermine inhibits the processing of the oligosaccharide portion of the influenza viral hemagglutinin. Biochemistry 22: 3975-3984.

56. BURKE, B., K. MATLIN, E. BAUSE, G. LEGLER, N. PEYRIERAS, H. PLOEGH. 1984. Inhibition of N-linked oligosaccharide trimming does not interfere with surface expression of certain integral membrane proteins. EMBO J. 3: 551-556.

57. BOSCH, V., R.T. SCHWARZ. 1984. Processing of $gPr92^{env}$, the precursor to the glycoproteins of

Rous sarcoma virus: use of inhibitors of oligosaccharide trimming and glycoprotein transport. Virology 132: 95-109.

58. ROMERO, P.A., R. DATEMA, R.T. SCHWARZ. 1983. N-methyl-1-deoxynojirimycin, a novel inhibitor of glycoprotein processing and its effect on fowl plague virus maturation. Virology 130: 238-242.

59. WARREN, L., C.A. BUCK. G.P. TUSGYNSKI. 1978. Glycopeptide changes and malignant transformation: a possible role for carbohydrates in malignant behavior. Nature 253: 457-460.

60. SANTER, U.V., M.C. GLICK. 1979. Partial structure of a membrane glycoprotein from virus transformed hamster cells. Biochemistry 18: 2533-2540.

61. COLLARD, J.G., W.P. VAN BEEK. J.W.G. JANSSEN, J.F. SCHIJVEN. 1985. Transfection by human oncogenes: concomitant induction of tumorigenicity and tumor-associated membrane alteration. Int. J. Cancer 35: 207-214.

62. HADWIGER, A., H. NIEMANN, A. KABISCH, H. BAUER, T. TAMURA. 1986. Appropriate glycosylation of the _fms_ gene product is a prerequisite for its transforming potency. EMBO J. 5: 689-694.

63. OSTRANDER, G.K., N.K. SCRIBNER, L.R. ROHRSCHNEIDER. 988. Inhibition of _v-fms_-induced tumor growth in nude mice by castanospermine. Cancer Res. 48: 1091-0194.

64. DESANTIS, R., U.V. SANTER, M.C. GLICK. 1987. NIH 3T3 cells transfected with human tumor DNA lose the transformed phenotype when treated with swainsonine. Biochem. Biophys. Res. Commun. 142: 348-353.

65. HUMPHRIES, M.J., K. MATSUMOTO, S.L. WHITE, K. OLDEN. 1986. Oligosaccharide modification by swainsonine treatment inhibits pulmonary colonization by B16-F10 murine melanoma cells. Proc. Nat. Acad. Sci. USA 83: 1752-1756.

66. HUMPHRIES, M.J., K. MATSUMOTO, S.L. WHITE, K. OLDEN. 1986. Inhibition of experimental metastasis by castanospermine in mice: blockage of two distinct stages of tumor colonization by oligosaccharide processing inhibitors. Cancer Res. 46: 5215-5222.

67. HUMPHRIES, M.J., K. MATSUMOTO, S.L. WHITE, R.J. MOLYNEUX, K. OLDEN. 1988. Augmentation of

murin natural killer cell activity by swainsonine, a new antimetastatic immunomodulator. Cancer Res. 48: 1410-1415.
68. DENNIS, J.W. Effects of swainsonine and polyinosinic:polycytidylic acid on murine tumor cell growth and metastasis. Cancer Res. 46: 5131-5136.
69. KINO, T., N. INAMURA, K. NAKAHARA, S. KIYOTO, T. GOTO, H. TERANO, M. KOHSAKA, H. AOKI, H. IMANAKA. 1985. Studies of an immunomodulator, swainsonine. II. Effect of swainsonine on mouse immunodeficient system and experimental murine tumor. J. Antibiot. 37: 936-940.
70. WALL, K.A., J.D. PIERCE, A.D. ELBEIN. 1988. Inhibitors of glycoprotein processing alter T-cell proliferative responses to antigen and to interleukin 2. Proc. Nat. Acad. Sci. USA 85: 5644-5648.
71. BERNACKI, R., M.J. NIEDBALA, W. KORYTNYK. 1985. Glycosidases in cancer and invasion. Cancer Metast. Rev. 4: 81-102.
72. HORI, H., Y.T. PAN, R.J. MOLYNEUX, A.D. ELBEIN. 1984. Inhibition of plant N-linked oligosaccharides by castanospermine. Arch. Biochem. Biophys. 228: 525-533.
73. SZUMILO, T., G.P. KAUSHAL, H. HORI, A.D. ELBEIN. 1986. Purification and properties of a glycoprotein processing α-mannosidase from mung bean seedlings. Plant Physiol. 81: 383-389.
74. CHRISPEELS, M.J., A. VITALE. 1985. Abnormal processing of the modified oligosaccharide side chains of phytohemagglutinin in the presence of swainsonine and deoxynojirimycin. Plant Physiol. 78: 704-709.
75. YAMASAKI, Y., H. KONNO. 1985. Three forms of α-glucosidase from suspension-cultured rice cells. Agric. Biol. Chem. 49: 3383-3390.
76. McGEE, C.M. D.R. MURRAY. 1986. Comparative studies of acid glycosidases from three legumes. Ann. Bot. (Lond.) 57: 179-190.
77. LIZOTTE, P.A., J.E. POULTON. 1988. Catabolism of cyanogenic glycosides by vicianin hydrolase from Squirrel's Foot fern (Davalia trichomanoides Blume). Plant Physiol. 86: 322-324.
78. SAKURAI, N., Y. MASUDA. 1978. Auxin-induced changes in barley coleoptile cell wall

composition. Plant Cell Physiol. 19: 1217-1223.

79. WALLNER, S.J., D.J. NEVINS. 1974. Changes in cell walls associated with cell separation in suspension cultures of Paul's Scarlet Rose. J. Exp. Bot. 25: 1020-1029.

80. NAGAHASHI, G., P.M. BARNETT, S.I. TU, J. BROUILLETTE. 1986. Purification of primary cell walls from corn roots: inhibition of cell-wall associated enzymes with indolizidine alkaloids. In Regulation of Carbon and Nitrogen Reduction and Utilization in Maize. (J.C. Shannon, D.P. Knievel, C.D. Boyer, eds.), Am. Soc. Plant Physiologists, Rockville, Maryland, pp. 289-293.

81. ROSENFIELD, C.L., P. MATILE. 1979. Glycosidases in pear pollen tube development. Plant Cell Physiol. 20: 605-613.

82. MÜLLER, L. 1986. Microbial glycosidase inhibitors. In Microbial Products II. (H. Pape, H.J. Rehm, eds.), VCH Verlagesellschaft, Weinheim, pp. 532-567.

83. SAMULITIS, B.K., T. GODA, S.M. LEE, O. KOLDOVSKI. 1987. Inhibitory mechanism of acarbose and 1-deoxynojirimycin derivatives on carbohydrases in rat small intestine. Drugs Exp. Clin. Res. 13: 517-524.

84. RHINEHART, G.L., K.M. ROBINSON, A.J. PAYNE, M.E. WHEATELY, J.L. FISHER, P.S. LIU, W. CHENG. 1987. Castanospermine blocks the hyperglycemic response to carbohydrates in vivo: a result of intestinal disaccharidase inhibition. Life Sci. 41: 2325-2331.

85. NASH, R.J., K.A. FENTON. A.M.R. GATEHOUSE, E.A. BELL. 1986. Effects of the plant alkaloid castanospermine as an antimetabolite of storage pests. Entomol. Exp. Appl 42: 71-77.

86. CAMPBELL, B.C., R.J. MOLYNEUX, K.C. JONES. 1987. Differential inhibition by castanospermine of various insect disaccharidases. J. Chem. Ecol. 13: 1759-1770.

87. BLANEY, W.M., M.S.J. SIMMONDS, S.V. EVANS, L.E. FELLOWS. 1984. The role of the plant secondary compound 2,5-dihydroxymethyl-3,4-dihydroxypyrrolidine as a feeding inhibitor for insects. Entomol. Exp. Appl. 36: 209-216.

88. DREYER, D.L., K.C. JONES, R.J. MOLYNEUX. 1985. Feeding deterrency of some pyrrolizidine, indoolizidine and quinolizidine alkaloids towards pea aphid (Acyrthosiphon pisum) and evidence for phloem transport of the indolizidine alkaloid swainsonine. J. Chem. Ecol. 11: 1045-1051.
89. BLANEY, W.M. 1974. Electrophysiological responses of the terminal sensilla on the maxillary palps of Locusta migratoria to some electrolytes and non-electrolytes. J. Exp. Biol. 60: 275-293.
90. SIMMONDS, M.S.J., W.M. BLANEY. 1984. Some effects of azadirachtin on lepidopterous larvae. In Proceedings of the 2nd International Neem Conference. (H. Schmutterer, K.R.S. Ascher, eds.), GTZ Eschborn, pp. 163-180.
91. BLANEY, W.M., M.S.J. SIMMONDS, S.V. LEY, P.S. JONES. 1988. Insect antifeedants: a behavioral and electrophysiological investigation of natural and synthetically derived clerodane diterpenoids. Entomol. Exp. Appl. 46: 267-274.
92. FLEET, G.W.J., A.N. SHAW, S.V. EVANS, L.E. FELLOWS. 1985. Synthesis from glucose of 2,5-dideoxy-1,5-imino-L-fucitol, a potent α-L-fucosidase inhibitor. J. Chem. Soc. Chem. Commun. 841-842.
93. INOUYE, S., T. TSURUOKA, T. ITO, T. NIIDA. 1968. Structure and synthesis of nojirimycin. Tetrahedron 23: 2125-2144.
94. HARRIS, C.M., M.J. SCHNEIDER, F.S. UNGEMACH, J.E. HILL, T.M. HARRIS. 1988. Biosynthesis of the toxic indolizidine alkaloids slaframine and swainsonine in Rhizoctonia leguminicola: metabolism of 1-hydroxyindolizidines. J. Am. Chem. Soc. 110: 940-949.

Chapter Twelve

BIOSYNTHESIS OF ALKALOIDS USING PLANT CELL CULTURES

MEINHART H. ZENK

Lehrstuhl für Pharmazeutische Biologie
Ludwig-Maximilians-Universität
D-8000 München
West Germany

INTRODUCTION

With an estimated 7000 individual structures thus far isolated from plant material, alkaloids are the most diverse of all low molecular weight nitrogen containing compounds. The amount of nitrogen immobilized in these structures varies drastically but, for instance, in cell suspension cultures of *Berberis stolonifera* which produce up to 3 g of protoberberine alkaloids per liter of medium,[1] 15% of the nitrogen supplied in the medium can be found in alkaloids. Plants in general in their natural environment are nitrogen limited; therefore the sacrifice of valuable nitrogen fixed into alkaloids which generally have no apparent turnover but rather are dead end products is justified only by the fact that these metabolites serve important ecochemic functions for the differentiated plant.[2] The extreme diversity of alkaloids and their use as pharmaceuticals[3] and stimulatory agents (i.e., caffeine, nicotine) of considerable commercial value make this group of compounds an interesting subject for studying their formation within the living system. At this stage of development of plant biochemistry, the elucidation of biosynthetic pathways for secondary compounds has become possible. Such studies were greatly assisted by the use of plant cell suspension cultures as an experimental system. It has become an art in the past decade to chemically differentiate cells which are morphologically dedifferentiated, that is, to induce them to produce secondary compounds. In those instances where this goal was realized, it has become possible to clarify whole chains of metabolic reactions leading from primary to secondary metabolites. Thus, success has been achieved for the enzymology of flavonoids,[4] and the biosynthesis of indole[5] and isoquinoline alkaloids.[6] The use of cell cultures has several major advantages over differentiated higher plants: a) Suspension cells have compressed into their growth cycle of about two weeks all the synthetic capacity of a differentiated plant which may require a vegetative period of 4 to 5 months. As a result, enzyme activities are normally a factor 10^2 to 10^3 higher in suspension cells than in differentiated tissue. b) Undifferentiated cells are cultivated under strictly defined conditions on synthetic media, thereby eliminating problems encountered with seasonal variations. c) Cultured cells are grown as single cells or as small aggregates and therefore represent uniform, transferable material in which problems of translocation are minimized. d) The sterile conditions used in the cultivation process precludes any microbial activity, which may lead to false conclusions in non-sterile, whole plant systems. e) For the reasons given above, rates of incorporation of labeled precursors using cell cultures are much higher than with field grown plants. Cell cultures therefore are an excellent source for precursor feeding experiments.

The most serious problem in using cell cultures for the objectives mentioned above remains, however, the successful chemical differentiation of the culture under study. This remains a problem which up to now has only been solved empirically.

This report will concentrate on a group of alkaloids which we have studied during the past 8 years, the benzylisoquinoline alkaloids. An enormous number of alkaloids in this group are derived from the central intermediate (S)-reticuline; for review see Reference 7.

In the past 12 years we have established approximately 170 cell cultures of members of the families Berberidaceae, Menispermaceae and Papaveraceae, some of which produce isoquinoline alkaloids in substantial amounts. All the major types of alkaloids can be found in these cultures, even morphinandienone-type compounds.[8] One of the major goals set for benzylisoquinoline biosynthesis was elucidation of the (S)-reticuline pathway at the enzymatic level.

REVISION OF THE RETICULINE PATHWAY

On several occasions[9,10] we have pointed out the problems encountered in finding the first alkaloidal intermediate after the condensation of dopamine with either 3,4-dihydroxyphenylacetaldehyde or 3,4-dihydroxyphenylpyruvate. This question seemed resolved after we isolated an enzyme which catalyzed the stereospecific condensation of dopamine with 3,4-dihydroxyphenylacetaldehyde but not with 3,4-dihydroxyphenylpyruvate. This enzyme was purified, characterized and was named after what we thought at that time to be physiological substrate: (S)-norlaudanosoline synthase.[11,12] At that time we had already observed that not only the dihydroxylated phenylacetaldehyde but also the mono-substituted p-hydroxyphenylacetaldehyde served as substrates.[11,12] However, even these results which established that the aldehyde instead of the pyruvate is the true biosynthetic precursor of the benzylisoquinoline alkaloids, could not completely resolve the statement of I.D. Spenser[13] that

> "the two C_6 - C_2 units derived from tyrosine which are incorporated into the alkaloids under discussion differ from one another."

Tyrosine, being an immediate precursor of L-DOPA, is incorporated into both halves of isoquinoline alkaloids, i.e., the isoquinoline or "upper" portion and the benzylic or "lower" part, while DOPA and dopamine are only incorporated into the isoquinoline half.[13,14] There still remained this unresolved observation. A first hint as to what was occurring was obtained when, in feeding experiments it was found that labeled tyramine was incorporated effectively into the benzylic moiety of the protoberberine molecule, jatrorrhizine, produced by the cell culture. This may be a rather unique characteristic of the species employed: Berberis canadensis, since several other Berberis species (B. stolonifera, B. wilsoniae var. subcaulialata, B. koetineana, B. henryana) incorporate tyramine only (90 - 99%) into the upper, isoquinoline half of the target molecule. It was R. Stadler in our laboratory who found the solution to the problem. Investigating the biosynthesis of bisbenzylisoquinoline alkaloids by feeding ^{13}C and ^{14}C labeled (S)- and (R)-coclaurines to B. stolonifera cell cultures which contain good quantities of these dimers,[15] he observed an exceedingly high incorporation of (S)-coclaurine into the protoberberine fraction of this culture. This incorporation observed only with the (S)-isomer led to a more elaborate study.[16] If, indeed, the 4'-monohydroxylated (S)-coclaurine is the precursor of advanced isoquinolines such as protoberberine alkaloids instead of the dihydroxylated norlaudanosoline, then both (S)-coclaurine and (S)-reticuline should have the same isotopic ratio if a plant producing both compounds is supplied with doubly labeled L-tyrosine (2,6-^{3}H:ring ^{14}C; ratio 7.4:1). Annona reticulata, the original plant from which reticuline was isolated and characterized[17] and after which it was named, was used as the experimental system. When mature leaves were worked up for (S)-coclaurine (ratio 6.3:1) and reticuline (5.5:1), the isotopic ratio found in both of these compounds agreed well with the theoretical value expected (5.5:1) calculated on the basis that one tritium atom of dopamine is removed during the condensation of the amine with the phenylacetaldehyde. Within experimental limits, these results demonstrate that (a) both benzylisoquinoline derivatives, the dihydroxylated reticuline as well as the monohydroxylated coclaurine, are formed from tyrosine; (b) no degradation of tyrosine occurs in this plant, and (c) both intermediates have the same biogenetic precursor. Furthermore, labeled (S)-[6-O-$^{14}CH_3$]-coclaurine was specifically and well (2.6%) incorporated into (S)-reticuline as proven by degradation of the molecule; specifically, no scrambling of the methyl group occurred. In contrast, (R)-coclaurine containing the ^{14}C label in the same (6) position was not incorporated into (S)-reticuline, indicating that no racemization occurs during this transformation, and that the monohydroxylated (S)-derivative is the true precursor. If it is assumed that (S)-coclaurine is the biosynthetic precursor of (S)-reticuline, identical ratios of the C_6-C_2 units derived from tyrosine should be found in the two halves of both the coclaurine and reticuline molecules. Conducting this experiment using L-[ring-U-^{14}C]-tyrosine and leaves of *A. reticulata* it was shown[16] that coclaurine contained a ratio of isoquinoline to benzyl portion of 1:4.90 while reticuline yielded a ratio of 1:4.71. Within experimental error, this experiment shows that both compounds must have one and the same biogenetic origin. Surprising was the fact that the majority of label from tyrosine was clearly found in the lower (benzylic) half of coclaurine and reticuline. This is in contrast to previous observations using different plant material.[13,14] This discrepancy was resolved when leaf tissue of *A. reticulata* was shown to contain an exceedingly large pool of dopamine (ca. 35 mM or 3% d.wt.), while only trace quantities of tyramine were observed in the same tissue. This finding explains the low incorporation of tyrosine into the isoquinoline portion of both molecules. Labeled dopamine derived from tyrosine will be highly diluted by the endogenous dopamine pool prior to incorporation into norcoclaurine. This dilution phenomenon plausibly explains the higher levels of radioactivity found in the benzylic halves of coclaurine and reticuline.

Since (S)-reticuline is firmly established as a common precursor to all major benzylisoquinoline alkaloids and (S)-reticuline is derived from (S)-coclaurine, then (S)-coclaurine should be incorporated into all reticuline-derived alkaloids. Indeed a first experiment[16] demonstrated the intact incorporation of (S)-[6-O-$^{14}CH_3$]-coclaurine into the protoberberine jatrorrhizine (5.9%), the benzophenanthridine macarpine (4.6%), and the morphinan alkaloid thebaine (5.7%) using either cell cultures of *B. stolonifera*, *Eschscholtzia californica*, or 5-day-old seedlings of *Papaver somniferum*. In each case label from the precursor was located at the correct position of the final metabolite. Repetition of the feeding experiments with the labeled enantiomer, (R)-coclaurine, showed absolutely no incorporation into these types of reticuline-derived alkaloids with the exception of the bisbenzylisoquinoline alkaloid berbamunine where (R)-coclaurine was incorporated (1.9%) into the (R)-half of this molecule exclusively.[18] In order to prove unequivocally the specific incorporation not only of coclaurine but also of norcoclaurine, the now suspected first alkaloid of the reticuline pathway, ^{13}C-labeled precursors were fed (30 - 100 μg of the labeled material) to each poppy seedling through the root system. Typically 200 seedlings were worked up after a metabolic period of 48 h in continuous light (2000 Lux, neon) at 20°C and 83% humidity.[19] It is known that seedlings under these conditions contain measurable amounts of thebaine.[20] This alkaloid was isolated and typically 150 g of thebaine was enough to conduct the ^{13}C-NMR analysis. While L-[2-^{13}C]-tyrosine labeled both C-9 and C-16 of the thebaine molecule (41% ^{13}C enrichment), [8-^{13}C]-tyramine labeled only C-16 of this morphinandienone (42%). (R,S)-[1-^{13}C]-Norcoclaurine, as well as coclaurine labeled in the same position, clearly and unequivocally labeled the C-9 of thebaine (each ca. 20% ^{13}C enrichment). In addition, (S)-[N-$^{13}CH_3$]-reticuline was extensively (55%) and specifically (N-$^{13}CH_3$ of thebaine) incorporated into thebaine. These experiments were extended in that (S)-[1-^{13}C]-norcoclaurine was fed to cell cultures of *B. stolonifera*, *Pneumus boldo*, *E. californica* and intact plants of *Macleaya cordata* as well as *Argemone hispida*. In each case the specific incorporation of the precursor into the target molecules which represented the protoberberines, aporphines, benzophenanthridines, and pavines was demonstrated.[21]

These experiments, in our opinion, demonstrate beyond any doubt that (S)-norcoclaurine and not norlaudanosoline, as previously assumed, is the true first alkaloidal intermediate opening up the (S)-reticuline pathway. Supporting this finding is the fact that norlaudanosoline has never been isolated as a natural product from any plant source, while norcoclaurine also called desmethylcoclaurine or higenamine has been isolated on several occasions.[22-26] Norcoclaurine has to stand now as the first and central intermediate in the biosynthesis of a vast majority of isoquinoline alkaloids (Fig. 1). The question remains, however, how the two building blocks of norcoclaurine (i.e., dopamine and 4-hydroxyphenylacetaldehyde) are formed. The amine is

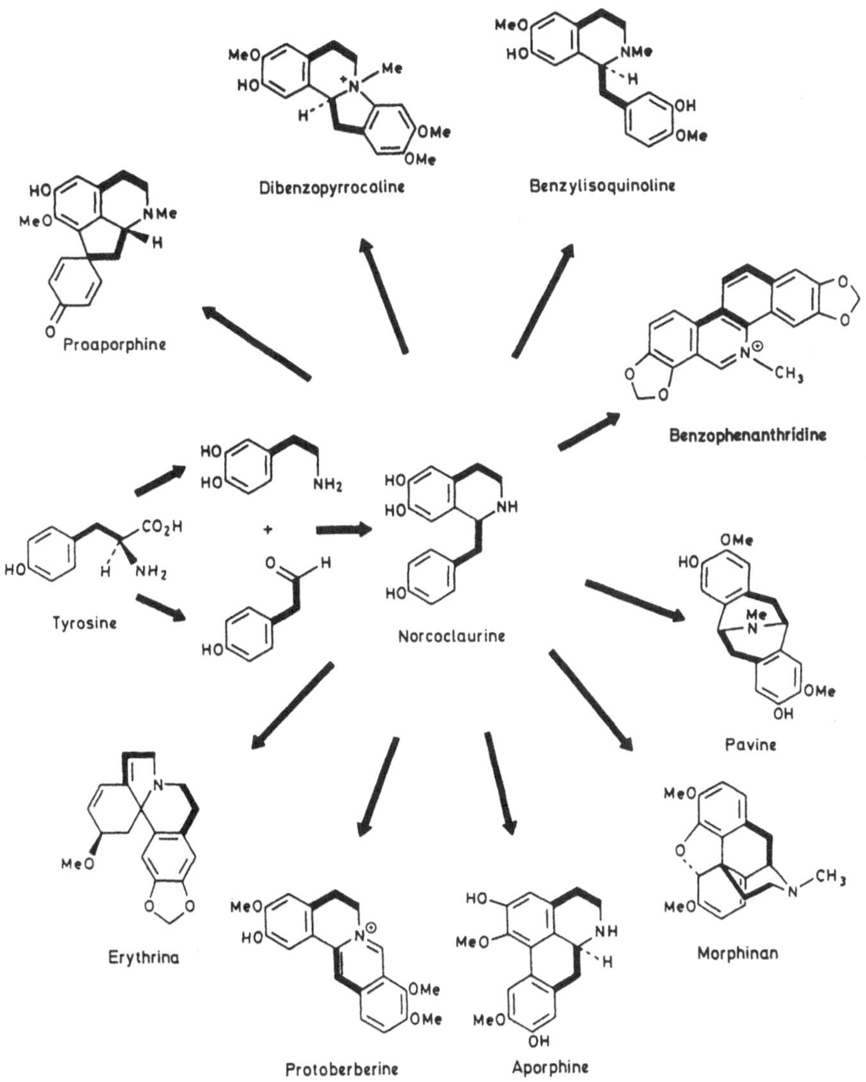

Fig. 1. Biosynthesis of various benzylisoquinoline alkaloids from (S)-norcoclaurine.

undoubtedly formed from tyrosine by hydroxylation to yield DOPA and subsequent decarboxylation of this 3,4-dihydroxy amino acid. The necessary enzymes have been found in isoquinoline-containing cell cultures of *B. stolonifera*, and the literature is discussed in this paper.[27]

The more difficult question to answer is how the aldehyde building block is formed. In *B. stolonifera* tissue, high levels of transaminase have been found which can explain the ease of transamination of tyrosine to p-hydroxyphenylpyruvate. Since, however, in most cell cultures thus far tested (see above) tyramine is not incorporated into the benzyl portion of the reticuline derived alkaloids,[13] a simple amine oxidase activity acting on tyramine to furnish p-hydroxyphenylacetaldehyde is precluded. The only other possibility seems therefore a p-hydroxyphenylpyruvate decarboxylase, an enzyme which was recently discovered and partially characterized in plant tissue.[27] This enzyme has a pH optimum of 6.5 and a molecular weight of ca. 30 kD. The K_M for the 4-hydroxylated phenylpyruvate is reasonably low (0.7 mM) and the reaction product was identified to be the expected aldehyde by forming the 2,4-dinitrophenylhydrazone derivative. This decarboxylase is present in *Berberis* tissue in about a 10-fold higher concentration than the respective amine oxidase. The suggested metabolic pathway leading to both building blocks of norcoclaurine is shown in Figure 2. There is little doubt that the amino acid L-tyrosine in isoquinoline alkaloid containing tissue is formed from prephenate via arogenate as shown for *Sorghum* seedlings.[28] If this is true and if the tyrosine pathway leads exclusively through arogenate, then the aromatic amino acid transferase taking tyrosine to p-hydroxyphenylpyruvate will play a specific role for channeling tyrosine into the aldehydic component.

(S)-Norcoclaurine has now been firmly established as the first alkaloid of the reticuline pathway. However, (S)-norlaudanosoline, if channeled into the metabolism of plants, is also incorporated into the various benzyliso-quinoline structures[29,30] obviously using non-specific enzymatic tracks. The previously proposed metabolic route based on the assumption that (S)-norlaudanosoline[10,31] is an intermediate is shown in Figure 3. This sequence

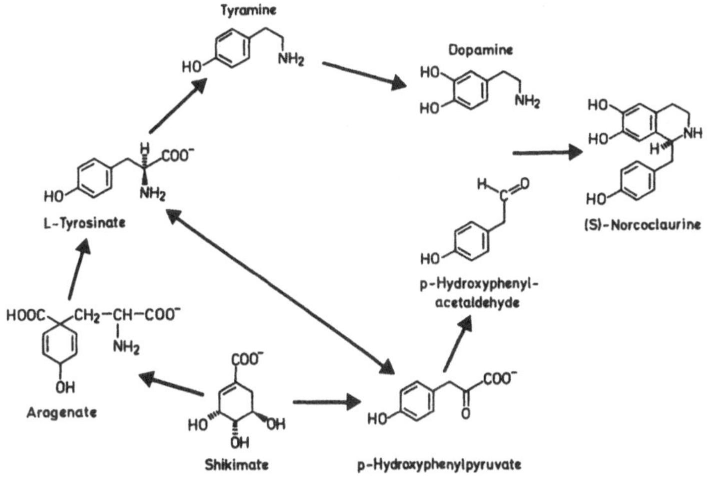

Fig. 2. Formation of (S)-norcoclaurine from L-tyrosine.

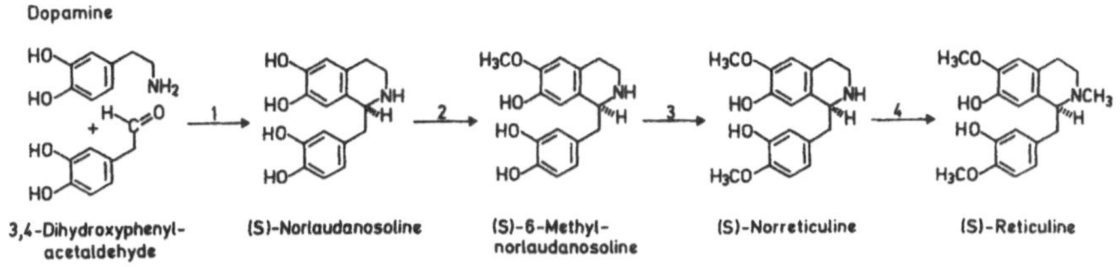

Fig. 3. Previous pathway leading from (S)-norlaudanosoline to (S)-reticuline.

involved three enzymatic steps: first, 6-O methylation; secondly, 4'-O methylation; and finally, N-methylation of norlaudanosoline. Norcoclaurine, possessing one hydroxyl group less than required for the reticuline structure, has therefore to undergo an additional hydroxylation reaction. The substrate for this hydroxylation will determine the steps of the pathway. One has to assume that norcoclaurine will first be 6-O-methylated to yield coclaurine which is both an excellent precursor for reticuline derived alkaloids (see above) and has been detected frequently in plant material as shown by radioenzymatic determination and by reports in the literature.[32] Both the (R) and (S) epimers have been isolated as well as the racemic form.[33] Since N-methylcoclaurine is the precursor for bisbenzylisoquinoline alkaloids[18] and since this compound is also excellently and specifically incorporated into the various types of reticuline derived alkaloids (Stadler and Zenk, unpublished results), we have to assume that N-methylation occurs prior to 3'-hydroxylation. Appropriate N-methyltransferases, isolated and characterized from _B. koetineana_, show that (S)-coclaurine is the preferred substrate over (S)-norreticuline (Frenzel and Zenk, unpublished). N-Methyl-(S)-coclaurine is, in turn, hydroxylated by a rather non-specific phenolase which transforms tyramine, tyrosine, N-methylcoclaurine and coclaurine to the corresponding 3,4-dihydroxy derivatives.[32] One has to assume, therefore, that methylation at the 4'-O-position of 3'-hydroxyl-N-methylcoclaurine is the final step in formation of (S)-reticuline. The enzyme catalyzing this step has been discovered, purified and characterized.[34] It is surprisingly stereo- and regiospecific in that only compounds with (S)-configuration serve as substrates and only the 3',4'-dihydroxy (C-ring) substrate is methylated while the 4'-hydroxy compound (N-methylcoclaurine) is not. Interestingly, (S)-3'-hydroxy-N-methylcoclaurine is the preferred substrate yielding (S)-reticuline after enzymatic methylation while the other possible substrate, (S)-6-O-methylnorlaudanosoline (yielding norreticuline), is only 10% as active.[34] This fact also indicates that N-methylation occurs prior to 4'-O-methylation.

Up to now, in all metabolic schemes for isoquinoline biosynthesis an intermediate, norreticuline, has been considered because of its high rate of incorporation into reticuline and compounds derived from reticuline.[31] On the other hand norreticuline, a chemically stable compound, has never been isolated from any plant source, strongly suggesting that it may not be a natural product and therefore not involved in the reticuline pathway. We presently assume that norreticuline is not an intermediate in the pathway (Fig. 4) leading from tyrosine to reticuline. The five enzymes required for this sequence are in hand, and some of their properties are given in Table 1. However, one intriguing feature of this pathway is its apparent non-specificity in that both norcoclaurine

Table 1. Select properties of the five enzymes involved in (S)-reticuline biosynthesis.

No.	Name	Cell Culture Species	Purification Factor (x-fold)	pH Optimum	Molecular Weight (kD)	Stereo-specificity
1	Norcoclaurine synthase	Eschscholtzia tenuifolia	40	7.8	15	S
2	6-O-Methyltransferase	Berberis koetineana	80	7.8	40(50)*	R and S
3	N-Methyltransferase	Berberis koetineana	28	7.4	60	R and S
4	3'-Hydroxylase	Berberis stolonifera	homogeneous	6.0	60	R and S
5	4'-O-Methyltransferase	Berberis koetineana	400	8.4	40(50)*	S

*Difference in molecular weight determination between SDS-PAGE and gel permeation HPLC, respectively.

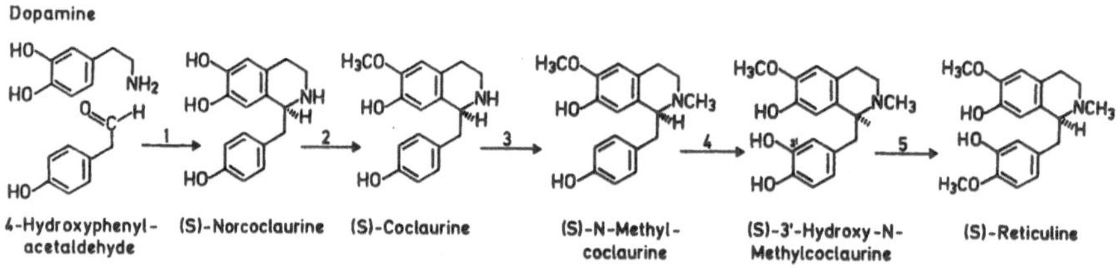

Fig. 4. Revised pathway beginning with the central intermediate, (S)-norcoclaurine and leading to (S)-reticuline.

and norlaudanosoline are transformed into reticuline. In light of our recent findings, however, we are convinced that norcoclaurine is the physiological and natural precursor, while norlaudanosoline is not.

As early as 1910 Winterstein and Trier,[35] in a far-sighted biosynthetic speculation, assumed norlaudanosoline to be the first isoquinoline intermediate. Their proposal, which was expanded by R. Robinson,[36,37] influenced Battersby and colleagues[29,30] to conduct experiments involving labeled norlaudanosoline as precursor. Proof is given here, however, that norcoclaurine is the true precursor instead of norlaudanosoline. If one compares Figures 3 and 4, one can now see that there is no common intermediate in the two pathways but that both lead to reticuline.

TWO METABOLIC PATHWAYS TO BERBERINE

The final metabolite of these pathways is berberine, a useful antimicrobial, gastric and antiinflammatory agent. Berberine is produced by cell cultures of Coptis japonica on a commercial level and the yield obtained (7 g/1 medium) is the highest so far recorded with plant cell cultures.[38] Berberine is the prototype of the protoberberine alkaloid family. Jatrorrhizine, which is a protoberberine alkaloid derived from berberine, is the final metabolic product in Berberis cell cultures.[1] The biosynthesis of berberine has been completely clarified at the enzyme level using highly purified enzymes from Berberis cell cultures.[10]

The conversion of (S)-reticuline into berberine involves 4 enzymes. The first step involves the conversion of the N-methyl group of reticuline into the so-called berberine bridge to form scoulerine. This compound is specifically methylated at the former 3'-hydroxyl group of (S)-reticuline to form (S)-tetrahydropalmatine which, in Berberis is oxidized to form the protoberberine, columbamine. In the final reaction step, the 3-O-methyl group of columbamine is oxidized to form the methylenedioxy group of berberine. The best characterized enzyme of this pathway is (S)-tetrahydroprotoberberine oxidase (STOX), a flavin enzyme which aromatizes ring C of a great variety of tetrahydroprotoberberines.[39,40] In this reaction, 1.5 moles of oxygen are taken up for each mole of substrate oxidized and one mole each of H_2O_2 and H_2O are formed. All four enzymes catalyzing this sequence including STOX have been isolated and characterized from Berberis species belonging to the family Berberidaceae. They are located in a specific class of vesicles whose only role seems to be the synthesis of protoberberine alkaloids.

After we had published our results on STOX,[39] a similar enzyme was described from C. japonica,[41,42] a species belonging to the closely related family Ranunculaceae which is placed together with the family Berberidaceae in the order of Ranunculales. This enzyme specifically catalyzes the oxidation of (S)-canadine ((S)-tetrahydroberberine) to berberine. The enzyme was named (S)-tetrahydroberberine oxidase[41,42] and, to avoid confusion with STOX ((S)-tetrahydroprotoberberine oxidase) we shall call it canadine oxidase (COX). This enzyme catalyzes the oxidation of canadine by the removal of four hydrogen atoms and the formation of 2 moles of H_2O_2. No alkaloidal intermediates could be detected in the reaction. Therefore the suggestion was made that COX should be classified as a new kind of oxidase.[42] Since COX from Coptis seemed to be absolutely specific for canadine, these results were at variance with the pathway we had worked out in the Berberis species. To resolve this discrepancy, the pathway for berberine synthesis in Coptis was reinvestigated.[43] As in the case of Berberis, the terminal four enzymes involved in the formation of berberine in Coptis were localized in specific vesicles having a density of $\rho = 1.14$ as determined by sucrose density centrifugation.[43] Subsequently a comparison of

Table 2. Comparison of the properties of purified oxidases from Berberis and Coptis.

Name	Plant Source	pH Optimum	Molecular Weight (kD)	Co-factor	Substrate Specificity	Per mole of Substrate, mole: O_2 Up-take	Per mole of Substrate, mole: H_2O_2 Pro-duced	Particle Bound
(S)-Tetra-hydroproto-berberine oxidase	Berberis wilsoniae var. sub-caulilata	8.9	105 (2 x 53)	Flavin	Broad specificity towards tetrahydroprotoberberines and (S)-norbenzylisoquinolines	1½	1	Yes
(S)-Canadine oxidase	Coptis japonica	8.8*	58* (2 x 28)	Fe*	(S)-Canadine only*	2*	2*	Not determined
		8.7	150 58	Fe	(S)-Canadine (S)-Stylopine (S)-Tetrahydro-columbamine	2	2	Yes

*Data from Reference 42.

STOX and COX was made. The physical-chemical properties of both enzymes were rather similar (Table 2), except that STOX was a flavin enzyme while COX contained iron. An important difference was found, however, in the substrate specificity of the two enzymes. While STOX aromatizes tetrahydroprotoberberines and also converts norreticuline into 1,2-dehydronorreticuline, the *Coptis* enzyme was inactive towards such substrates. Antibodies directed against STOX did not cross react with COX. COX also has a much narrower substrate specificity than STOX, although it is not specific towards canadine as originally reported.[41,42]

If it is true that canadine is oxidized in *Coptis* to berberine by COX, then the methylenedioxy group has to be synthesized prior to the oxidation of ring C by the enzyme. To test this possibility, [3-O-C^3H_3]-columbamine was incubated with vesicular extracts of *Coptis*, but no labeled berberine was detected. *Coptis*, however, transformed (R,S)-[3-O-C^3H_3]-tetrahydrocolumbamine to canadine with high efficiency. This new enzyme, which was named (S)-canadine synthase, has a completely different substrate specificity than the previously discovered berberine synthase.[44]

We can now compare the reactions for berberine formation in *Berberis* and *Coptis* as follows: The berberine pathway begins when the central intermediate (S)-reticuline enters smooth vesicles with a density of $\rho = 1.14$. In both cases, (S)-reticuline is transformed to (S)-scoulerine which is subsequently methylated at the 9-O-position in the presence of S-adenosyl-L-methionine to yield (S)-tetrahydrocolumbamine. At this point, the pathways diverge. In *Berberis*, the tetrahydro derivative is first oxidized by STOX to columbamine which in a final step is further oxidized to yield the methylenedioxy group specific for berberine. Thus, the methylenedioxy group is introduced at the level of the quaternary protoberberine, a compound which can no longer penetrate the vesicle membrane. In contrast the berberine pathway in *Coptis* vesicles involves first the formation of the methylenedioxy ring at the tetrahydroprotoberberine level to yield (S)-canadine, followed by oxidation of the C-ring to yield berberine. It should be pointed out that this pathway gives the tetrahydroprotoberberine derivative possessing the 3,4-methylenedioxy

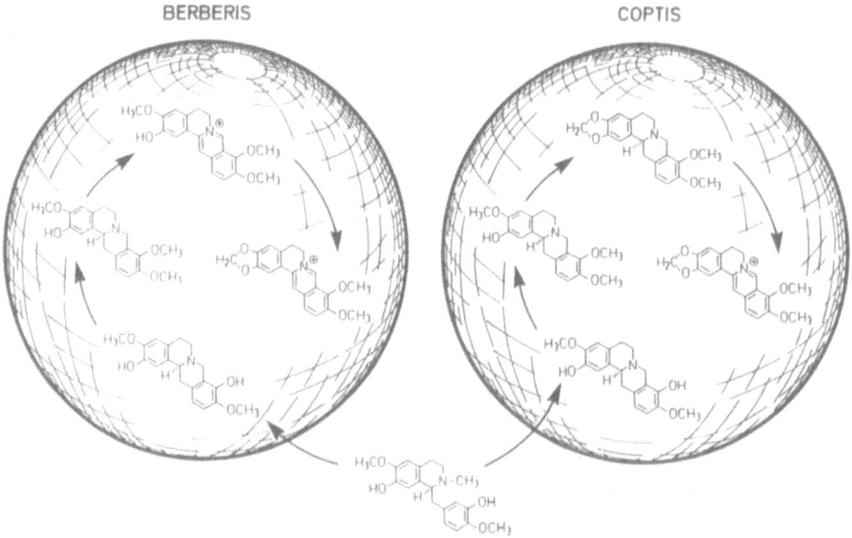

Fig. 5. Two alternative biosynthetic routes from (S)-reticuline to berberine realized in vesicles of Berberis and Coptis cells.

bridge the chance to diffuse out of the vesicle for further metabolism (for instance, N-methylation). If the molecule stays in the vesicle it will be oxidized to the protoberberine molecules of the berberine type.

As shown in Figure 5, the Berberis and Coptis systems are fundamentally different. The transformation of (S)-reticuline to berberine in Berberis involves C-ring oxidation initially with subsequent methylenedioxy ring formation and 2 H_2O_2 molecules for each (S)-reticuline consumed are formed. In Coptis the reverse order of oxidations is observed and 3 moles of H_2O_2 are formed for each mole of substrate entering the smooth vesicle. The fact that there are two alternative pathways for the same secondary plant product should serve as a warning to us not to generalize pathways unless the individual enzyme reactions have been elucidated in each case and for each species investigated.

THE PROTOPINE PATHWAY

If a molecule of (S)-reticuline enters smooth vesicles (density of $\rho = 1.14$) occurring in the cytoplasm of many isoquinoline alkaloid containing plants,[45] it is fated to become a tetrahydroprotoberberine alkaloid. This, however, does not mean that the carbon skeleton of the compound containing the berberine bridge (carbon atom 8) is a final end product. This is only the case if the compound is further oxidized in the C-ring and becomes a protoberberine molecule; the latter is prevented from diffusing out of the vesicle due to its positive charge. This quaternary alkaloid will ultimately end up in the vacuole when the vesicle membrane fuses with the tonoplast membrane and the contents of the vesicle are released into the vacuole where, as far as we know, no further metabolism occurs. (S)-Scoulerine, the entry product formed by action of the berberine bridge enzyme[46-48] on (S)-reticuline, can be further substituted, as we have seen above, at least to canadine, within the vesicle. At the tetrahydro stage all of these alkaloids are free to diffuse out of the vesicle to be further modified. We assume that conversion of (S)-scoulerine via (S)-cheilantifoline to (S)-stylopine occurs in the cytoplasm. This tetrahydroprotoberberine molecule carrying two methylenedioxy groups is bound to undergo important modifications outside the vesicle.

It has been known for a long time that very diverse compounds such as benzophenanthridine, phthalide isoquinolines and protopines as well as rhoeadine alkaloids are derived from tetrahydroisoquinoline molecules.[7] The key reaction leading to all these compounds is the N-methylation of the tetrahydroprotoberberine skeleton.[49] This reaction takes place again outside the vesicle since the N-methyltransferase clearly was not associated with the vesicle.[50] Since substitution of the nitrogen atom of the tetrahydroprotoberberines prevents aromatization of the C-ring, further oxidation to form quaternary protoberberines is no longer possible. The molecules are thus channeled away for alternate metabolism.

The N-methyltransferase was recently discovered in several members of protoberberine-containing cell cultures.[50] It was named S-adenosyl-L-methionine:

(S)-7,8,13,14-tetrahydroberberine-cis-N-methyltransferase. The enzyme was purified 40-fold from cell suspension cultures of *Corydalis vaginans*; it has a pH-optimum of 8.0 and a molecular weight of 72 kD. The *Corydalis* enzyme is specific for (S)-canadine and (S)-stylopine. The (R)-enantiomers, as well as (S)-scoulerine, (R,S)-tetrahydropalmatine, (R,S)-tetrahydrojatrorrhizine, (R,S)-columbamine, (S)-norreticuline and (S)-reticuline, are not metabolized. The enzyme therefore acts exclusively on tetrahydroprotoberberines of (S)-configuration containing a methylenedioxy bridge at carbon atoms 2 and 3. Since, however, benzophenanthridine, rhoeadine, protopine as well as phthalide alkaloids with substitution patterns corresponding to tetrahydropalmatine do exist in nature (e.g., sanguilutine, alpigenine, muramine, cordrastine) it is to be expected that N-methyltransferases will be found in the future with different substrate specificities.

The product of the N-methyltransferase reaction using (S)-stylopine as substrate was determined to have a cis-configuration since the enzymatically formed product showed good incorporation into protoberberine and sanguinarine when fed to callus cultures of *Fumaria capreolata*. This experiment showed the product formed by the N-methyltransferase is biologically active. Since it had previously[49] been unequivocally demonstrated that only the cis-N-methyl derivatives of tetrahydroprotoberberines can be metabolized into the protopine and benzophenanthridine type skeleton, there is no doubt that the product of the N-methyltransferase reaction possesses cis-N-methyl configuration. This N-methyltransferase has an important regulatory option in that it determines the alkaloid flow between quaternary protoberberines and protopine-derived (and phthalide) isoquinoline alkaloids. Furthermore, it is known that benzophenanthridine alkaloids[49] and rhoeadine alkaloids[51] are derived from protopine. Thus, this oxidized alkaloid again occupies a crucial position in the biosynthesis of alkaloids with different structural types. Protopine-type alkaloids are one of the most widely distributed groups of alkaloids found in the plant kingdom.

One important question remained, namely how is the oxidation of C-atom 14 of the N-methyltetrahydroprotoberberine brought about. Several oxidation systems have been tried in vain. No reaction was observed with ^{14}C-labeled N-methylstylopine as a substrate in reaction mixtures designed to detect α-ketoglutarate dependent dioxygenases, phenolases or peroxidases. Since, however, cytochrome P_{450} dependent hydroxylases active in secondary product metabolism have been shown to be localized in microsomal preparations[52] and since several reactions are known,[53] microsomal fractions of several cell cultures were tested.[54] As substrates (S)-[8,14-^{3}H]-_cis_-N-methylcanadine or the correspondingly labeled stylopine were prepared enzymatically. If any hydroxylase activity were present, tritium from the 14-position should be released into the aqueous medium of the test system. Microsomal preparations from plants not containing isoquinoline alkaloids were inactive in this assay. However, cell-free preparations from cultured cells of _Fumariaceae_ and _Papaveraceae_ showed good activity in the presence of NADPH. The product of the reaction from large scale incubation mixtures was isolated and unequivocally identified as protopine by mass spectral analysis. The hydroxylating activity showed a broader substrate specificity than the N-methylase described above in that all important N-methyltetrahydroprotoberberines were oxidized. The enzyme complex was solubilized from the microsomal membranes by sodium cholate and the protein purified further on an affinity column to yield an enzyme purified 100-fold with an overall yield of 35%. The pH optimum of the microsomal hydroxylase is 8.5. The UV spectrum of the $Na_2S_2O_4$ reduced enzyme shows peaks at 420, 450 and 480 nm suggesting a cytochrome P_{450} enzyme, which fact was verified by the use of typical inhibitors of cytochrome P_{450}-systems such as ketoconazole, metyrapone etc. The enzyme complex was named (S)-_cis_-N-methyltetrahydroprotoberberine-14-hydroxylase (Fig. 6). It is the first monooxygenase discovered in benzylisoquinoline alkaloid metabolism. The formation of this protopine, ubiquitously found in the family _Papaveraceae_, therefore involves 15 enzymes starting with the primary metabolite L-tyrosine. It is the longest biosynthetic sequence known at the enzyme level in secondary plant metabolism.

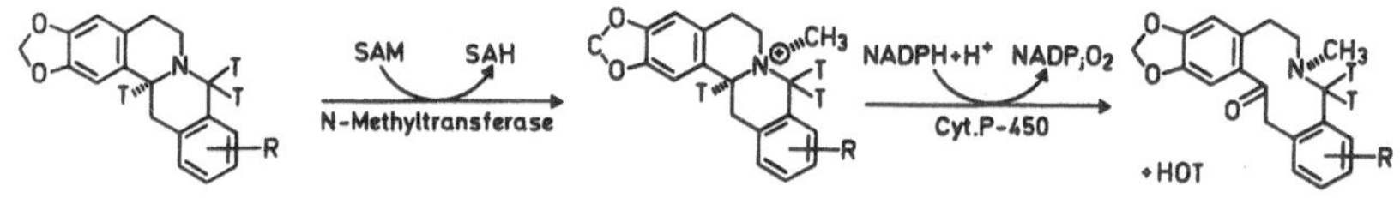

Fig. 6. Enzymatic steps leading from stylopine to protopine.

METABOLIC DEGRADATION OF BENZYLISOQUINOLINE ALKALOIDS

One of the drawbacks in the elucidation of metabolic pathways is the fact that biogenetic hypotheses are based on the assumption of precursor-metabolite relationships based on our general knowledge. This situation is obtained both in experiments involving the feeding of precursors to intact living systems as well as work on enzymes. Both metabolites and enzymes can only be searched for on the basis of a hypothetical metabolic grid. In order to obtain insight into the channeling of certain distance precursors into final products, the metabolism of reticuline in B. stolonifera was studied by "in vivo" ^{13}C-NMR.[50] Cells were charged with [N-$^{13}CH_3$]-reticuline (δ = 40.82) at a temperature of 30°C and, after a period for metabolism, were transferred into NMR tubes and examined. Under these conditions (S)-reticuline was completely taken up by the cells and metabolized to such an extent that the expected resonance for this precursor had almost completely disappeared. As expected and predicted, high concentration (12 - 15%) of the precursor into the protoberberines jatrorrhizine and columbamine was observed as shown by a strong signal at δ = 144.60 corresponding to C-8 of those compounds. Furthermore, a resonance at 43.43 ppm was detected in the cells which was assigned unequivocally to magnoflorine. This compound is present in the cultures to only a small extent (25 mg/l medium) but is highly labeled under these conditions.

The strongest signal, however, was seen at 47.64 ppm and corresponded to a new metabolite of reticuline. This compound was purified as a yellow, blue fluorescent compound. It was identified as the simple isoquinoline, pycnarrhine and was labeled in the N-CH_3 group. This compound also occurs endogenously in small quantities in

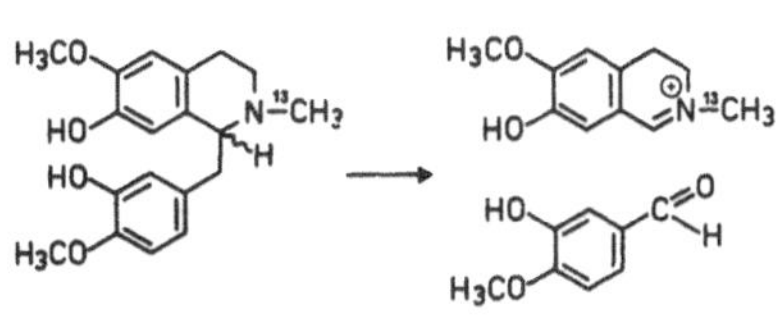

Fig. 7. Non-stereoselective degradation of reticuline to the simple isoquinoline alkaloid, pycnarrhine, as analyzed by *in vivo* ^{13}C-NMR.

cell cultures of *B. stolonifera*. Further investigation[55] lead also to the isolation of isovanillin which is not normally found in the *Berberis* cell cultures (Fig. 7). It is assumed to originate from the benzyl portion of reticuline by oxidative cleavage. This process most likely involves the action of peroxidase as was experimentally shown.[55] It is tempting to speculate that the H_2O_2 necessary for this peroxidative reaction is supplied by the metabolism of (S)-reticuline inside the smooth vesicles (Fig. 5). As noted above at least two moles of H_2O_2 are produced for each most of (S)-reticuline converted to protoberberine alkaloids. Degradation of reticuline into pycnarrhine and isovanillin is not restricted to the (S)-enantiomer; (R)-reticuline also undergoes the same cleavage. This reaction is also exhibited by cells which do not produce benzylisoquinoline alkaloids, such as *Nicotiana tabacum* and *Rauvolfia serpentina*. This non-stereo-selective degradation of reticuline may be a phenomenon of the elevated temperature (30°C) used for incubation of the plant cells. At lower temperatures (23 ± 2°C) under which these cell cultures have been grown for 10 or more years, this degradative phenomenon is *not* observed. In no case during extensive investigations on the metabolism of (R)- and (S)-coclaurine or reticuline using several plant species was this degradation ever observed (R. Stadler and M.H. Zenk, unpublished). We therefore assume that this degradation may be related to the heat-shock phenomena observed in plants and cell suspension cultures; such phenomena may possibly lead to an activation of peroxidases and thus to non-specific degradation of intermediates in a general defense mechanism of plant cells.[56]

Pycnarrhine has been isolated as a natural product from *Pycnarrhena longifolia* and *Corydalis ophiocarpa* as well as other species.[57] We assume that, in these species, simple isoquinoline alkaloids are derived from benzylisoquinoline alkaloids in a manner independent of the temperature effect as was experimentally shown for *C. ophiocarpa*.[55] Previously it has been shown that phthalideisoquinoline alkaloids were also degraded to simple isoquinoline alkaloids when fed to *Corydalis* callus cultures.[58] Experiments *in vivo* using N-methylcoclaurine and ascorbic acid oxidase, a common plant enzyme, resulted in the formation of the simple isoquinoline alkaloid corypalline; the latter is obtained by reduction of pycnarrhine.[59] On the basis of all these experiments, we can assume that some classes of simple isoquinolines may have their biogenetic origin in a cleavage reaction of common benzylisoquinolines.

In the future ^{13}C-NMR of intact plant cells may prove an extremely useful technique with which to study metabolic reactions of secondary metabolism. Prerequisites will be cell suspension cultures yielding high concentration of metabolites, chemical synthesis of specific ^{13}C-labeled precursors, and high resolution NMR instrumentation. Through the combination of these techniques, a new and efficient approach to the study of biosynthesis will evolve.

CONCLUSION

From the above we may conclude that the combination of precursor feeding techniques, *in vivo* ^{13}C-NMR experiments and enzyme studies will eventually lead to the establishment of the correct biosynthetic sequences for plant products. Very little research has been done in the past on the purification and characterization of these enzymes of alkaloid biosynthesis. As noted, they catalyze some of the most remarkable reactions, transformations which in some cases cannot be readily achieved by known chemical reactions. A new subject may also emerge from this research, namely the cytology and compartmentation of the reactions of secondary pathways. As we have seen above, there are specific vesicles most likely derived from the smooth endoplasmic reticulum which are the sites of highly specific enzymes catalyzing only the formation of protoberberine alkaloids. The oxidation of N-methyltetrahydroprotopines by microsomal cytochrome P_{450}-dependent monooxygenases is yet another example. It was a surprise in our work to see that the long established pathway leading from tyrosine to (S)-reticuline had to be modified. The fact that the 5 enzymes involved in reticuline synthesis were rather non-specific was an unexpected result. Long assumed precursors such as the four hydroxylated norlaudanosolines were transformed by these enzymes when fed into the pathway and this led to a misinterpretation of results. The trihydroxylated norcoclaurine is the newly established precursor, consistant with experimental data which, until now, have been difficult to interpret.

It is the duty and pleasure of our generation to work out the pathways which lead to natural products. We must isolate, characterize and purify to homogeneity the enzymes involved in secondary metabolism. This will lay the groundwork for the isolation of the genes involved in these pathways. We can envisage the transfer of genes from higher plants to bacteria and we can anticipate that these enzymes, or maybe even small chains of plant metabolic reactions, will be expressed in microorganisms. By a combination of biomimetic organic syntheses and fermentations involving these cloned genes, we will see new applications for plant biotechnology.

ACKNOWLEDGMENTS

I would like to thank the members of this laboratory for their enthusiastic, hard work. My thanks are also due to Dr. T.M. Kutchan for her linguistic help in the preparation of this manuscript. Research was sponsored by grants of the Deutsche Forschungsgemeinschaft, Bonn, through SFB 145 and by the Bundesminister für Forschung und Technologie as well as Fonds der Chemischen Industrie.

REFERENCES

1. HINZ, H., M.H. ZENK. 1981. Production of protoberberine alkaloids by cell suspension cultures of Berberis species. Naturwissenschaften 67: 620.
2. HARBORNE, J.B. 1982. Introduction to Ecological Biochemistry. Academic Press, London, New York, 2nd Edition, 278 pp.
3. FARNSWORTH, N.R. 1984. The role of medicinal plants in drug development. In Natural Products and Drug Development. (P. Krogsgaard-Larsen, S.B. Christensen, H. Kofod, eds.), Munksgaard, Copenhagen, pp. 1-30.
4. EBEL, J., K. HAHLBROCK. 1982. Biosynthesis. In The Flavonoids, Advances in Research. (J.B. Harborne, T.J. Mabry, eds.), Chapman and Hall, London, New York, pp. 641-680.
5. STÖCKIGT, J. 1986. Enzymatic biosynthesis of indole alkaloids: ajmaline, sarpagine, vindoline. In New Trends in Natural Products Chemistry, 1986, Studies in Organic Chemistry. (Atta-ur-Rahman, P.W. Le Quesne, eds.), Elsevier Science Publishers B.V., Amsterdam, Vol. 26, pp. 497-511.
6. ZENK, M.H., M. RUEFFER, T.M. KUTCHAN, E. GALNEDER. 1988. Biotechnological approaches to the production of isoquinoline alkaloids. In Application of Plant Cell and Tissue Culture. Ciba Foundation Symposium No. 137, Wiley, Chichester, pp. 213-227.
7. CORDELL, G.A. 1981. Introduction to Alkaloids. John Wiley-Interscience, New York, pp. 330-517.
8. TANAHASHI, T., M.H. ZENK. 1985. Isoquinoline alkaloids from cell suspension cultures of Fumaria capreolata. Plant Cell Rep. 4: 96-99.
9. ZENK, M.H. 1985. Enzymology of benzylisoquinoline alkaloid formation. In The Chemistry and Biology of Isoquinoline Alkaloids. (J.D. Phillipson, M.F. Roberts, M.H. Zenk, eds.), Springer-Verlag, Berlin, Heidelberg, New York, Tokyo, pp. 240-256.
10. ZENK, M.H., M. RUEFFER, M. AMANN, B. DEUS-NEUMANN, N. NAGAKURA. 1985. Benzylisoquinoline biosynthesis by cultivated plant cells and

isolated enzymes. J. Nat. Prod. (Lloydia) 48: 725-738.
11. RUEFFER, M., H. EL-SHAGI, N. NAGAKURA, M.H. ZENK. 1981. (S)-Norlaudanosoline synthase: The first enzyme in the benzylisoquinoline biosynthetic pathway. FEBS Lett. 129: 5-9.
12. SCHUMACHER, H.M., M. RUEFFER, N. NAGAKURA, M.H. ZENK. 1983. Partial purification and properties of (S)-norlaudanosoline synthase from Eschscholtzia tenuifolia cell cultures. Planta Med. 48: 212-220.
13. SPENSER, I.D. 1968. The biosynthesis of alkaloids and other nitrogenous secondary metabolites. In Comprehensive Biochemistry. (M. Florkin, E.H. Stotz, eds.), Elsevier, Amsterdam, Vol. 20, pp. 231-413.
14. HOLLAND, H., P. JEFFS, T.M. CAPPS, D.B. MACLEAN. 1979. The biosynthesis of protoberberine and related isoquinoline alkaloids. Can. J. Chem. 57: 1588-1597.
15. CASSELS, B.K., E. BREITMAIER, M.H. ZENK. 1987. Bisbenzylisoquinoline alkaloids in Berberis cell cultures. Phytochemistry 26: 1005-1008.
16. STADLER, R., T.M. KUTCHAN, S. LOEFFLER, N. NAGAKURA, B.K. CASSELS, M.H. ZENK. 1987. Revision of the early steps of reticuline biosynthesis. Tetrahedron Lett. 28: 1251-1254.
17. GOPINATH, K.W., T.R. GOVINDACHARI, B.R. PAI, N. VISWANATHAN. 1959. Konstitution von Reticulin, einem neuen Alkaloid aus Annona reticulata Linn. Chem. Ber. 92: 776-779.
18. STADLER, R., S. LOEFFLER, B.K. CASSELS, M.H. ZENK. 1988. Bisbenzylisoquinoline biosynthesis in Berberis stolonifera cell cultures. Phytochemistry 27: 2557-2565.
19. LOEFFLER, S., R. STADLER, N. NAGAKURA, M.H. ZENK. 1987. Norcoclaurine as biosynthetic precursor of thebaine and morphine. J. Chem. Soc. Chem. Commun., 1160-1162.
20. WIECZOREK, U., N. NAGAKURA, C. SUND, S. JENDRZEJEWSKI, M.H. ZENK. 1986. Radioimmunoassay determination of six opium alkaloids and its application to plant screening. Phytochemistry 25: 2639-2646.
21. STADLER, R., T.M. KUTCHAN, M.H. ZENK. 1988. (S)-Norcoclaurine is the central intermediate

in benzylisoquinoline alkaloid biosynthesis. Phytochemistry, in press.

22. KOSHIYAMA, H., H. OHKUMA, H. KAWAGUCHI, H.-Y. HSU, Y.-P. CHEN. 1979. Isolation of 1-(p-hydroxybenzyl)-6,7-dihydroxy-1,2,3,4-tetrahydroisoquinoline (demethylcoclaurine), an active alkaloid from Nelumbo nucifera. Chem. Pharm. Bull, 18: 2564-2568.

23. KOSUGE, T., M. YOKOTA. 1976. Studies on cardiac principle of aconite root. Chem. Pharm. Bull. 24: 176-178.

24. KOSUGE, T., M. YOKOTA, H. NUKAYA, Y. GOTOH, M. NAGASAWA. 1978. Studies on antitussive principles of Asiasari Radix. Chem. Pharm. Bull. 26: 2284-2285.

25. WAGNER, H., M. REITER, W. FERSTL. 1980. New drugs with cardiotonic activity. I. Chemistry and pharmacology of the cardiotonic active principle of Annona squamosa L. Planta Med. 40: 77-85.

26. LeBOEUF, M., A. CAVÉ, A. TOUCHÉ, J. PROVOST, P. FORGACS. 1981. Isolement de L'Higénamine a Partir de L'Annona squamosa; Interet des Résines Adsorbantes Macromoléculaires en Chimie Végétale Extractive. J. Nat. Prod. (Lloydia) 44: 53-60.

27. RUEFFER, M., M.H. ZENK. 1987. Distant precursors of benzylisoquinoline alkaloids and their enzymatic formation. Z. Naturforsch. 42c: 319-332.

28. CONNELLY, J.A., E.E. CONN. 1986. Tyrosine biosynthesis in Sorghum bicolor: Isolation and regulatory properties of arogenate dehydrogenase. Z. Naturforsch. 41c: 69-78.

29. BATTERSBY, A.R., R. BINKS. 1960. Biosynthesis of morphine: Formation of morphine from norlaudanosoline. Proc. Chem. Soc., 360-361.

30. BATERSBY, A.R., R. BINKS, R.J. FRANCIS, D.J. MACCALDIN, H. RAMUZ. 1964. Alkaloid biosynthesis. Part IV. 1-Benzylisoquinolines as precursors of thebaine, codeine and morphine. J. Chem. Soc., 3600-3610.

31. BROCHMANN-HANSSEN, E., C.H. CHEN. R. CHEN, H.C. CHIAN, A.Y. LEUNG, K. McMURTREY. 1975. The biosynthesis of 1-benzylisoquinolines in Papaver somniferum. Preferred and secondary

pathways, stereochemical aspects. J. Chem. Soc. Perkin Trans I, 1531-1537.

32. LOEFFLER, S. 1988. Norcoclaurin: das zentrale Zwischenprodukt in der Biosynthese der Benzylisochinolinalkaloide. Dissertation der Fakultät für Chemie und Pharmazie der Universität München.

33. JOHNS, S.R., J.A. LAMBERTON, A.A. SIOUMIS. 1967a. 1-Benzyl-1,2,3,4-tetrahydroisoquinoline alkaloids from Alseodaphne archboldiana (Allen) Kostermans. Aust. J. Chem. 20: 1729-1735.

34. FRENZEL, T., N. NAGAKURA, N.H. ZENK. 1989. S-Adenosyl-L-methionine: 6-O-methyllaudanosoline 4'-O-methyltransferase a regio and stereospecific enzyme of the (S)-reticuline pathway. Phytochemistry, in press.

35. WINTERSTEIN, E., G. TRIER. 1910. Die Alkaloide. Gebrüder Bornträger, Berlin, p. 307.

36. ROBINSON, R. 1917. A theory of the mechanism of the phytochemical synthesis of certain alkaloids. J. Chem. Soc. 111: 876-899.

37. ROBINSON, R. 1955. The Structural Relations of Natural Products. Clarendon Press, Oxford, 150 pp.

38. FUJITA, Y., M. TABATA. 1987. Secondary metabolites from plant cells - pharmaceutical applications and progress in commercial production. In Plant Tissue and Cell Culture. (C.E. Green et al., eds.), Alan R. Liss, New York, pp. 169-186.

39. AMANN, M., N. NAGAKURA, M.H. ZENK. 1984. (S)-Tetrahydroprotoberberin oxidase the final enzyme in protoberberine biosynthesis. Tetrahedron Lett. 25: 953-954.

40. AMANN, M., N. NAGAKURA, N.H. ZENK. 1988. Purification and properties of (S)-tetrahydroprotoberberine oxidase from suspension-cultured cells of Berberis wilsoniae. Eur. J. Biochem. 175: 17-25.

41. YAMADA, Y., OKADA, N. 1985. Biotransformation of tetrahydroberberine to berberine by enzymes prepared from cultured Coptis japonica cells. Phytochemistry 24: 63-65.

42. OKADA, N., A. SHINMYO, H. OKADA, Y. YAMADA. 1988. Purification and characterization of (S)-tetrahydroberberine oxidase from cultured Coptis japonica cells. Phytochemistry 27: 979-982.

43. GALNEDER, E., M. RUEFFER, G. WANNER, M. TABATA, M.H. ZENK. 1988. Alternative final steps in berberine biosynthesis in Coptis japonica cell cultures. Plant Cell Rep. 7: 1-4.
44. RUEFFER, M., M.H. ZENK. 1985. Berberine synthase, the methylenedioxy group forming enzyme in berberine synthesis. Tetrahedron Lett. 26: 201-202.
45. AMANN, M., G. WANNER, M.H. ZENK. 1986. Intracellular compartmentation of two enzymes of berberine biosynthesis in plant cell cultures. Planta 167: 310-320.
46. BÖHM, H., E. RINK. 1975. Zue Rolle eines Reticulinumwandelnden Enzyms (Berberin-Brücken-Enzym) im pflanzlichen Benzylisochinolin-Stoffwechsel. Biochem. Physiol. Pflanz. 168: 69-77.
47. RINK, E., H. BÖHM. 1975. Conversion of reticuline into scoulerine by a cell-free preparation from Macleaya microcarpa cell suspension cultures. FEBS Lett. 49: 396-399.
48. STEFFENS, P., N. NAGAKURA, M.H. ZENK. 1985. Purification and characterization of the berberine bridge enzyme from Berberis beaniana cell cultures. Phytochemistry 24: 2577-2583.
49. TAKAO, N., M. KAMIGAUCHI, M. OKADA. 1983. Biosynthesis of benzo[c]phenanthridine alkaloids sanguinarine, chelirubine and macarpine. Helv. Chim. Acta 66: 473-484.
50. RUEFFER, M., M.H. ZENK. 1986. S-Adenosyl-L-methionine: (S)-7,8,13,14-tetrahydroberberine cis-N-methyltransferase, a branch point enzyme in the biosynthesis of benzophenanthridine and protopine alkaloids. Tetrahedron Lett. 27: 5603-5604.
51. RÖNSCH, H. 1986. Rhoeadine alkaloids. In The Alkaloids. (A. Brossi, ed.), Academic Press, New York, London, Vol. 28, pp. 1-93.
52. MADYASTHA, K.M., T.D. MEEHAN, C.J. COSCIA. 1976. Characterization of a cytochrome P-450 dependent monoterpene hydroxylase from the higher plant Vinca rosea. Biochemistry 15: 1097-1102.
53. RIVIERE, J.-L., F. CABANNE. 1987. Animal and plant cytochrome P-450 systems. Biochimie 69: 743-752.

54. RUEFFER, M., M.H. ZENK. 1987. Enzymatic formation of protopines by a microsomal cytochrome P-450 system of Corydalis vaginans. Tetrahedron Lett. 28: 5307-5310.
55. JENDRZEJEWSKI, S. 1989. Detection of reticuline degradation in Berberis suspension cultures by ^{13}C in vivo NMR. Phytochemistry, in press.
56. NOVER, L. 1984. Heat-shock Response of Eucaryotic Cells. Springer-Verlag, Berlin, 82 pp.
57. MENACHERY, M.D., G.L. LAVANIER, M.L. WETHERLY, H. GUINAUDEAU, M. SHAMMA. 1986. Simple isoquinoline alkaloids. J. Nat. Prod. (Lloydia) 49: 745-778.
58. IWASA, K., M. KAMIGAUCHI, N. TAKAO. 1987. Biotransformation of phthalideisoquinoline alkaloids by Corydalis tissue culture. Arch. Pharm. (Weinheim) 320: 693-697.
59. LUNDSTRÖM, J. 1985. The occurrence of simple isoquinolines in plants. Op. cit. Reference 9, pp. 47-61.

INDEX